HENRY GEORGE'S WRITINGS ON THE UNITED KINGDOM

RESEARCH IN THE HISTORY OF ECONOMIC THOUGHT AND METHODOLOGY

Series Editors: Warren J. Samuels and Jeff E. Biddle

RESEARCH IN THE HISTORY OF ECONOMIC THOUGHT AND
METHODOLOGY   VOLUME 20-B

HENRY GEORGE'S WRITINGS ON THE UNITED KINGDOM

EDITED BY

KENNETH C. WENZER
Tahoma Park, MD, USA

2002

JAI
An Imprint of Elsevier Science

Amsterdam – London – New York – Oxford – Paris – Shannon – Tokyo

ELSEVIER SCIENCE Ltd
The Boulevard, Langford Lane
Kidlington, Oxford OX5 1GB, UK

First edition 2002

Library of Congress Cataloging in Publication Data
A catalog record from the Library of Congress has been applied for.

British Library Cataloguing in Publication Data
A catalogue record from the British Library has been applied for.

ISBN: 0-7623-0793-5
ISBN: 0743-4154 (Series)

⊗The paper used in this publication meets the requirements of ANSI/NISO Z39.48-1992 (Permanence of Paper).
Printed in The Netherlands.

For Sophia, who has given me
so much joy and, of course,
wisdom through the pain

CONTENTS

SECTION III: ENGLAND

SECTION IV: NATIONALIZATION, COMPENSATION, AND SOCIALISM

PREFACE

My goal has been to collect and preserve Henry George's writings pertaining to the United Kingdom (including Ireland). It is primarily a corpus of hard-to-find literature that merits attention for scholars since it reveals largely unknown aspects of George, especially to Americans: one full of surprises. The introductory essay presents a short exposition of pertinent British history and ideas with an explanation of George's relationship with these two islands and the political milieu with which he was confronted. It serves as background reading to the selectons in the main text. A second part describes his influence on the Fabians, probably his most enduring legacy in England. And the last paragraphs touch upon his most cherished desire to see in operation there and elsewhere – a cooperative commonwealth: one not far removed from philosophical anarchism, which he readily acknowledged as kindred.

Spelling has been standardized and Americanized except for proper nouns. (e.g.: "Although he labored for the good of the Labour Party he did not support the nationalization policies of the Land Nationalisation League while his defense of the National Defence Programme was not part of his program.") Archaic hyphenated words have been updated but capitalization has been retained, and in some cases added for consistency. Formatting has been altered at times for purposes of modernization, especially with excessively long paragraphs which have been shortened. Sections are divided topically and then chronologically subcategorized. Indecipherable words taken from handwritten notes are marked by a blank space within brackets and footnotes by George or others will be indicated by their initials in parentheses. Annotation is presented at the end of each section. The reader is encouraged to explore Henry George and/or the history of Great Britain further in any number of accessible secondary works; many pertaining to the former are contained in the bibliography.

Gratitude is extended to the Lincoln Institute of Land Policy and the Robert Schalkenbach Foundation for their generous financial support. This book would have been unthinkable without Ted Gwartney, Nan Braman, Sonny Rivera, and Mark Sullivan of the Robert Schalkenbach Foundation. Lily Griner, Patricia Herron, Terry Saylor's interlibrary loan staff, and Nancy Caldwell's periodical division of McKeldin Library of the University of Maryland (College Park) have, as usual, extended their expertise in so many ways. Lorin Evans of Washington Apple Pi has also been indispensable with his computer help.

Thanks must be given to Warren Samuels for permitting this book to be part of his series *Research in the History of Economic Thought and Methodology*. Marvin Adams, Patricia Barnes, David Borak, Barbara Bull, Edward Dodson, Evelyn Fazio, Judith Graebner, Fred Harrison, Leo Hecht, Jack Schwartzman, Thomas West, Anne Wilcox, Simon Winters, and Joan Youngman have also been of inestimable value in either their suggestions, friendship, or both. My cats Oliver and Sophia have immeasurably lightened the workload at the most propitious times.

FROM THE VISIONARY *UTOPIA* OF THOMAS MORE TO THE COOPERATIVE COMMONWEALTH OF HENRY GEORGE

Kenneth C. Wenzer

Sir Thomas More, founder of the secular tradition of utopianism, was a prodigious scholar and noted statesman of whom Desiderius Erasmus wrote that: "He will be . . . celebrated in the letters of all the learned . . . as a portent rather than a man."[1] More embodied the Renaissance quest of adventurous learning. It was a time of bold explorations and the first excitement of the printing press had not worn away. When in 1516, More's *Utopia* in Latin came out in Louvain, only two years passed since publication of *The Prince* by Niccolo Machiavelli, another primer for governance.

Utopia, according to Joyce O. Hertzler, "one of the earliest expressions in England of the consciousness of social wrong," lays the foundation for all subsequent visionaries recoiling from social injustice and seeking a perfected world order: It therefore stands as a monument and its name has lent itself to this tradition.[2] Francis Bacon's *New Atlantis*, James Harrington's *Oceana*, Thomas Campanella's *City of the Sun*, William Morris' *News from Nowhere*, Samuel Butler's *Erewhon*, and Henry George's *Progress and Poverty* are but only a few of the most noted examples. More has been regarded even as a precursor of communism: a Thomas More room in the Kremlin attests to that. More was an enemy to the martial ambitions of princes, the lust of gold, gross social and economic inequalities, crude penal codes, avariciousness of religious orders,

Henry George's Writings on the United Kingdom, Volume 20-B, pages 1–313.
Copyright © 2002 by Elsevier Science Ltd.
All rights of reproduction in any form reserved.
ISBN: 0-7623-0793-5

and onerous laws and taxation. But his humanistic wedding of pure religious values and ethics with the Greek emphasis on virtue and reason was the real triumph.

What sparked More's writing of *Utopia* was the contrast between the wealth of the privileged landowning few and the misery of the English multitude. The amassment of private wealth, especially in the monopolization of land, was in his opinion the main scourge of mankind. During his time the enclosure movement was growing, for the most part allowing for a greater concentration of much acreage, expelling tenants and even small owners so that sheep could graze in peace. Prices rose; wages sank. The landless were forced into the cities and into lives of crime.

More's declaration speaks to such conditions as have brought privation throughout time:

> for when every man draws to himself all that he can compass, by one title or another it must needs follow, that how plentiful soever a nation may be, yet a few dividing the wealth of it among themselves, the rest fall into indigence ... I am persuaded that till property is taken away there can be no equitable or just distribution of things, nor can the world be happily governed: for as long as that is maintained, the greatest and far best of mankind will be still oppressed with a load of cares and anxieties.[3]

In Utopia, all property is held in common – notably the land, the source of all wealth. Agriculture is therefore the single occupation all citizens hold in common. All evils will then vanish, especially the class structure and the money economy.

From 1642 to 1646 and from 1648 to 1649, Civil Wars rent the England More had wished to redeem from the appetite for land. The Stuart monarch Charles I by his arbitrary policies, particularly in raising taxes and his reintroducing Catholic practices, had angered many Anglicans as well as other Protestants. His Royalists were pitted against the Parliamentarians led by Oliver Cromwell and his Roundheads.

During this period the enclosure movement continued to displace rural folk. Poverty made for history's first social welfare law, the Poor Law of 1619. Meanwhile, a class of commercial entrepreneurs was buying land for sheer profit and did not adhere to the ancient "customs of the manor" that had governed relations with the tenants. Landownership, as a general rule, dictated partisanship during the Civil Wars, pitting the older landowning west, primarily Anglican and Royalist, against the mercantilist east, most usually Presbyterian and Roundhead in inclination. So divergent views on the pulpit and politics, superimposed on patterns of landownership, ripped England apart.

These like other eras of social and political unrest inflamed religious passions. Anabaptists, Ranters, Quakers, some of them preaching the Apocalypse, called for repentance and spiritual rebirth. The popular Levellers (from leveling the fences raised on the common lands), the first political party in existence, proposed a democratic platform, a society composed of small landowners, and other reforms. Among these political insurgents were the Diggers of 1649 and 1650, who defined as no group of men had done before the main source of all social, political, and economic problems – the unjust ownership of the land.

Prior to the Civil Wars peasant riots had already broken out over the enclosures, the oppressive rack rents,[4] loss of ancient rights, and poverty. Petitions from those days complain of fields left untilled and the flight to cities, victim to landlord greed. And there were attempts in this period of injustice to hasten the coming of a utopia into the earthly world. One of the most eloquent spokesmen was Gerrard Winstanley, a Digger. The earth and all its fruits belong to everyone, he proclaimed; from the control of the land come all inequalities and social problems.

> And let all men say what they will, so long as such are rulers as call the land theirs, upholding this particular property of mine and thine, the common people shall never have their liberty, nor the land be freed from trouble, oppressions and complainings; by reason thereof the Creator of all things is continually provoked.[5]

The Diggers believed that they had the right to dig into the soil anywhere; for labor, and labor alone, was the only honorable source of a decent living. People must work together in harmony and share the products according to a just need. Not only land, but all personal property, according to Winstanley and his followers, was to go. The result would be a classless and moneyless commonwealth. All secular and religious authority except that of Reason meant corruption.

But like countless other seekers of a regenerated world, the Diggers were doomed. According to one historian, the final outcome of the Civil War

> was not the enlargement of English freedom. What in fact it accomplished was to bring the land under the new capitalistic system. It favored the great owner: it made the balance of class power under which the process of enclosure was completed by a Parliament of landlords. In the Act of Settlement of 1662 it virtually revived serfdom by tying the laborer to his parish, and with uncanny prescience made ready a proletariat for the mills of the industrial revolution. The King, indeed, was beaten at Marston Moor and Naseby, but the vanquished party in the Civil War were the peasants who charged behind Fairfax and Cromwell.[6]

The enclosure movement, continuing into the eighteenth century, made for greater productivity and spurred many advances in husbandry and stock breeding on the larger farms dominated by an aristocracy. It also altered the social life of the village, forcing the small farmer into hired labor and devastating the peasantry. A population uprooted from the land moved to the cities for relief.

But a productive agriculture on the enclosed lands, along with a militant and successful financial merchant class, a unitary internal market economy, and a technological inquisitiveness most visible in the application of steam power fueled by coal, gave rise to an industrial revolution based on the exploitation of the land. These singular changes were to make England at the beginning of the nineteenth century the leading power on the globe and the hub of a world-wide economy.

During this period England was a turmoil of dissenting religious groups, among them Methodists, Moravians, Muggletonians, and Antinomians. Evangelism, revivalism, millenarianism; religious schisms and communitarian experiments proliferated. Seeking freedom of conscience in self-isolated communities of the regenerates, such conclaves bred a "slumbering radicalism," declare some historians, that bordered on anarchy. Buffeted by the industrial revolution, the enclosure movement, social dislocation and poverty, popular working class movements espousing egalitarian ideas arose while revolutionary ideas awakened among intellectuals, including many former ministers.

The established church, the monarchy, and the aristocracy, the long and settled foundations of England, were shaken by the French Revolution and the Napoleonic Wars. As craft skills of the artisans gave way to the standardization and dehumanization of the factories and the cities packed with migrants from the countryside, urban problems grew in extension if not in severity. The old aristocracies were being replaced by a new bourgeois elite, and ancient superstitions and codes of honor were giving way to materialism, but of a kind hopeful that everything could be made better. The old personal relationships, however harsh, built up over the ages between the tenant and the lord were ripped up and replaced by an impersonal monetary exchange. Spiritual were giving way to worldly hopes of a better future: technology, science, and politics were to build it.

In the nineteenth century, social and political change hastened the growth of different varieties of socialism together with a more sedate reformism only for people who had money and aspired to reach a social position equal to that of the aristocracy. Coming into the chambers of deputies were not farmers and workers but physicians, lawyers, merchants, and manufacturers, committed consciously or not to the doctrine of *laissez-faire* and its reduction of existence to cold, steady, methodical acquisitiveness. Monopolization of natural treasures and all its benefits grew apace with industry and commerce. All wealth is founded ultimately on land, and land remained the most prestigious of property, the mark of old manners and old family. But now its seekers were industrialists who, having turned land into steam and brick and iron, wished to retire as leisured country gentlemen.

In 1830 a growing prosperity fostered revolution in Europe and a reform impetus in Great Britain. Then in 1848 revolution rocked the continent and fueled the growth of different varieties of socialism espousing change in the name of the masses. A more sedate liberalism spoke to the triumphant middle classes; but it was a liberalism only for those who had money and aspired to reach a social position equal to that of the aristocracy. But the acquisition of land with all its benefits still was the most singular feature of wealth even with the growth of industry and commerce. Later in the century, advocates of free trade and inter-nationalism would contradict their economic doctrines by supporting national-ism, militarism, and crusading missions for the acquisition of colonies for prestige and new markets. The contradiction is only apparent. Today the forces of globalization are simultaneously supporters of political and military domina-tion of what during the Cold War became termed the Third World.

Ever since a lad Henry George loved the sea and its adventures. When fame embraced him in his adult life he would primarily visit English-speaking coun-tries, where his single-tax system was received most openly. In the British Isles, in fact, he was more famous than at home. Something about crossing the waters invigorated George's radicalism and contentiousness. Here is George looking far beyond his times:

> It seems startling to think of woman suffrage becoming an accomplished fact in Great Britain within two or three years. And yet this is one of the things which the present political situation may bring about . . . The idea certainly frightens many of the . . . politicians. With the present electorate they feel certain of regaining power as soon as an appeal to the country is made. But if the women are called in they fear that the new vote may swamp them.
>
> It is certainly more than possible that in the very next general election women may vote as well as men, and that when the next Parliament assembles representatives of the sex who up to this time have only been permitted to look down on the Commons through the bars of a cage may take their seats on the benches of the House.
>
> The bill is not likely to pass at this session, but it begins to look as though the British Empire were within measurable distance of a resolution that, with a woman on the throne and a majority of women in the electorate, would put the ultimate political power in the hands of women.
>
> That the women in their new found power might do some foolish things is probable. But it is certain that they cannot do more foolish things than the men have done. And though at first they may stand in the way of some reforms that are on the verge of accom-plishment, it is certain that they will ultimately call for larger reforms.[7]

Victoria, beginning in 1837, had been queen of the British people and from 1876 empress of India. Marriage to Albert of Saxe-Coburg produced nine children who were to wed with other royal houses. During Victoria's reign, Great Britain's commercial and industrial growth as well as colonialization; astute diplomacy, and a strong navy gave the island kingdom international supremacy.

This nation had passed laws – some of them from landed conservatives repelled at the crudeness of the liberal and industrial and commercial magnates – that were intended to ameliorate domestic problems, most noticeable among the dislocated and poor in the cities. Such measures reflected a compound of genuine humanity with a fear that in the absence of reform these unenfranchised lower classes, especially the workers, would rise up. But despite legislation that eroded the bases of laissez-faire, half of English land was owned in 1870 by only 7,400 people. Then there was the depression of thirty years that began in 1873, coupled with a series of bad harvests. Britain became more dependent on imported foods and was losing her industrial superiority to the United States and Germany. Even though there was industrial peace, since the standard of living was not upset too much, new unions wielded power and new visions of reordering society surfaced at this time. The groundwork was thus laid for Henry George's meteoritic rise to popularity, and he, in turn, influenced the prevailing spirit.[8]

Although Victoria provided a symbol of much that was dynamic and good in imperial Britain, Henry George would think otherwise. A letter of 1882 announces, "We'll topple Mr. British Crown before we are done."[9] An article, "The Queen's Jubilee," from *The Standard* contains some reasons for George's disgust.

> Victoria Guelph is . . . a greedy, grasping, narrow-minded, commonplace woman, who never did a useful thing in her life unless to serve the purpose of a legal fiction that might just as well have been served by a wax figure from Madame Tussaud's show . . .
>
> The English throne is the capstone of a social pyramid of many ranks, each of which is interested in abasing itself before those who are above, in order that it may in return enjoy the abasement of those who are below. And in inculcating this habit of servility, in accustoming the public mind to look upon the useless incumbent of a throne as a gracious benefactor, and in confounding ideas of duty and patriotism with personal devotion to a family, no class is so active and so efficient as the professed ministers of Christ. [T]his heathenish adoration of a human creature prevails in Great Britain [because] . . . Christ, according to the religion that is taught . . . is not the friend and deliverer of the poor, but the patron of the rich. He is a guardian of game preserves and mining royalties and city ground rents; a protector of the smugly respectable, who consider the honor paid to one's betters as honor paid to Him. Almighty God, the people are virtually taught, has so ordered this world that while a few roll in luxury the great mass of its people can only get a poor living by the hardest toil, and large numbers cannot even get that, but must live, if they live at all, on the crumbs that fall from rich men's tables; but He has considerately provided another world, in which things will be ordered more equally and to which such of the poor will be admitted as have in this life conducted themselves lowly and reverently toward their betters and not quarreled with the existing order.[10]

Among all the classes George despised, his contempt ran especially deep for the aristocracy, its wealth resting on land exploitation. "Everywhere, in all times, among all peoples," *Progress and Poverty* asserts, "the possession of land is the base of aristocracy, the foundation of great fortunes, the source of power. As said the Brahmins, ages ago – 'to whomsoever the soil at any time belongs, to them belong the fruits of it. White parasols and elephants mad with pride are the flowers of a grant of land.'"[11] George's letter of October 29, 1881 to Thomas Briggs declares,

> Surely the masses of the English people cannot understand the sort of government that they are maintaining here, and here that first principles of human liberty are being trodden under foot by an irrepressible dictatorship wielded in the interests of a panic stricken and maddened class.
>
> But out of this will come good. What is going on here makes it but the more evident that land monopoly has received its death wound.[12]

Outside his country and especially in the British Isles, where George was on the hard campaign trail from John O'Groats to Land's End, his rhetoric was the most impassioned. A speech entitled "Confiscation," presented in this collection gives a sense of it.

> This long, lingering dragging out of justice is bad on all sides. If you are going to do a thing do it quickly and be done with it. Now the trouble about compensation [for expropriation of the wealth from private landholding] is this: I really cannot see any way of compensating for an injustice without to a greater or less degree continuing that injustice . . .
>
> Private property in land. Why it is simply a form of slavery. [T]he man who owns the land . . . from which another human being must live is his master, and his master even to life or to death. Under the state of things which we see here, and . . . all over the civilized world . . . [where] we are imposing it . . . are there not slaves? Are not the working masses of these civilized countries slaves, just . . . [like] chattel slaves, only they do not know their particular master. When a man, without doing a thing, can draw from the earnings of the community the results of labor, when he can have a palace, and yachts, and horses, and hounds, and all the things that labor produce, is not the laborer necessarily robbed? Are not the fruits of his labor necessarily taken without any return to the man who gave the labor? Why in a paper last week, in which I read a column[-long] denunciation of myself and my proposals for "theft and confiscation" . . . there was an article headed "The White Slaves of England," in which it stated . . . that the condition of a large section of the English people was worse than that of any chattel slaves . . . [N]o southern slaveholder would have worked and kept his negroes as white men and women and children are worked and kept in this free England. I will take the annals of any system of chattel slavery, and for every horror that you produce, I will produce a double horror from the files of our papers.[13]

Although he would be an embarrassment to present-day "Georgists" and their support of private property in land, George was well received by his listeners for his ardent radicalism. A letter dated November 20, 1881, describes his reaction to the speech. "I had an immense audience, and had them wild with enthusiasm. It was all I could do to prevent being dragged around the streets at its conclusion."[14]

Many problems of interpretation regarding George's radicalism during his lifetime that have continued down to the present day stem from George himself. In a short but insightful chapter in *Henry George in the British Isles*, Elwood P. Lawrence succincly states the reasons for this dilemma.[15] George has been labelled not only as a land nationalizer but also as a socialist. George did speak on both platforms and showed how his views conformed to theirs. Yes, he was a land nationalizer, but rather than do it by a state takeover of the land, he would employ rent seizure. He looked forward to a cooperative society lightly ordered by a minimal state. If the single tax does not make for a decent reapportionment of wealth, then socialism or near anarchism will be the logical extension. The debate with H. M. Hyndman in the last section of this book illustrates this important point. The distinction between socialism and anarchism was more blurred at this time: a synonym for anarchism was libertarian socialism and for socialism itself, state socialism. Confiscation and a redistribution of wealth, George takes as a given. It follows that the land belongs to the entire people: a principle that would appeal to anyone on the left.

Under British rule absentee landowners rack-rented the Irish people; their economy was depressed; out of repression grew rebellion and a vision of independence. In face of poverty, and thanks to the introduction of the potato, the Irish population doubled in sixty years to eight millions in 1841. Primarily dependent on this food source for sustenance and at the mercy of landlords, the rural Irish lived precariously. The Great Famine, which began in 1845 and lasted for three years, immeasurably worsened life: 750,000 died and over a million emigrated. A recurrence of agricultural distress in the 1870s inflamed Irish nationalism and brought land war. The Act of Union of 1801 had brought Irish questions into British politics. Such leaders as Charles Stewart Parnell brought into the halls of Parliament the issue of Home Rule and land reform. Parnell also advised his countrymen to establish the first boycott. After some time in jail, he was released for political concessions. William E. Gladstone, the prime minister, took upon himself the mission of pacifying Ireland and conceded a number of reforms, including the Land Act of 1881, which legalized "fixity of tenure, fair rents, and free land sale." Gladstone's unsuccessful attempt to push two Home Rule bills brought down one of his ministries and splintered his Liberal Party. Ireland was left in a state of unrest until after World War I.

The Irish Land Question had put Henry George in the front ranks of American radicals, especially among the Irish. His denunciation of landlordism and its abuses as the cause of the Ireland's misery soon earned him fame there. First visiting the Emerald Isle in 1881 as representative of the *Irish World*, he became immediately embroiled in its politics. He even managed to get arrested twice as a suspicious person. The short stays in jail intensified his radicalism. "It does not seem to me," George wrote of the British presence, "that any fair-minded Englisman can visit Ireland, mix with people, and see how laws passed by an English Parliament are administered there, and how English power is used to bolster up a reckless and stupid class tyranny, without feeling indignation and shame."[16]

It was George's wife of Irish descent who brought him close to Ireland, but immediate family ties linked him to Scotland. In this land the nineteenth century witnessed a quickening of commerce and industry and also its share of misery. Land was being cleared by the aristocracy for its deer parks and sheep raising. Julia Bastian writes of this experience:

> Under the old Celtic tenures, the Klaan (children of the soil) had been the proprietors of the land they occupied. Whole counties, Sutherland for example, belonged to its inhabitants. Their chief was held to be their monarch and not a proprietor, and had no more right to expel them from their homes than the king across the border in England. But in 1807, the Duke of Sutherland ordered his agents to clear some 15,000 inhabitants from his land and to burn their homes. Year after year, more of his tenants were forced to abandon their snug farms and find new livelihoods elsewhere, or to seek a new life by emigrating to America. Such "clearances" continued with increasing severity and were not confined to Sutherlandshire. Other parts of Scotland witnessed the tearing apart of the ancient ties which for centuries had bound clansmen to their chiefs. In Buteshire, the Duke of Hamilton caused the land area rented by twenty-seven families to be converted into one farm, who were deprived of their homes and possessions. The same sorry story was reenacted in Argyllshire, Inverness-shire, Ross-shire, Perthshire and the Hebrides, to list but some of the affected areas. Everywhere men gave place to deer, so that Highland landlords might reap their golden harvests in the increase of their rent rolls from shooting and fishing rights.[17]

Many people had to flee their homes for the teeming cities that groaned under the onslaught of this exodus together with Irish immigration even though there was emigration to the New World. The potato famine had also taken its toll in Scotland. Another period of land clearances and evictions by rich landowners around mid-century depopulated entire areas and many crofts were abandoned. A virtual civil war ensued in the countryside. With bad harvests in the 1870s and other problems an agricultural depression set in. Local cottage industries declined, unable to compete with a growing international market.[18] This was the situation that George sought to address.

George's call for judicious reform appealed to many minds, throughout the United Kingdom burdened with land monopolization. Arthur J. Balfour, the Unionist Prime Minister, who declared to George's advantage: "As will be readily believed, I am no Socialist, but to compare the work . . . of such men as Mr. Henry George with that of such men, for instance, as Karl Marx, either in respect of its intellectual force, its consistency, its command of reasoning in general, or of economic reasoning in particular, seems to be absurd."[19] Winston Churchill wrote that the "unearned increments in land are not the only form of unearned profit ... but it is the principal form ... which is derived from processes ... positively detrimental to the general public." [20] Most important of all, however, was George's contribution to the Fabians and, through them, to the direction of British politics and society.

The Fabians condemned the ill effects of industrialism and urbanization and looked for a just reordering of proper relations in cities and industry under a new order espousing an evolutionary transition to socialism; Sidney and Beatrice Webb, along with George Bernard Shaw and other intellectuals, took more from Henry George and the English liberal Thomas Hill Green, than from Karl Marx. Although they did believe in the nationalization of industry and the control of production and distribution, they rejected material determinism and the class struggle, looking instead to the ballot box, legislation, education, and other legal methods. Socialism would be the natural development of tendencies already inherent in the capitalist system. The Fabians thus named themselves after a Roman general, Fabius Cunctator, who had fought Hannibal in the Second Punic War by delaying tactics, rather than outright battle. The Fabians thought of themselves as purifying classical liberalism by empowering the state in social and economic relations. Governmental control over production, distribution, exchange, and society, which was the Fabian version of socialism, would bring greater political and economic freedoms, and the individual would be more at liberty to develop his potential. The Fabians presented the state as an agent of an actively advanced good, as opposed to an instrument merely for the elimination of obstacles to the good. George Sabine contends that the goal was based

> on the conviction ... that liberty is impossible without a reasonable degree of security and that in consequence social security and stability are as much an object of political policy as liberty ... that a planned society can be a "far more free society" than a competitive one, because it can "offer those who work in it the sense, on the one hand, of continuous opportunity for the expression of capacity, and the power, on the other, to share fully in the making of the rules under which they work."[21]

Shaw, the Fabian playwright and critic, was an advocate of vegetarianism, antivivisectionism, and other radical causes. His imagination was roused by Henry George's gift of oratory and he became a socialist. He declared that it was George who had awakened his social conscience. His "attention was first drawn to political economy as the science of social salvation by Henry George's eloquence, and by his *Progress and Poverty*, which ... had more to do with the Socialist revival of that period in England than any other book." Later on, Shaw wrote of "our debt to Henry George. If we [Fabians] outgrew *Progress and Poverty* in many respects, so did George himself too, but nobody ever has got away, or ever will get away, from the truths that were the center of his propaganda . . ."[22] The first historian of the Fabians, Edward R. Pease, observes:

> to George belongs the extraordinary merit of recognizing the right way of social salvation . . . political institutions could be molded to suit the will of the electorate; he believed that the majority desired to seek their own well-being and this could not fail to be also the well-being of the community as a whole. From Henry George I think it may be taken that the early Fabians learned to associate the new gospel with the old political method.[23]

"But George's true influence," the economist J. A. Hobson writes,

> is not rightly measured by the small following of theorists who impute of landlords their supreme power of monopoly . . . The spirit of humanitarian and religious appeal which suffuses *Progress and Poverty* wrought powerfully upon a large section of what I may call typical English moralists . . .
>
> No doubt it is easy to impute excessive influence to the mouthpiece of a rising popular sentiment. George, like other prophets, cooperated with the "spirit of the age." But after this first allowance has been made, Henry George may be considered to have exercised a more directly powerful formative and educative influence over English radicalism of the last fifteen years than any other man.[24]

Central to George's contribution to the thought of the Fabians was his belief that private property in land was theft. George's taxation of land values was therefore adopted as part of their platform. But, unlike George, they believed that industrial and financial capital should be common property. An essential element in their program was the emancipation of land and capital from private ownership (with compensation) – they were to be owned by the community and managed by the state, especially at the local level, the most democratic.

Shaw believed that that single-tax proposal should absorb the economic rent not only of the land but of the entire country; then the government would be in control of all capital: it is to result in state capitalism. Sabine describes Fabian economics as "for the most part not Marxian but an extension of the theory of economic rent to the accumulation of capital, on lines already suggested by Henry George. Fabian policy was based on the justice and the desirability of recapturing unearned increment for social purposes."[25]

In the Fabian Manifesto (Tract no. 1) George's influence is quite obvious:

> That a life interest in the Land and Capital of the nation is the birthright of every individual born within its confines and that access to this birthright should not depend upon the will of any private person other than the person seeking it.
>
> That the most striking result of our present system of farming out the national Land and Capital to private persons has been the division of Society into hostile classes, with large appetites and no dinners at one extreme and large dinners and no appetites at the other.
>
> That the practice of entrusting the Land of the nation to private persons in the hope that they will make the best of it has been discredited by the consistency with which they have made the worst of it; and that Nationalization of the Land in some form is a public duty.[26]

"The Basis of the Fabian Society," used until 1919, calls for

> the reorganization of Society by the emancipation of Land and Industrial Capital from individual and class ownership, and the vesting of them in the community for the general benefit ... The Society accordingly works for the extinction of private property in Land and of the consequent individual appropriation, in the form of Rent, of the price paid for permission to use the earth, as well as for the advantages of superior soils and sites.[27]

The Fabians, writes J. S. Schapiro, provided a viable outlet for radical social ideas without dogmatism that could be readily adopted by a country.[28] They were important in the formation of the Labour Party and the trade union movement at the beginning of the twentieth century. Both the Labour Party and the Liberal Party adopted some form of land-value taxation at one time or another. The high watermark of Georgist influence in Great Britain occurred in 1909 when land-value taxation was incorporated in what became known as the "People's Budget." In 1918 the Fabians adopted Sidney Webb's progam. Called "Labour and the New Social Order," it demanded a national minimum wage, nationalization of industry, radical changes in national finance, and the use of surplus wealth for the nation. The Fabians also served in various ministries when the Labour Party came to power in 1923, 1929, and 1945. The government formed by the Fabian Prime Minister Clement Attlee in 1945 was their great triumph. It was during this period that the modern welfare state came to fruition. Socialized medicine, guaranteed subsistence, insurance, and the extension of public ownership of the coal mines, the railroads, utilities, and the Bank of England were among the many reforms they carried out. To a large degree, Fabian ideas brought about not only the welfare state but liberalism in the United States as well as Great Britain. Also to the credit of the Fabians is the establishment of the London School of Economics and the launching of the respected *New Statesman*.

For all practical purposes, the Great Britain of today is the grandchild of George, though the child of the Fabians. Many of her major industries and utilities are controlled by the state. Many public services are likewise dispensed by the government for the benefit of the community at large.

George would have applauded the use of ground rent for society, but he would have disapproved to extend it to other spheres of the economy save in the event that the single tax should not succeed in a fundamental distribution of wealth and power. He feared that a major realignment of property, at least in the absence of a prior trial of this more confined scheme, would eventually stifle free enterprise. Is it possible that one reason that Britain lags behind some other economies, especially in Europe, is that the goverment levies such high taxes? Has the state arrogated so much power as to interfere with the workings of society and the lives of individuals? *Progress and Poverty*, imagining a minimal state within a cooperative rather than a competitive society, looks for

> the ideal of Jeffersonian democracy, the promised land of Herbert Spencer, the abolition of government. But of government only as a directing and repressive power. It would at the same time, and in the same degree, become possible for it to realize the dream of socialism. All this simplification and abrogation of the present functions of government would make possible the assumption of certain other functions which are now pressing for recognition. Government could take upon itself the transmission of messages by telegraph, as well as by mail; of building and operating railroads, as well as of opening and maintaining common roads. With present functions so simplified and reduced, functions such as these could be assumed without danger or strain, and would be under the supervision of public attention, which is now distracted. There would be a great and increasing surplus revenue from the taxation of land values, for material progress, which would go on with greatly accelerated rapidity, would tend constantly to increase rent. This revenue arising from the common property could be applied to the common benefit, as were the revenues of Sparta. We might establish public tables – they would be unnecessary; but we could establish public baths, museums, libraries, gardens, lecture rooms, music and dancing halls, theaters, universities, technical schools, shooting galleries, playgrounds, gymnasiums, etc. Heat, light, and motive power, as well as water, might be conducted through our streets at public expense; our roads lined with fruit trees; discoverers and inventors rewarded, scientific investigations supported; and in a thousand ways the public revenues made to foster efforts for the public benefit. We should reach the ideal of the socialist, but not through government repression. Government would change its character, and would become the administration of a great cooperative society. It would become merely the agency by which the common property was administered for the common benefit.[29]

Throughout the course of his writings George indirectly addresses a particular dilemma that comes with cooperation. It is this: if working together is at the expense of individual initiative, it is not cooperation, for the individuals are not freely participating; and if initiative is at the expense of cooperation, it cannot be a creative initiative, drawing on the rich store of ideas and activities that society offers. George's solution is elegant. Hold in common the land and its resources: this will remind individuals that ultimately they are beneficiaries of a shared heritage from God. But let individuals freely work upon the land, its soil, water and ores, whatever their brain and muscle driven by their imagination can achieve. This must be effected in a spirit of morality and mutual concern, and the community ownership of land will evoke that spirit. Thus a balance of cooperation with competition is struck between man and man, and between man and God.

At the basis of all George's inquiries, then, is the relationship of people to the land. Human beings cannot survive without utilizing the bounties of the earth, whether it is agricultural pursuits for sustenance or various extractive processes for a more abundant life. People are to organize themselves for a more efficient working on the land, not only socially through cooperative endeavor but through progress in technology.

Russian thinkers, especially Lev Tolstoi, a self-avowed Georgist, drew their inspiration from the suffering peasants of Russia.[30] For the Russian people to treat as their own the land and its products would not mean only the conquest of poverty and the achievement of freedom and dignity in work. It meant over-turning a system in which the serfs belonged to the land and the land belonged to the gentry, before 1861 as serf owners and afterwards as legally and econom-ically bound to an onerous system that continuously threatened the means of subsistence. Among all peoples, Russians were most sharply aware of the evil that monopolization of the land entails. In the United States Henry George similarly counterpoised land monopolizaton and the ethos of cooperation.

"What I have done in this book," announces *Progress and Poverty*,

> if I have correctly solved the great problem I have sought to investigate, is to unite the truth perceived by the school of [Adam] Smith and [David] Ricardo to the truth perceived by the schools of [Pierre-Joseph] Proudhon and [Ferdinand] Lasalle; to show that *laissez faire* (in its full meaning) opens the way to a realization of the noble dreams of socialism; to identify social law with moral law, and to disprove ideas which in the minds of many cloud grand and elevating perceptions.[31]

The socialist ideal, George declares later in *Progress and Poverty*,

> is grand and noble; and it is, I am convinced, possible of realization; but such a state of society cannot be manufactured – it must grow. Society is an organism, not a machine. It can live only by the individual life of its parts. And in the free and natural development of all the parts will be secured the harmony of the whole. All that is necessary to social regen-eration is included in the motto of those Russian patriots sometimes called Nihilists – "Land and Liberty."[32]

Individual and social needs, furthermore, would be more wisely disbursed by a benign government, or "cooperative association – society,"[33] for "the best government is that which governs least."[34]

Many of George's putative followers have spurned the cooperative goals of his thinking, preferring individualism, right-wing religious fanaticism, petty monetary reforms, specious libertarianism, and acquisition of personal wealth freed from any tax outside of land; gross misunderstandings of George's inten-tions.[35]

NOTES

1. Desiderius Erasmus to Ulrich von Hutten, July 23, 1517. In: M. Campbell (Ed.), *The Utopia of Sir Thomas More*. The Classics Club (Roslyn, NY: Walter J. Black, 1947), p. 204.

2. Joyce O. Hertzler, *The History of Utopian Thought* (New York: The Macmillan Co., 1926), p. 130.

3. As quoted in Ibid., pp. 132–133.

4. Rack rent has two definitions: Rent equal to or nearly equal to the full annual rent of a property or, in this instance, the practice of exacting the highest possible rent.

5. George Woodcock, *Anarchism: A History of Libertarian Ideas and Movements.* (Cleveland, OH: The World Publishing Co., Meridian Books, 1967), p. 46.

6. Henry Brailsford, *The Levellers and the English Revolutio,* C. Hill (Ed.). (Stanford, CA: Stanford University Press, 1961), p. 452.

7. Henry George, "Henry George in England," *The Standard,* Apr. 6, 1889, In: Kenneth C. Wenzer (Ed.), *An Anthology of Henry George's Thought* (Vol. I) *The Henry George Centennial Trilogy* (Rochester, NY: University of Rochester Press, 1997), pp. 202–203.

8. Walter L. Arnstein, *Britain Yesterday and Today.* (Lexington, MA: D.C. Heath & Co., 1971), pp. 100–113 and 160–172.

9. Henry George, Letter to Patrick Ford, June 8, 1882, In: Wenzer, *An Anthology of Henry George's Thought,* p. 135.

10. Henry George, *The Standard,* June 25, 1889, in Ibid., pp. 133–134.

11. Henry George, *Progress and Poverty: An Inquiry Into the Cause of Industrial Depressions and of Increase of Want with Increase of Weatlh . . . The Remedy.* (1879; New York: Robert Schalkenbach Foundation, 1960), p. 296.

12. Henry George to Thomas Briggs, Oct. 29, 1881. In: Wenzer (Ed.), *An Anthology of Henry George's Thought,* p. 135.

13. Henry George, "Lecture in Birmingham, England," June 23, 1884, in Ibid., 88 and 95–96. See pages 313–333 for the complete text.

14. Henry George to Dr. [Edward Taylor?], Nov. 20, 1881, reel No. 2, Henry George Papers, Rare Books and Manuscript Division of the New York Public Library; Astor, Lenox, and Tilden Foundation. Hereinafter this citation will be abbreviated as HGP.

15. See the chapter entitled "Toward the Single Tax." In: E. P. Lawrence: *Henry George in the British Isles.* (East Lansing: Michigan State University Press, 1957), pp. 51–60.

16. As quoted in Ibid., p. 23.

17. Julia Bastian, Patrick Edward Dove. In: Kenneth C. Wenzer (Ed.), *An Anthology of Single Land Tax Thought* (Vol. III) *The Henry George Centennial Trilogy* (Rochester, NY: University of Rochester Press, 1997), p. 159.

18. See R.L. Mackie, *A Short History of Scotland.* (Edinburgh: Oliver & Boyd, 1962), pp. 266–289.

19. As quoted in Edward R. Pease, *The History of the Fabian Society* (London: The Fabian Society, 1925), p. 45.

20. As quoted in Vernon I. Saunders, In Defense of the Two-Rate Property Tax. In: Kenneth C. Wenzer (Ed.), *Land-Value Taxation: The Equitable and Efficient Source of Public Finance.* (Armonk, NY: M.E. Sharpe, 1999), p 275.

21. George H. Sabine, *A History of Political Theory.* (New York: Holt, Rinehart & Winston, 1961), p. 740.

22. As quoted in Anne Fremantle, *This Little Band of Prophets: The British Fabians.* (New York: Mentor Books, 1960), p. 31.

23. Pease, *The History of the Fabian Society,* pp. 20–21.

24. As quoted in Charles A. Barker, *Henry George.* (New York: Robert Schalkenbach Foundation, 1991), pp. 415–416.

25. Sabine, *A History of Political Theory,* p. 740.

26. Pease, *The History of the Fabian Society,* pp. 41–42.

27. Ibid., pp. 259–260.

28. J. Salwyn Schapiro, *Movements of Social Dissent in Modern Europe*. (Princeton, NJ: D. Van Nostrand Co., 1962), p.75.

29. George, *Progress and Poverty*, pp. 455–457.

30. See Kenneth C. Wenzer, *An Anthology of Tolstoy's Spiritual Economics* (Vol. II) *The Henry George Centennial Anthology* (Rochester, NY: University of Rochester Press, 1997).

31. George, *Progress and Poverty*, p. xvi.

32. Ibid., p. 321.

33. Henry George, Land Question. In: *Land Question and Related Writings*. (1881; New York: Robert Schalkenbach Foundation, 1982), pp. 83–84.

34. Henry George, American Republic. In: Kenneth C. Wenzer (Ed.), *Our Land and Land Policy*. (East Lansing: Michigan State University Press, 1999), pp. 119 and George, *Progress and Poverty*, pp. 454–457.

35. For a discussion of how these distortions came about the reader is encouraged to consult: Kenneth C. Wenzer, "The Degeneration of the Georgist Movement: From a Philosophy of Freedom to a Nickel and Dime Scramble". In: K. C. Wenzer & T. R. West, *The Forgotten Legacy of Henry George*. (Waterbury, CT: Emancipation Press, 2000), pp. 46–91.

SECTION I: IRELAND

TWO ARTICLES FROM THE *IRISH TIMES*
(November 15, 1881)[1]

Mr. Henry George, all the way from San Francisco, lectured last night in the Rotunda on Land and Labor, and exceeded even our own best eloquence in the volume and vastness of his phraseology. The Americans at home will hear from him with some surprise that they are to have a Land Question on the Irish model which will "lead them upward and onwards" until the "revolution sweeps the civilized world." A new broom this with a vengeance!

To illustrate his discovery of quite a novel principle, Mr. George referred to an Irish monopoly, not of land, but of house property; but he might have found a better example of monopolies in his own paradise, where the proprietor of the Great Bonanza far overtops the riches of the Pembroke Estate.[2] On another of his points a word may be said, that "not in the Southern States of America could there at any time be found Negroes that were housed and fed as so-called free people are in New York." Mr. George's views are very simple: starting with the assumption that everyone has the right to live, he easily concludes that everyone should be a landed proprietor. The baby of a duchess had no right to existence greater than the baby of any other woman, and did it not follow that their agrarian rights are the same? Mr. George has come over to teach us all about it. The Land Act has been thrust down the throats of the Irish people.[3]

But there has been no land for the laborers or their infants. "This is the keynote of the whole thing." The man to be considered was the man who had nothing but his labor: his is the true title to the soil. Spades are trumps. But Mr. George will not have peasant proprietors at all. They are "played out" along with the aristocracy. "What," he said, "did they want with peasants of any kind?" Evidently he is unfamiliar with the classic couplet: "A bold peasantry, their country's pride, when once destroyed can never be supplied."[4]

Land and Labor

A lecture was last night delivered on "Land and Labor" in the Round Room, Rotunda, by Mr. Henry George of the United States. There was an extremely large attendance. In the right hand corner of the gallery the St. Nicholas of Myra's Catholic Total Abstinence Band was stationed, and from the railings of the gallery there waved a large green banner emblazoned with a crownless harp. At the opposite end of the gallery the flag of the United States was suspended. The wall at the back of the platform was ornamented by an American and an Irish flag, hung side by side, the foldings, which touched, being looped together by the figure of a harp. Two similar harps were hung at the outer sides of the flags. The proceedings, which commenced shortly after eight o'clock, were characterized throughout by the greatest enthusiasm, the entire audience frequently rising waving their hats, and cheering loudly, while the flags in the gallery were waved in concert . . .[5]

Mr. Henry George, who was received with applause, said that for the first time in his life he rose to speak before an audience in a country where everyone held his liberty at the scratch of a pen. He did not want to be understood as saying anything against the Government. It was a very remarkable Government, and he had no doubt that it was a very good Government of its kind (laughter); of a kind rapidly approaching what was found in the dominions of the Czar. Contrasting the free institutions of the United States with those of Ireland, he said that in America it was a disgrace to a man to have been in jail, but in Ireland to have been imprisoned seemed to be an honor. (Cheers.) Between the United States and Ireland there was a very close connection, and each country exercised a strong influence on the other. America was democratizing Ireland (cheers) and the Irish land agitation would have a vast effect for good upon America. (Applause.) The Irish Land Question was making Americans think of an American Land Question, and leading them on and leading them upward, and if the people of Ireland carried on that movement, as he believed they would carry it on (cheers) until fully and truly the land of Ireland was the property of the people of Ireland (cheers); they would have led the van in a revolution that was destined to sweep the civilized world. (Loud cheers.)

They had come there that night, not merely to hear him, but to contribute their mite to the Society for the Protection and Aid of Political Prisoners: (loud cheers) a society which had among its active members Miss Anna Parnell (prolonged cheers) and whose president was an illustrious Englishwoman, Miss Helen Taylor. And who were the prisoners they intended to aid? Among them was the one-armed hero and patriot, Michael Davitt (cheers) and the real leader of nine-tenths of the Irish people, Charles Stewart Parnell.[6] (Loud cheers.)

Referring to his subject of his lecture, "Land and Labor," Mr. George said land was the passive factor, labor was the active factor. Without land none could live, and the man who absolutely owned the land on which his fellow men were dependent for their living was their master. Let a class own the land of a country, and that class would rule the rest of the people of that country, and the great mass of the people would have to toil and strive and die for them. All history, and especially the history of Ireland showed that.

The people of Ireland had begun an agitation against the grosser forms of that institution, but they must go further than the great majority of them had yet dreamed of going. In Dublin itself they had an instance of the evils of the system. The Earl of Pembroke annually drew an income of £70,000[7] from the land he possessed in that city, and that revenue necessarily came from the labor of the people. That sum represented the earnings of 14,000 laborers, and using other words the Earl of Pembroke might just as well have 14,000 slaves among the population. It made no difference whether a man was a chattel slave or whether his earnings were taken from him in some other way so long as he must work for another, and he (Mr. George) did not believe that in the Southern States of America could there at anytime have been found Negroes that were housed and fed as so-called free people were in Ireland. (Applause.)

Did it not follow from the equal right of every human being to live, seeing that life was dependent on land, that there necessarily existed an equal right to land? (Applause.) Did it not follow that the land of every country belonged by inheritance to the whole people of that country? (Applause.) Would anyone say that the baby of a duchess had a better right to live than the baby of any other woman? Then did it not necessarily follow that their rights to the land were and must be the same? (Applause.) They could never settle the Land Question until they came down to the firm basis of natural right and natural justice. (Cheers.)

The Land Act which had been thrust down the throats of the Irish people would not give satisfaction to the people, and even suppose it did reduce some of the rents. It would be but of slight good to the tenant farmers, and would give no help to the laborers. (Applause.) And that was the keynote of the whole thing. Until the condition of the laborer was raised the condition of other classes could never be raised. The man to be considered was the man who had nothing but his labor.

Let rents be reduced as much as they pleased, and this only would be accomplished. The new man who took the farm would pay less rent than was charged before, but would pay more for the tenant right, and the laborer, as the result, would be profited by the change very little, if he were profited at all. Lord Salisbury (hisses) had said in a recent speech that the difference between Mr. Gladstone[8] (groans) and Mr. Parnell (cheers) was that one professed to reduce rents 30%, and the other 50%, and that it was for the 20% difference that one put the other in prison. (Cheers and laughter.)

There really was a great deal of truth in that saying. The Land League had not gone very much further than Mr. Gladstone.[9] They thought it in Ireland very radical to demand a peasant proprietary. But what did they want with peasant proprietary? What did they want with peasants of any kind? What they wanted was not peasants on one side and landowners on the other – what they wanted was free and equal and independent citizens. (Cheers.) Peasant proprietary and aristocracy were played out. (Cheers.) Democracy was the hope of the people – not merely political democracy, but a social democracy as well

in which the equal right of every citizen was fully recognized, his equal right
to obtain the full value of his labor, his equal right to his native soil, and to
all natural opportunities. Those Tories[10] who said Mr. Gladstone was a
Communist were right. The land belonged either to the landlords or to the
people. If it belonged to the landlords then any interference whatever with their
ownership was wrong. ("Hear, hear.") If the land belonged to the people there
should not have been any halfway work about it. (Applause.) It was time the
present system had been done away with, and he believed the revolution had
already commenced.[11]

TRAVELS IN THE WEST OF IRELAND
(Probably sometime in 1882)[12]

Ireland is a beautiful country even when it rains, for the eye everywhere rests
on living green, which gives to the aspects of Nature a wonderfully soft and
finished appearance, but when the sun does shine, as it did on the afternoon
[I] left Galway, [13] lighting up[14] the landscape[15] and tingeing with gold the green,
it is a lovely[16] country indeed.

The road was lively too with the country people coming from market.[17]
Some could boast a donkey cart, into which frequently two or three people
would be piled, but the majority were women, in the costume of the country
– short red petticoats and bare legs, who carried on their backs baskets
containing the returns for the poultry, eggs, etc., they had taken to market,
which in most cases seem to be wool, which they spin by hand and then weave
into the coarse cloth which forms the usual clothing or the handknit socks which
again they take to market. Whatever these people may be they are certainly not
lazy. Many of them trudge for twenty miles or more to market with produce
which cannot bring more than a few shillings, walking in many cases all night,
and then after having disposed of their little stock and laid in their little supplies
walk back again.[18]

Except the people returning from market nothing else was visible on the
roads, save a car carrying passengers which we had noticed soon after leaving
Galway and which steadily followed us at a little distance. I asked our driver
to quicken his pace, and the car behind immediately did the same; then to go
slow, and the car behind slowed down too. We walked and they walked, and
when we started into a quick pace again the car behind us started too. Then
we came to a full stop as though to admire the landscape with the Aroon Islands
looming there in the foreground, and the car behind us stopped too. And so we
went along, but do what we might the car behind us insisted on keeping its
place.

Finally turning suddenly[19] as though to stop at[20] a little cabin we had passed we came face-to-face with our follower who had stopped his car the moment he saw our intention. He was evidently not a detective, but a constable in plain clothes, and clearly asking of his [ ], for he blushed like a girl, and then grinned all over his face as we talked to him. He had evidently been detailed to follow us to the next police barracks for this is a country in which there are no telegraph wires. And so having looked into the miserable little cabin, which was however [ ] as good as the average, got a drink of water and put something into the hands of the half-naked little children who crowded to the door to look at us. Then on we went, our shadow keeping us in sight.

We passed the home place of Mr. Blake, the agent who was killed so tragically near [ ], and in a little enclosure near by his tomb. Whatever may have been the good qualities of the man, one does not hear them from the peasantry. He was hard, grasping, and tyrannous by common repute, and as he rests by the roadside I fear that it is more curses than blessings [that] the marble above him provokes from passersby.

Spiddle is a poor little fishing village, consisting mainly of a couple of rows of squalid cabins. There is however a neat little church, but the curate who ministers in it was engaged in having confessions, and we did not see him. There is also at Spiddle an institution maintained by some sort of a Protestant proselytizing society, which sends forth Bible readers and gathers in destitute children to bring them up as Protestants. The children, a group of whom were playing in the grounds while a Bible reader was resting himself in the sun against a fence, were well-clad and chubby, but they are regarded by the people with feelings of pity and horror, and any sacrifice will be undergone by parents or relatives rather than permit a child to enter an institution where as it seems to them the body is cared for at the expense of the soul.

That there is a police barracks in Spiddle goes without saying, for every little hamlet in Ireland has its police barracks, and it is the best building in the place will also be understood by anyone who has traveled in this part of the country. We drove up at the door of what here goes for a shop, and our shadow stopping at the barracks a few yards from it, and joining the group of policemen who were engaged in watching us. We did not bother ourselves with their [ ] but finding that no horse could be had, as the four horses to be had had gone to Galway[21] for the market, and would not return till late in the evening, and our driver disclaiming that his horse could not go further until he had had a long rest, we set off to have a swim, picking up a bright lad who could speak English though with some hesitation and something of a foreign accent rather than a brogue to show us a good place.

The lad seemed to me eminently bright and intelligent, though his raiment was little better than a picturesque assortment of rags. His father,[22] an old man whom we saw sitting by the roadside had been blinded some years before by an accident, was dependent he said upon: "What the good God sent;" he himself worked when he could, but it was not often he could get work to hire. He could neither write nor read, [but] a bright lad; with us one can tell what latent mental power was growing up. I speak of him only because he is a type of very many.

We went down on a massive[23] breakwater and the Government is constructing here as a harbor of [ ] for fishing craft. We went off the stone steps on the inside of the breakwater and had a most enjoyable swim in the clear, warm water. But before we went in our guide told us that he thought he could get us a horse and cart to[24] go in with, as a man who was draining sand for the breakwater had a cart to which he could put his horse after his day's work and the breakwater was done. We suggested that a horse who had been working in a sand cart all day was not a[25] very desirable animal to take a cart ride at night; but the boy said that this was a good horse and a run after his work would be nothing to him, for the man gave him plenty to eat. So we told him to go ahead, and when he came out, and before we had dressed, as six o'clock came, we saw a man take his horse out of a sand cart, and leaping on his back gallop off at a pace that could have won many a stretch, and before we got back to the village again, he was there with a cart for us.

ENGLAND AND IRELAND: AN AMERICAN VIEW
(*The Fortnightly Review*, 1882)[26]

The area of Ireland is only 32,000 square miles; its population somewhat over five millions. Yet in time of profound peace its Government requires some 15,000 military constables and 40,000 picked troops, to say nothing of the navy constantly encircling its shores. And whatever else may be the marks of bad government are to be seen in Ireland today: a declining country, a discontented people; the resort to all the powers of despotism on the one side, contempt of authority and defiance of law on the other; arbitrary arrests, vindictive punishments, searches, seizures, proclamations and suppressions, the garrisoning of hamlets, the patrolling of roads, burnings, maimings, and assassinations – at once the white terror and the red terror.[27] Surely it is time thoughtful Englishmen began to ask themselves what it is they are trying to do in Ireland.

The common belief among Englishmen seems to be that they are trying to do a righteous and benevolent thing – to keep order among a turbulent people while redressing real grievances; and coercion is regarded as the necessary incident to a kindly act – the tying of a patient mad with pain in order that an operation for his relief may be performed. But so far from making coercion less hateful to Irishmen, this sharpens its sting. To be coerced is bad, but to be coerced upon the assumption that coercion is good for you is worse still. Nothing could be more irritating to a high-spirited people than the assumption of superiority that blends with so much that Englishmen intend for the expression of kindly feeling towards Ireland, and it naturally produces that indignant revulsion which Englishmen often take for Irish gratitude.

The assumption of race differences that do exist is, in fact, responsible for much misunderstanding. The belief that Ireland is discontented and turbulent because it is the nature of the Irish race to be discontented and turbulent, stops further inquiry into the causes of discontent; the notion that the restiveness of Irishmen under English rule is the restiveness of a lower civilization under the impact of a higher, suffices to prevent any examination of the character of that rule.

The majority of Englishmen do not begin to realize the badness of the government they maintain in Ireland; still less have they the remotest idea that the people of Ireland may have passed the point when even such a measure of self-government as prevails in England could satisfy them. In fact, nothing better shows why one people should never attempt to govern another people than the utter ignorance of Ireland that prevails in England.

But the Government of Ireland is not the government of one people by another people. It is worse. It is English force that is holding Ireland; but it is a small privileged class who, by the aid of this force, are the real governors of Ireland.

Here is the real reason that, after the lapse of centuries of political connection, Ireland has never been really incorporated with the British nation, but yet remains, in greater part, a conquered province, held by sheer force, and, given but the chance, as ready to rebel as ever.[28]

The Normans conquered Saxon England.[29] They were ravenous and brutal. They despised the people they had conquered, and were hated in return. Yet after a time the two peoples blended, and formed the English nation.

But supposing Normandy had been as much greater and stronger than England, as England is greater and stronger than Ireland, and that the Conqueror and his successors had remained in Normandy, looking upon England merely as a conquered and barbarous province, hardly to be visited once in a reign; looking upon those among whom the English lands had been parcelled as his civil garrison, just as the Norman soldiers maintained there to support them were his military garrison: Supposing, too, that the Reformation[30] had affected one country without affecting the other, and that advantage had been taken or religious differences to mark more clearly the gulf between conquerors and conquered, and to increase the power of the one, and intensify the degradation of the other by atrocious penal laws – What would have been the result? The result would have been that, though as in Ireland one language might have supplanted the other; and the blood of conqueror and conquered have become thoroughly mixed; and, after a time, the penal laws have been relaxed or repealed – A ruling class would have formed, which relying upon Norman strength to secure its dominance, and engrossing all power and emolument, would be clearly marked off from the body of the people. Under such circumstances England would today be as restless and as turbulent as Ireland, and the masses of the English people would hate the Norman union as bitterly as the people of Ireland hate the English union.

Irish landlords have been sufficiently characterized by English writers. The name, wherever the English language is spoken, has become synonymous with recklessness, cruelty, and extortion. Yet it is by and for this class and their dependents that Ireland has been and is governed. England had not governed Ireland, does not govern Ireland; England but supplies the force and bears the shame.

The history of English dominance in Ireland is a history of misgovernment blind and cruel. This Englishmen readily admit. What they largely fail to see is that, irrespective of bitter traditions, the realities of the present are enough to make Ireland restive. Yet I think no right-minded Englishman can go to Ireland and mix with the people (not exclusively with the landlord and official class, as most Englishmen who go there do) without feeling that only a race of slaves could be quiet under the Government maintained in Ireland – without what was left of his English pride consoling itself with the belief that, were the Irish people English, Ireland would be ten times as turbulent.

The masses of the Irish people have no more control over the Government under which they live than they have over the process of the suns. The suffrage is restricted in England, but, grouping borough and county franchise, it is much more restricted in Ireland. And to those who have the franchise it amounts to little more than a means of occasionally showing their feelings or making a protest. This is not merely so at present, when anyone chosen by the people may be locked up by the Government; it is so irrespective of the Coercion Act.[31] The nonpayment of members, much as it tells against the proper representation of the English masses, tells even more powerfully in Ireland. For while an Irish member must travel farther and submit to a greater interruption of his ordinary business than an Englishman, the real political division in Ireland is more nearly a division between those who have means and those who have not, and the popular party in Ireland can find proper representatives only with extreme difficulty.

But representation in Parliament, whether better or worse, does not give the Irish people control of their own affairs; for the Imperial Parliament is not, like the American Congress, a general legislature having power only to make general laws, applying alike to all parts of the country. It makes local laws as well: and in Ireland not a gaspipe can be laid or mile of railroad built without permission of the Imperial Parliament. And as the Irish members are in the minority, and that section of them most in sympathy with the masses of their people in a woeful and detested minority, it is really the English and Scotch members who make laws for Ireland that they do not make for their own kingdoms. I am told that in regard to the Land Bill of the last session – a measure of the first importance, relating exclusively to Ireland – the Irish members were not even consulted.[32] So little control over their own affairs does their representation in Parliament give the people of Ireland that they do not value it, save as a means of making a protest or gaining a concession by annoying the representatives of the rest of the country.

Over the branches of government the Irish people have, if that be possible, still less control. All judicial officers, from the mere honorary magistrate to the highest judge, are appointed mediately or immediately by the Government. This is substantially as it is in England; but in Ireland it means something much worse than in England, as the governing class is sharply marked off from the rest of the people, and between them are class animosities. The High Sheriffs of Counties are appointed in the same way, and they practically have the appointment of the Grand Juries, which, in addition to the presentment of indictments, assess local rates and make grants in compensation for injuries to property. The schools are under control of one central board, appointed by [the] Government; the prisons of another, and so on.

The ratepayers elect some members of the Local Poor Law Guardians;[33] but this is little more than an empty privilege, for the additional votes given on account of rate payments accrue to the benefit of the landlords, so that some landlords have thirteen, or even fifteen votes. But in addition to this advantage given the landlords as to the elective guardians, the returning-officer names from the magistrates (all landlords) highest rated, a number of *ex-officio* guardians equal to those elected. And, further still, over all these Boards of Guardians is a Government Bureau, called the Local Government Board, consisting of the Chief Secretary and two or three other Castle officials,[34] who have absolute power to review their proceedings, disallow their grants, dismiss their employees, or even set the whole Board aside and appoint others in their places. The incorporated cities have Mayors and Councillors elected by the ratepayers on a very restricted franchise, who control certain little matters of municipal regulation and finance. But these Mayors and Corporations (even of the city of Dublin) have no control whatever over the municipal police. The direct and entire control of all the police in Ireland is in Dublin Castle.

As to the police outside the metropolis – the Royal Irish Constabulary – nothing better shows the real character of the Government of Ireland. They are anything but constables of the Anglo-Saxon type. They are a standing army of occupation, carefully disciplined and drilled to prevent them from having any sympathy with the people among whom they serve, and carrying, not the staff of the peace officer, but the rifle and bayonet of the soldier. The rank and file are recruited from the sons of tenant-farmers, to whom the high pay, the good uniform, and life of ease offer great temptations. The commissioned officers – Inspectors and Sub-Inspectors as they are called – are taken from the landlord or bureaucratic class.

When a farmer's son enlists in the Royal Irish Constabulary he separates from family and friends as effectually as though he had enlisted for foreign service. For never while he remains in the force can he serve in the district in which he is born or brought up. After a time spent in drilling in the constabulary headquarters in Dublin, he is sent to some different part of the island. Here he lives in barracks, and wears a uniform on duty and off. He is encouraged to learn the habits and characters of the country people about where he is stationed, in order that he may act the spy and informer, but is kept from any such association with them as might lead to friendship or sympathy, and if there are any signs of this he is moved to another locality. After good conduct – which of course means conduct agreeable to his superiors – for a certain number of years, he is allowed to marry. But he can never again serve in the district in which his wife was brought up, or in which her family reside, without the special leave in writing of the Commanding General of Constabulary. He must still live in barracks. He is permitted to keep his children until they are thirteen years old; but then they must be sent away lest through them he should imbibe sympathies with the people.

Thus is created a gendarmerie who in their native country are almost as completely isolated from the people as would be Swiss, or Hessians, or any of those foreign mercenaries in which despotic governments have always delighted.[35] There are, in addition to nearly 40,000 regulars, some 15,000 of these troops in Ireland – more than the whole standing army of the United States when they had 30,000,000['s] of people and an enormous Indian frontier; half as many as the whole standing army of the United States is now. It must furthermore be remembered that this constabulary does duty mainly in the country, and among a people remarkably moral and religious, where ordinary crime is almost unknown. Their helmets and bayonets are to be see at every railway station; in every little hamlet are their barracks; and they are to be met by twos and fours prowling along the most unfrequented roads. A good Government could have no need for such a force.

The spirit of the Royal Irish Constabulary is the spirit of the whole administration. Whenever in Ireland a man goes into Government employment he becomes a member of the official class, and is cut off from the people. How he may conduct himself towards the people makes little difference. Their respect and friendship can do him no good, and may do him harm. The public sentiment to which he must defer is the public sentiment of the ruling class. For continuance in place and for promotion he must look to his official superiors. So long as they consider him a good and faithful servant, he may be to the people with whom he comes in contact as insolent and as brutal as he pleases, and in many cases it seems that the more insolent and brutal he is, the better a servant he is considered by those who alone have power to reward or to punish him.

There being little commerce or manufactures in Ireland, the growth of an independent middle class has been slow and small. The landlord-official class being the only class that had money to spend, patronage to bestow, and social recognition to give, had controlled the professional and trading classes. Subserviency to it has been the road to success and the badge of respectability, and has produced slavish habits of thought. There are more lions and unicorns and similar heraldic devices over the shops in the principal Dublin streets than can probably be seen in the same space anywhere else in the world. To get an invitation to a Castle ball is said to be a well-to-do shopkeeper's life-ambition, while the privilege of attaching the initials "J.P." to one's name is valued as the proverbial dog is supposed to value an extra tail:[36] While for the physician not only does the best and most reputable practice lie among the landlord-official class, but there are an extraordinary number of places connected with prisons, workhouses, hospitals, and examining boards, which in a poor country of high official salaries, such as Ireland is, constitute the overpowering prizes of the profession.

As for the Bar, its relations with the present movement are worth dwelling on a moment, as they illustrate both the character of the Government and the character of the movement.

What more natural than to find among the active leaders of a movement so sweeping and strong that in the greater part of Ireland it can, in spite of the restrictive franchise, elect whom it pleases to such offices as are open to election – numbers of young and ambitious lawyers. But there are none. This seems at first the more strange because Irish movements heretofore have been largely recruited from the Bar. Patriotism – the writing of ringing verses and the making of fiery speeches – used to be the orthodox way of attracting the attention of the Government and getting bribed into loyalty with good places; and there are more than one of the present higher Irish judges who thus commenced their career. But now the feeling of the Bar is bitter against the Land League and all its works. Nor is this, I think, because the Land League is peculiarly wicked. There is a readier explanation.

In the first place, no Irishman can become a barrister in Ireland. He must go to England and keep two years' terms in the English Inns of Courts, the business of which consists of eating dinners.[37] The time that must be spent, and the money required, amounting to what any but the rich would consider a small fortune, operate to keep out from the Bar all but young men in sympathy with the dominant class. And once within the Bar, not only does practice depend upon subserviency to this class and its representatives the judges, but to ambition and cupidity the strongest temptations are offered. To say nothing of the comfortable places made by that Special Providence of the Irish Bar, the Land Act, there is said to be one official position for every three practicing barristers. Naturally, therefore, every barrister is striving for a place. And when the barrister becomes a judge, while he is independent for life of the people, there is still a keen sense of favors to come from the Government. For beyond one step rises another step, with still larger pay, and above the highest is the vision of a peer's coronet and a stall in St. Patrick's Cathedral – all to be won only by thorough identification with the views and wishes of "the Government."

But the governing class of Ireland has cared for the English connection only as furnishing the power necessary for the maintenance of its own position. It has in part become at times restive under this connection, and has had no serious objection to a little sentimental patriotism, if it did not indeed look back regretfully to the good times preceding the Union, when the English Government had to buy through a Parliament of Irish landlords every measure it wanted. Home Rule, or even absolute separation, did not directly threaten the landlords. Smith O'Brien was a landlord, wanted to make his revolution with sacred respect for vested rights. Isaac Butt used to urge in favor of Home Rule that it would secure Irish landlords from being overwhelmed by the agrarian wave that sooner or later must arise in England.[38] But the present movement is primarily a social movement; it directly menaces, not the English connection, but the landlord class. They repel it instinctively, and the sentiment of the landlord class dominates the Bar.

It is difficult to imagine a Government more demoralizing and more irritating than this class government of Ireland. How contemptuous is its spirit towards the people, [and] whoever reads the Dublin Castle organs may see; how brazenly despotic, [for] anyone who reads in the same papers the reports of the doings of constables, and magistrates, and judges may also see. The Irish Government is, in short, a vast system of repression, and espionage, and corruption, run in the interests of a particular class, disgraceful to the English nation whose power maintains it, and degrading to the people who are compelled to live under it. A purely foreign Government, even though as repressive and as tyrannical, would not be as corrupting; for just as this Government takes the peasant's son, and, bribing him with an easy life and high pay, turns him against his own people, so does it, through all the walks of life, tempt men to forget their country and join the ranks of its oppressors.

At the head of the Irish Government stands nominally the Lord-Lieutenant. His functions seem to be to sign such documents as the Chief Secretary directs, to preside at banquets, and hold levees; and no one in Ireland ever talks of him except in such connection. The real Viceroy is the Chief Secretary. He is the true head of the whole centralized administration, and has in his hands almost unlimited power over it.

It is bad enough that the government of a country of five million people, with the power of sending to jail whoever it chooses, should be turned over by a lot of overworked Cabinet ministers to one of their number, just as the first lady of the bedchamber might turn over to the second lady of the bedchamber the sweeping and dusting of her Majesty's apartments. But the Chief Secretary is only a nominal ruler. Were he a man who knew things by intuition, who had as many eyes as a fly and as many hands as some of the Hindu deities, who could get along without sleep and be in several places at once, he might run the bureaucratic machine; but being, no matter how good and conscientious, only a man built on the ordinary pattern, the bureaucratic machine runs him.

This is the mysterious thing that is called "the Government" in Ireland. Who it is or what it is, nobody seems to know; but it is spoken of vaguely as "the Castle." This bureaucracy, which constitutes the real government of Ireland, is of course animated by the spirit of the landlord class, who, having been masters of Ireland, have filled up all branches of the administration.

It is hard for one who has imbibed traditions of English liberty to realize that there can exist an English-speaking community such a condition of things as exists in Ireland today. Of course, all the world knows that in Ireland over five hundred men are held in prison on suspicion, and that an English official may at any hour of the day or night send to jail whom he pleases. But wantonly and recklessly as the Coercion Acts have been used, the worst tyranny and oppression take place under the ordinary law. Talk to an intelligent Irishman in Ireland about what the Government can or cannot do, and he will tell you the Government can do anything. And anyone who reads the daily records of the Irish papers will come to very much the same conclusion. In fact, it is difficult to see what the Coercion Act was really needed for. Men and women are being daily sent by the county magistrates to jail – and to a worse punishment than that

permitted by the Coercion Act, for they are treated as common felons – under
an old statute of Edward III,[39] which has been dug up for the occasion, and by
virtue of which the magistrates send anyone to jail of whom they are suspicious,
and who will not or cannot give bail. Mr. Parnell or Mr. Dillon might have been
as easily sent to jail in this way as Father Feehan, or Captain Dugmore, or the
ladies of the Land League.

The advocates of "stronger measures for Ireland" talk about martial law; but
unless by martial law they mean a revival of the atrocities of the yeomanry of
the last century[40] – mean indiscriminate hanging and shooting, and the sending
of men into penal servitude on suspicion – martial law could hardly be any
worse, and if administered by English officers would probably be much better,
than the "magistrates' law" that prevails in Ireland now.

When the police can seize a gentleman and rifle his pockets in the market-
place; when a magistrate can point his finger at a peaceable citizen and order
him to be taken to the police barracks and searched; when he can send for
shopkeepers and tell them that if they refuse to sell goods to such and such
persons he will commit them to jail; when the police force an entrance into
private houses where two or three ladies are meeting, and insist upon remaining
to see what they do; when a dozen young fellows found walking together, and
suspected of being on their way to a Land League hunt, can be committed
without bail on the charge of "marching in a manner calculated to terrify her
Majesty's subjects," and boys are imprisoned for "whistling with derision;"
when newspapers can be suppressed by detectives; when a policeman can search
a shop, and carry off what he pleases without exhibiting any warrant or making
any payment; when letters can be opened and mail matter seized; when the
police can enter any peasant's cabin in which they see a light after ten o'clock;
and when they suspect a man, rouse him up two or three times a night to make
sure he is at home – it is rather hard to say what additional power martial law
can give. And even as to beating and shooting, it is to be observed that the
respectable citizens who were wantonly clubbed by the Dublin police at the
Phœnix Park and on the evening following Mr. Parnell's arrest never got any
redress.[41] In the one case an "Act of State" was pleaded in lieu of proceedings;
in the other those who dared prosecute were wearied out by having their cases
postponed and postponed. Practical immunity has been granted to the police
who bayoneted to death a young girl at Belmullet.

These things are justified by the plea that they are necessary to preserve
order. But they do not preserve order, as the advocates of strong measures
now admit in their demand for stronger [sic]. Is this to be wondered at?
Is Ireland a country where grapes may be gathered from thorns and figs from
thistles?

Take any intelligent man who knows anything of human nature and of history; tell him that there is [on] the moon, or [on] the planet Mars, a country governed as Ireland is governed, and he will tell you that it must be a turbulent country.

For in any place or time the enforcement of law and the preservation of order must rest on public opinion. Let the constituted authority become corrupt and inefficient, and what in the United States is called "lynch law" will spring up. Let it become tyrannous and arbitrary, and it will be hated and despised, and will have no power beyond the reach of the policeman's arm. For when the idea of legality is divorced from the idea of Justice, the strength and virtue of law is gone.

Whoever will consider what is being attempted to be done by law in Ireland will not wonder that the law fails. The true province of law is narrow. It is to maintain order; to secure from violence person and property. It may well be doubted whether legal interference as between man and man can ever be carried farther than this without doing more harm than good; but at any rate it is well-settled, both in theory and practice, that it should go no farther than the collection of debts and the enforcement of contracts by pecuniary penalty. And it is equally well-settled that law should never attempt to punish anything less than overt acts of violence or criminal fraud, or at most the direct and specific incitement thereto. The reason of this is clear. The moment punitive law is extended beyond these bounds an endless field for tyranny and abuse is opened; the freedom of speech and of action which is necessary to healthy social and political life is destroyed; peaceful methods of withstanding usurpation, of bringing about reform, or of adapting institutions to national growth and social progress are interdicted; and the hatred and contempt into which the adminis-tration of the law inevitably falls, palsy its legitimate functions. Here is the line of the long battle between the spirit of despotism on one side, and the spirit of liberty on the other, which runs all through English history from the field of Runnymede to the trades-union cases.[42] And when, under the British flag, men are being dragged to prison upon "reasonable suspicion of having encour-aged diverse persons to incite other persons to intimidate certain persons from doing what they had a legal right to do," is it not enough to make one wonder whether [the] Magna Charta is yet read in England, or the name of John Hampden remembered?[43]

What social, or political, or religious reformer has there ever been against whom such a charge as this would not lie? – What reform movement which, to the satisfaction at least of the interests that would suffer from it, could not be charged with just such constructive intimidation as this? As for the trades unions, whose battle has been fought and won in England, they are one and all based on precisely the same kind of intimidation which the Land League has advocated; for whatever individuals may have done, the Land League as a body cannot be fairly charged with advocating more than passive resistance and nonintercourse. And when it was attempted to suppress these unions by the law, as it is now attempted to suppress the Land League, did not outrages of the same kind occur?

The advice, incitement, or combination to pay no rent, or to take no farm from which a tenant has been evicted, are of precisely the same character as the advice, incitement, or combination to strike, or to take no place from which a union man has been discharged. In the one case, as in the other, the only legitimate point for legal interference is the point of violence. The landlords will say, as the masters said, that if the law does not interfere terrible things will happen. But both reason and experience show that worse things happen from legal interference with anything short of violence, or the direct incitement thereto; for not only is there no stop to the principle when it is once admitted, but it is certain that there can be no effective combination of this kind unless some bitter injustice is felt by a large class. Such combination is not easy. It cannot be made effective except under tremendous pressure, or carried past a very moderate success. And if the law be applied to its suppression, either the wrong will remain unredressed, still to fester, or the struggle will be made more desperate.

As for "boycotting," to attempt to prohibit and punish that, as is now attempted in Ireland, is so clearly to carry law into a province where it can do but harm, that it should not need discussion. "Boycotting" is not an invention of the Land League, nor a thing peculiar to Ireland; it is known wherever the social state exists. To suppress it by law is as hopeful as to control thought by law. In the main it answers a need and serves good social purposes. It may be abused, and has been abused in Ireland, even from a Land League standpoint, but the abuse is incident to excitement, which all these repressive measures serve to heighten. And certainly the lesson of acting together is one sorely needed by the Irish tenants. In their terrible eagerness to bid against one another has lain the power of the landlords to extract such monstrous rack rents.

There can be no doubt that there has been much exaggeration in the recitals of outrages, which have done so much to excite passion in England. The landlords, credulous as to everything relating to the Land League, have been themselves deluded. They have been deluded partly in the way Mr. Herbert Gladstone[44] pointed out, by peasants who desired an excuse for withholding rent they could not or did not want to pay, and partly by the desire to obtain damages from Grand Juries or to make favor. And beyond all this, there has been in the political demand, and in the commercial demand (for the outrage report has been paid for more readily than any other item of news from Ireland), a constant temptation to make everything an "outrage," and even to manufacture outrages out of whole cloth.[45]

Nevertheless, there are real outrages many and cruel. But it must not be forgotten that there is beneath this Irish movement a terrible reality. It springs from famine.[46] It has in it the desperation of men literally at bay. Large numbers of the poorer tenants of Ireland cannot pay their rents, were they ever so anxious. Their only hope of escaping eviction is to stand together and to have it understood that their holdings, should they be evicted, shall not be taken.

Here, combined with the suspicion sown throughout Ireland by arbitrary arrests and the secret bribes known to be offered to informers, is the main cause of the outrages. And, without justifying outrages, are English and Americans such a peaceable race that they can justly look upon Irish outrages, occurring under the circumstances that they do, as reason for condemning a whole people or a great party?

No matter how high or just its aims, every struggle that arouses passion and into which force enters is sullied by violence. Was not the triumph of Christianity over paganism marked by outrages? Has the Gospel never been preached by the sword? Was not the Reformation everywhere stained by brutalities, and cruelties, and vandalisms, and the effort of the older faith to keep its place accompanied by butcheries and persecutions? Were not Tories tarred and feathered, their houses burned, their goods taken and destroyed during the American Revolution by those whom the majority of [the] Anglo-Saxon race now revere as patriots? Or if on either side of the Atlantic there yet be those who regard the side of King George [III][47] as the side of right, was there not called on in its aid even the tomahawk, and scalping knife, and torture stake of the savage? Did not the army of God and Holy Church that won us [the] Magna Charta have its camp followers, just as the army that broke the back of the American slaveholders' rebellion had its "bummers?"[48]

Every large movement includes men of all kinds. All the good people never get on one side, nor all the bad on the other. Did any great issue stir the people of London as the people of Ireland are stirred, the roughs who are constantly committing as brutal outrages as ever heard of from Ireland would be on the popular side, and would doubtless give to many of their outrages political pretext or color[?][49] But it would be unfair on that account to condemn either the whole people of London or the popular cause. And so is it unjust to condemn, on account of outrages, the people of Ireland or the Land League movement? Agrarian outrages are no new thing in Ireland. They have marked every period of distress and repression. And with or without the Land League no one who knows anything of Irish conditions or Irish history could expect the present period to be exempt from them.

The Land League preached passive resistance. In its instructions to its orga-
nizers it urged them to discourage violence by every means in their power, and
this was always the effort of those best entitled to speak in its name. In this
Michael Davitt was specially earnest and anxious, and to the day he was sent
to Portland Prison exerted his great influence. The Land League was an open
organization, and a very loose one. The men in its lead could not have afforded
to countenance murder or outrage, and to have done so would have set against
it the clergy, whose power, especially among the classes from the Land League
drew its strength, is very great, and would have killed the movement. But,
further than this, the generative idea in this Land League movement is not the
idea of violence, but of moral force.

"Spread the light," the watchword of the radical or "socialistic" section, means
spread the truth, disseminate ideas; and the doctrine of "the Land for the People"
is to them as applicable and as needed in England as in Ireland; in America as
in Europe. And it is instructive to note how, both in individuals and in popular
movements, the ideas of a fundamental error in the organization of society
displaces those narrower notions that look for remedies to the employment of
force and to mere political change. For with a recognition of the truth that the
enslavement of the masses all over the world is due to the same cause, comes
also a recognition of the truth that the only thing that can emancipate them is
that intellectual quickening and moral awakening that will lead to a more just
and healthy organization of society.

In speaking thus of the Land League I am speaking of the Land League
proper, and of that peculiar leaven which distinguishes this from previous Irish
movements. The truth is that what is spoken of as the Land League embraces
most divergent elements, from the man who would not take off his coat for the
social question, but its bearing on the political question, to the man who cares
nothing for the political question, save as it bears on the social question. Just
as any attempt to prohibit any worship save that of the Established Church
would in England unite all other denominations from the Catholic to the Quaker,
so repression and misgovernment have in Ireland welded together divergent
elements. Or rather, to draw a more exact parable, just as all other denominations might in such case unite under the lead of the strongest, so had it been
in Ireland. Land Leaguism proper is a social movement, but under its banner
have united Nationalism, Democracy, Fenianism, Ribbonism, the "cupidity," as
the landlords call it, of the tenant-farmer, and the vague discontent of the
laborer.[50]

To suppress the open organization of the Land League, cautious from the
very fact of its openness, was at once to prevent the legitimate expression of
strong feelings, while greatly exasperating them, and to remove a check from

the more violent, while the ruthless use that has been made of legal and extralegal authority has intensified hatred of that authority.

The truth is, that the whole strength and activity of the Government of Ireland is directed, not to repress disorder, not to punish outrage, not to give the possession of land to those whom the law declares to be its owners, but to compel the payment of rent and to break up and punish any combination, direct or indirect, for its nonpayment. It is to effect this purpose that every resource of the ordinary and extraordinary law is being strained. Men of the highest character are dragged to prison on suspicion – not of intimidation, but of giving countenance and moral support to nonpayment of rent; ladies are hampered and bullied and sent to felons' cells, not because anyone imagines they are inciting to outrage and murder, but because the work of charity in which they are engaged destroys the deadly fear of eviction, by which Irish landlords have extorted their rack rents; and to the same end police terrorism is invoked and draconian sentences imposed by maddened Dogberries.[51]

Law, to command more than forced obedience, must be impartial; but law in Ireland is but a weapon in the hands of one of the parties to a great social struggle. The landlords may freely write, talk, meet, combine, boycott, do what they please for the protection of their interests; but the popular party are gagged, dispersed, imprisoned. A great movement, stirring the Irish people as only at long intervals and under great provocations any people are stirred, is driven in. Is it any wonder that there is lawlessness and outrage, that evidence cannot be obtained, and that juries will not convict unless they are packed?

That criminals cannot be detected in Ireland does not prove that the Irish are peculiarly a lawless people, but that among them has been used for purposes that outrage the moral sense. The Irish horror of the informer has become traditional during generations in which priest and patriot have been hunted by the bloodhounds of the law. And today this feeling is being intensified. In countries where the constable chases only the thief and the murderer, every bystander will join in the hue and cry; but where constables drag off to prison those whom the people must love and honor, he who flies from the constable, even though he be thief or murderer, finds help and concealment. This is only human nature.

Nor yet does the failure to find juries to convict in cases which have any tinge of agrarian or political complexion prove that the jury system is not suited to Ireland. The Anglo-Saxon jury is not an invention for the surer punishment of crime; it is a device to prevent the enforcement of law when not sanctioned by public opinion. Let judges charge as they please, this always has been the real power and the real usefulness of the jury. And it is this that has made it a safeguard of popular liberties. Many criminals have escaped, even in the teeth of law and evidence, by reason of the jury system; but on the whole when it has prevailed, social life has been freer and purer. For when juries habitually fail to do justice, the fault lies deeper than any judicial method. And to give up the jury because in certain cases convictions cannot be had, is to abandon those principles that have made the Anglo-Saxon race what it is, and to adopt the theory of despotism.

This is the choice that in regard to Ireland the people of England must make – full liberty or the most ruthless, brutal despotism; there is no halfway course.

Here, in a few words, is the situation in Ireland. A privileged class in whose hands is all the machinery of Government, and who have long been accustomed to look upon the rest of the people as their serfs, find making way against them a social revolt, which their rapacity has provoked. And nothing is more bitter, more cruel, and more unscrupulous, and at the same time more blind, than a privileged class threatened with loss of power.

This class has had the ear of the English Legislature, and through the English press of the English people, and all things relating to Ireland have been seen in England through the medium of their prejudices and their fears, and they have led English statesmen into the blunder of treating a revolution as though it were a petty conspiracy, and so accelerating what they thought to crush. Surely it is time that English statesmen and the English people should seriously ask themselves what they are trying to do in Ireland.

Why should the people of England let the people of Ireland settle their own affairs? Why should England take upon herself the cost, the trouble, and the danger of trying to govern Ireland? Is this effort to keep one set of Irishmen under the feet of another set of Irishmen to the profit or the strength or the glory of England? On the contrary, Ireland is to England today an expense, a weakness, and a disgrace.

A connection is possible between the two countries that would be to mutual advantage; but the present connection is plainly a curse to both.

LECTURE BY HENRY GEORGE: THE IRISH LAND QUESTION; WHAT "THE LAND FOR THE PEOPLE" MEANS; HEARTY RECOGNITION OF HIS VIEWS
(June 10, 1882).[52]

At eight o'clock on Saturday evening the 10th of June, a very large, orderly, and respectable gathering had assembled in the Round Room of the Dublin Rotunda to hear Mr. Henry George's exposition of the Land Question. The charges of admission ranged from 4s. to 6d. and the proceeds were devoted to the advancement of the Political Prisoners' Aid Associations. American flags were prominently displayed, and two bands were in attendance, discoursing appropriate national airs. There was such enthusiasm manifested that made one think that the people were indifferent to any terrors the forthcoming Coercion measure had. At twenty minutes past eight, Mr. George ascended the platform, and received a great ovation, one that has rarely been accorded to even the most popular Irishmen. Dr. Kenny was called upon to preside, amid loud applause.

There was present Miss Anna Parnell and several prominent ladies of the Land League. . . .[53]

The Chairman, who received a very warm reception from the audience and he did not think it was necessary, thought custom demanded, to introduce Mr. George, who had the strongest sympathy with their cause, and that alone was quite sufficient to secure for him the warmest welcome. (Applause.) But the distinguished lecturer had made his mark [on] the age, and wherever there were thinking men, his name was well known. (Applause.) Wherever men occupied themselves with the solution of problems which affected the position of the people his words received marked attention and whether they would agree in the ideas that he would put forward or not he (the Chairman) should say that he thoroughly and entirely agreed with them. (Applause.)

[The Secretary then read letters from the following: names and comment.][54]

Mr. Henry George then came forward and was received with ringing and repeated cheers. He said he was afraid that tonight he could hardly give them anything that was worthy to be called a lecture. Nevertheless, when he was asked to speak [ ] he could not refuse, and he was glad to come among them for the sake of the associations for whose benefit the gathering was met – the Political Prisoners' Associations of Ireland. (Applause.) He was glad of it and he would go out of his way to do them service for their cause, and he would go further than that for themselves. These Political Prisoners' Associations (and he did not seem to mince his words) represented, he was sure really and truly, the organization that was now under a ban. At any rate, they were associations of earnest patriotic men, and he was delighted to see they had continued together to form a central association. He hoped that that was but the prelude to similar organizations all over the country. ("Hear, hear.") The men of Ireland have yet he thought something to learn in the way of associations – something to learn in the way of those practical organizations, that formed only the surest sound basis for political action – organizations that would bring the people together, and that would bring men together in little knots, that they might discuss, that they might think and that they might plan, and so unite their forces. (Applause.)

He was in Paris a while ago, and he met there a distinguished Irishman, a well-known citizen of their town. They called him "Pat" for short: Pat Egan.[55] (Applause.) And one of the things he said to him was: "Well, there is one reason I have faith in this movement beyond, and it is this – that they have not been quoting much poetry." (Laughter.) He (Mr. George) thought there was sense in that in the way Mr. Egan meant it. No one loves poetry better than he (Mr. George) did. No one better appreciates the force of the sentimental than he did, but after all, if people must do anything they must think, and in what he would have to say to them tonight he should not attempt to win their applause. What he wanted to do if he could was to appeal to their intelligence. ("Hear, hear.") It was thought after all that conquered everything what existed; what man did, or what was done, must have been preceded by thought, and as a prelude to all right action, there must be right thought first. "As an American saying goes," he said: "First make sure you are right, then go ahead. (Loud applause.)

That all over the world the majority of mankind are enslaved or oppressed, that the long history of the human race, has been a history of the oppression of the masses, is due to the fact that the masses of men do not think. Did they ever see one of those great strong dray horses, driven by a little bit of a man, in physical power not able to cope with him. Yet that horse was enbridled, strapped, and driven. Why? What is the difference between the man and the horse? The difference was that the man thinks, and the horse does not. And so it was throughout all the world. The masses of men were enslaved today, simply because the masses of men did not think. Partly it was their misfortune; partly it was their fault. But whether it be fault or whether it be misfortune, there could be no redemption for any people while they coolly and calmly set themselves down to look at the situation – to see what is best to do, what they want, and then have to do it. (Applause.) And surely if there was any country in the world today where patriotism called upon her children for their best thought it was this country of theirs. (Applause.)

What could be sadder than the situation of this fair isle? Here you have thousands of men, women, and little children, who are being driven from their homes. Here they have from a land that could easily maintain ten, fifteen, aye twenty millions of people, people going in thousands to seek a living across the Atlantic. You have outrage, you have assassination, you have almost everything that is bad.

He did not want to talk politics tonight but he thought among other things that they had about the worst Government that existed today. (Long and continued applause.) But there was a breathing space. The new Lord Lieutenant and the new Chief Secretary seemed to him to be wiser and better men than those who had proceeded them. ("Hear, hear.") But so long as the character of their Government depended upon the accident of character of the men, who are in those positions, what security had they? (Applause.) It was no resting place.

There should be no resting place until the people govern themselves (Loud applause.) And if they were true to themselves, if steadily, calmly, and firmly they went forward the time would come, the time must come, when they would find that there was no power in physical force, no power in bayonets that could keep underfoot five millions of people who knew what they wanted and who determined to get it. (Applause.)

But he did not want, as he said before, to talk to them about the political question but about the social question. He believed with Michael Davitt (tremendous applause), that the social question was larger than the political question – that the social question includes the political question because political institutions were ultimately determined by the individual character of people. (A voice: "Physical force.")

A hundred years ago there was a struggle in two parts of the British Dominions. We on our side of the Atlantic won independence. (Applause.) You here in Ireland after fruitless struggles were trodden down again. Why the American Colonies won their independence was simply because of the independence of the people. ("Hear, hear.") If the contest that led to American independence had devolved upon the Southern States where slavery existed the arms of Great Britain would have carried all before them. It was those self-respecting independent, sturdy farmers of the Middle and Eastern States, that fought that war, and if they had here in Ireland social institutions that produced the same traits of character, the result with them would have been the same. (A voice: "It is coming;" applause.)

All over the world show him the country in which the social condition of the masses was low, and he would show them a country of tyranny and oppression. Look at Egypt, the point on which the eyes of the world are concentrated.[56] What is the struggle there? Nothing more than a choice of masters. The people of Egypt! There were no people in Egypt. These poor fellows, downtrodden and ground down with taxes, struggling but to make the very barest and scantiest living could have no thought beyond that. So with India,[57] so with every country in which similar social conditions exist, and democratic institutions good as they might be in themselves, were of no use and would avail nothing, unless there was also something like equality of social conditions. Take Mexico, nominally a Republic.[58] What has she been? Nothing but a prey to one dictator and military chief after another. So with nearly all the other South American Republics, and this truth they could also see exemplified in his own country.

They in America had gone as far as they go in political democracy. With them every man was equal before the law. With them every man could vote, and every native-born citizen could become President; yet it was very plain to whatever person chose to look that their social conditions were becoming such, as to create a great gulf between the rich and the poor, and political democracy would avail them nothing until these existing conditions were gone, so as to enable every man to make for himself an independence and thereby create the true independence of the nation. (Cheers.)

No one would yield to him in his love of political freedom, in his faith in democratic institutions. Yet after all, were the choice presented to him, he believes he would be acting the noble part so far as he was concerned, if he put his lot under the most despotic Government on the face of the earth today, with an opportunity of making a very good living, than to put it under the freest institutions if he had to bend, and scrape, and crawl for a living. Considering all that had been said of political tyrannies, he would ask them what tyranny was so galling, so humiliating, and so oppressive as that tyranny of poverty and of want that made men cowards and slaves. (Loud applause.)

He had believed for two or three years past that the men of Ireland were destined to play an important part in the history of the world at this epoch. ("Hear, hear.") He had believed that this had commenced here with the institution of the Land League. (Loud applause; a voice: "We have proved that.")

No, they had not proved it yet. (Applause.) That remained for them to prove. He had faith, however, in their proving it. He believes that there had been commenced here in Ireland, a great Social Revolution that wants no [ ] here. He believes there came to the front in this Irish Land Question, the greatest question of modern times; (applause) and his faith now at last was in a fair way of being justified. Now at last the [ ] departure has been taken. Now at last with Michael Davitt's Liverpool speech (loud and continued applause) the standard has been raised, the standard of "the Land for the People." (Loud applause.) The land for the whole people, for every one of them, man or woman, rich or poor, down to the humblest child.

And now this question came up before them, for it was with them, the men of Dublin, and the men of Ireland, and not with that Parliament over there, would rest the ultimate solution of this question. If they knew what they wanted, if they would make up their minds what they wanted, and if calmly, soberly, [and] carefully they would think aright in this question, they finally would win it, must win it, and in winning it for themselves they would do much more than that – they would start a movement that would run around the globe. Once more they would bring Ireland to the front, once more after the lapse of all these hundreds of years, they might say that Ireland again has become the beacon light of the world. (Applause.)

"The Land for the People." This was a cry that has rung out before this. They were words that had been spoken from all their platforms, but after all there were many men, and even now there were many men who did not give it its full and true meaning. "The Land for the People!" Why, they had only recently been told that "the Land for the People" meant that the tenant-farmers should have the privilege of buying out the landlords. He rejoiced that at last the true standard has been lifted, the true interpretation made, and that the ringing words went forth that "the Land for the People" meant the law for the whole people of Ireland. (Applause.)

He wanted to talk to them tonight about this, and before he proceeded to balance against and compare with each other, the system of Peasant Proprietary as advocated by some, and the system of Land Nationalization[59] as advocated by others he wanted, if he could, call their attention to two or three general facts, and to ask them to follow him through some reasoning that may be a little abstract in character, but that was nevertheless very simple.

The first thing he would like [that] they would ask themselves was: "When they talked of land, what did they mean?" He asked that question because a great many people talked of land and wrote of land, and seemed always to have in their minds the idea that it was nothing but agricultural land ("hear, hear") and they thought and wrote as though the Land Question was a question which related exclusively to farmers and their landlords. Nothing can be more delusive. Let them think for a moment – "what was land?" So far as they were concerned, so far as they could see there were but two things in this universe – æther and matter and so far as they could come in contact with it, or use it in any way the whole material universe was comprised under that term "Land." Land! Something that relates merely to farmers! What man or woman here in the world, what human being had ever existed or could exist independently of land? (A voice: "Nobody.")

What work could be performed and what production could there be without land? Take a man away from land and where would they put him? Up in the interstellar spaces. They would have no man. His flesh and blood, body and bones, where were they drawn from? From the soil. Without land no man could live no matter what his occupation. Without land no avocation no matter what it may be could be carried on. When they fully grasped that idea they would begin to appreciate the importance of the Land Question.

The Land Question! The Labor Question! What were they but different names for the same thing. All production was the result of the union of two things, of human exertion and of the raw material of the universe which they called Land. Now then, that being the case, give to some man, or class of men in a community the ownership of the land from which, on which, and by which, the whole people must live, and the few must be the masters of the many. ("Hear, hear.") The Almighty has deemed that nothing they desired could be produced without labor. ("Hear, hear.") "By the sweat of his brow man shall earn his bread" was the law which revealed religion made clear to them.[60] But this did not require revelation. It was as apparent to anyone who wants [to] reflect on it as was the sun at noon day. What was there – think of any desirable article – what was [ ] that could be procured without human labor? Why, in the Garden of Eden, if its occupants did not make any exertion, they would starve. The fruits of the earth must be gathered and the fishes of the sea must be caught. The Almighty had given them the power of conscious, voluntary, [and] intelligent exertion. Then He had provided for them a world that would respond to that exertion. Let, therefore, any one class in a community take possession of that raw material, to which labor must be applied for the production of wealth and the satisfaction of human desire[s], and without labor by themselves they can demand and will receive the proceeds of the earnings of those who do labor. ("Hear, hear," and applause.)

Without land, labor could produce nothing, and therefore without access to land, the man who would use his labor for its legitimate purposes – the supplying of his own wants and those of his family – was helpless. He could do nothing for himself. He must supply his labor for somebody who had the raw material; and where the pressure of competition was strained, he could only get employment upon his terms. Take agricultural avocations. Here were a number of farm laborers who had no land of their own. They could produce nothing – they could earn nothing save by uniting their labor with the raw material of the universe. Therefore, they must go if the land belonged to an individual, and ask him to give them leave to toil, and to tax his toil. They must either go and ask him to rent them the land paying him for it, and give him his demand out of the proceeds of their labors, or they must hire themselves to him and let him tax the proceeds of their labor. Unless they do either, they must starve, and they, the men of Ireland, knew what the consequences were of either alternative.

Down in the country districts, in as rich a country as he (the lecturer) had ever seen they would find the wealth [taken away from the] producers [who were] living in the most abject cabins and living on the most miserable fare, and they would see the children of these men growing up ragged and ignorant. And what would they find on the other side – the man who did nothing drawing from *their* labor £10,000, £20,000, or £30,000 a year, living in a magnificent castle that looked down upon their miserable hovels, and taking the proceeds of their toil, over to London, or Paris, or Italy and living on it there, he did not know how. (Loud applause.)

These workers had no option. They must either work on the landlords' terms, or those of the middlemen, the farmers, or they must starve. And if they are willing under these circumstances, as they must be, to work, then the competition arising from one man pushing another, crowding wages down to the lowest level. What was the result? Did it stop there? No!, not at all, when the agricultural laborers' wages gave but a bare living, what would be the wages of the working men, of some other kind? The wages of the blacksmith, the shoemaker, or the clerk were held down by the same competition, and did not this enter too, into the life of their cities and affect their masses. All other avocations depended upon it, and when wages in the lower or agricultural stratum came down to that point where the laborer made but a bare living, wages would be in other avocations, very near the same point. (Applause.)

Was not that the reason and no other why wages here in Dublin in the trades of mechanics of all occupations were so low? Was not that the reason why the new countries were the countries where laborers were best paid, the countries to which capital flowed, and laborers emigrated? As a matter of fact, wages ought to be higher in the old countries than the new, in the countries of dense populations than in the countries where the population was sparse, because in the former, human labor is more effective. In old countries they had whatever remained of the work of the men who had gone before you. In these countries of dense population they had all the advantages that came from close associations. A dozen men acting together would individually get a great deal more than the man who worked singly, and where the spaces between them were not large, their labor would be more effective . . .[61]

But to get an honest living one must go the new country where the land was got cheap and could not be monopolized. Now that being the case with wages and when the social conditions of the masses of the people, no matter what be their occupations depend upon their freedom of access to the land, what was the true solution of the Land Question? Supposing here in Ireland they went to work, and turned the tenants into owners; supposing they made these tenant-farmers the proprietors of their farms. What then? They would have done something no doubt. They would give to these tenant-farmers something that the landlords have, but what good would that do to the agricultural laborer? (A voice: "None;" applause.) What advantage would it be to the mechanic of the city? None whatever; none to the men who stood outside, and who had nothing but their hands to support them. Their conditions would be but little, if in any way improved, while they were really the men to be considered. If they raised the base, they raised the whole social edifice, but if they raised any other portion, they [are still] the lowest class in the same condition it was in before. ("Hear, hear.")

He was not talking theory, he was talking facts. Go into the countries of Europe where there had been made a minute division of land, and an examination of Peasant Proprietary in these places would show that under the conditions of this the nineteenth century, it could not exist in this country. In these countries, they would find laborers at the lowest wages. In Belgium, for instance, they would find wages lower than they were in England, and he believed lower than they were in Ireland.

The same thing could be seen over in his country. They started with very fair conditions. They started with a whole continent of virgin soil. They started with institutions that gave every man who wanted it an opportunity of getting a farm for himself, and with all that, had they solved the Social Question? (A voice: "No.")

Not at all. Over in his country today one of the greatest strikes that had ever agitated the world was going on there.[62] In the United States today there were men and women working for the merest pittance. There was destitution and there was want, not so much it was true as might be seen on this side; but certainly very much more than there should be. And as they grew older and as the soil would be taken up, these conditions would come more and more into prominence.

They should go further than Peasant Proprietary in this country. (Applause.) If there should be given to labor its full reward, and if they would establish their institutions upon a basis that would stand they should go farther than that. They should make the land the property of the whole people, so that every child [who] came into the world, so that every man who lived in the country should have his full and equal right acknowledged and secured. (Applause.) He wanted to say that if they could establish such a thing as Peasant Proprietary, if they could turn over to the tenants what was the property of the landlords, and which he thought was not practicable – nothing of the kind was proposed.

The Tories and the English House of Lords – the great landholders of England, aye, and the landlords of this country, were now in favor of a measure of this kind. They wanted to extend purchase clauses in such a way, as to diffuse land property among the farmers and they were very wise in so doing, indeed, because by getting a number of people interested in the ownership of land, they would constitute a buffer to them, and be a bulwark to them, so that when a man got up and said: "It is unjust that the 'Duke of This' or the 'Lord of That' should have so much land and others so little," they could say: "Oh, you want to rob the poor farmer of the land he has got after so much toil." When people would ask: "What right has this duke or that lord to get £30,000 or £40,000 a year from their estates, when others had not an inch of land," they would say: "Oh, you are a Communist. You want to turn these poor people out of possession of the soil that they won by so much toil and sacrifice."

He believed it would have the effect of postponing the final settlement of the question – to make it very much more difficult. And then how long would the thing last? All the tendencies of the times were pointing to [the] concentration [of landownership]. The people who said if you were to divide up the things of this world equally today, after a year there would be inequality, were perfectly right. If they were to take all the property that existed in the world or in any one country, if divided equally among the people, in a very little time there would be inequality just as there is now.

The thing to aim at was not an exact division of land, they did not each want the same quantity of land or the same quantity of any other property. Equality decrees not that, but simply that each one should have equal opportunities, and the true solution of the social question, of the Land Question, or the labor question, call it what they please, was that solution, . . . which would enable every man to get that which he fairly earned. (Applause.)[63] And if every man got that which he fairly earned, no man would get more than was fairly earned by him; but that was not the state of things existing today.

On the contrary, look at Ireland, look at England, look at the United States, look at any country of modern civilization. Would they find the producers of wealth the men who enjoyed most wealth? No. If they wanted to see the men faring sumptuously they must look for the men who did no work at all. The men who lived poorest, were as a general rule, those who worked the hardest. That was not as it should be; that was not what the Creator ordained, when He made the necessity of labor the preliminary to enjoyment.

Now how could equal justice be secured? Why not divide up the land piece by piece. Nothing would be gained were they to do so, because with the growth of population inequality would soon begin again. But it was not necessary to do anything of that kind. It was merely necessary to take for the benefit of the whole people those profits coming from land, which were more than sufficient to reward the exertions of those who labored or those who expended their capital in the promotions of these exertions. That was the plan proposed by Michael Davitt. (Applause.) And that was the true plan. I did not believe there was any other plan. There was an estate that belonged of right to the whole people of the country. They, the men of Ireland; they, the whole people of Ireland owned by Justice and by natural right the whole soil of Ireland. (Applause.) They and none others. And if they took for themselves and applied to their common uses the income that arose from it, every man would be put upon an equal level. The laborer would get simply what he earned, if nothing more, while the "dog in the manger" would get nothing. (Laughter and applause.) That simple plan he had spoken of was Michael Davitt's plan, but it was not the plan of Michael Davitt alone. He did not believe it could be said to belong to any man. It was the old commonsense plan, that had been acted upon time out of mind. It was the plan that had occurred to man after man as he had thought of it. And it was the plan that must occur to every man who thought of it. It was the plan that in a new country 6,000 miles away from here, he saw in operation. He wished to read to them something that was new to him until a few weeks ago and something that possibly may be new to them and that was the Declaration of one of the best patriots of 1848, James Fenton Lalor. (Loud applause.) It was his Declaration of his beliefs published in the *Irish Felon* (applause) of 4th June, 1848.[64]

[Here insert two quotations of J.F. Lalor; [ ] applause during delivery.][65]

Could there be a truer, or fuller enunciation of the principles which Michael Davitt enunciated than that put forth by one of their foremost men in 1848. (Loud applause). Let him read again to them from another Irishman. He did this to show them that, though one of their papers (laughter) called him an American Communist, the ideas he was putting before them, while they were the ideas he firmly held, were good Irish ideas, put forward by eminent Irishmen (applause), and they were good Christian ideas as well. What he held in his hand was a letter of the most Reverend Dr. Nulty.[66] (Tremendous cheering.) Dr. Nulty said: . . .[67]

And so he said to them, men of Ireland, if they consented to any settlement of this Land Question short of the full true settlement that the land was the common property of the whole people – that the land of Ireland belonged to the whole people of Ireland; and that the humblest child born in Ireland had to it his equal right.

They would be doing not only an injustice, and a wrong but they would be committing a blasphemy. Was it not blasphemy of the worst kind today that the Creator had intended and instituted such a state of things as could be seen in this country today? Such a state of things as they might see in this very city. When they went out into the streets after leaving here – What would they see? Poor girls stalking around to make a living by dishonor, and poor little children, as he had seen them creeping up at twelve o'clock at night to the good ladies of the Land League. (Tremendous cheering.)

Mr. Crinion: Another cheer for the Land League ladies. (Great cheering.)

[Mr. George continued.][68]

They were right, men of Dublin, to cheer for the Land League ladies and he hoped they would do something more. Since he came to Ireland and had seen what these women have done, he had become a convert to women's suffrage. And when the men of Ireland came to make their own laws and decreed what they should vote, he advised them to take the women and give them a vote too, as these ladies have shown what women could do. (Applause.)

But what he was saying was if [that] they consented to any settlement of the Land Question that would leave out those ragged little children of those streets, they would be doing as Bishop Nulty said, not only a wrong – they would be consenting to a blasphemy. Were not they, Dr. Nulty asked, children of God just as much as the heir of the proudest duke, for had they not come into the world by the will of God? And had He not provided the material for their sustenance? These were the persons to look to, the lowest, poorest class, and when they had secured them freedom and fair play; when they had secured to them their free rights they would have secured everything. (Applause.)

He believed this struggle should go on – that it could not stop short of that. That movement which he felt was not merely a local movement. The very same impulses were felt all over the world today. In England, on the continent of Europe, and in the United States men were everywhere beginning to ask themselves what was the use to them of all this progress, if still they should toil and strain their energies for a bare living – if still there should be paupers and tramps, and barefooted children. And the fight was commenced that could not stop until popular rights were secured – not merely political rights – something more than that. They had over in his country full political rights, but these rights would amount to nothing unless they would go farther. They had there to secure to every man not merely his right to vote, but his right to the material of the universe, his equal right to the bounty of his Creator. Else all that had been done was [in] vain; else the Republic had not be[en] established. (Applause.)

And in the way pointed out by Michael Davitt, in the way pointed out by Bishop Nulty, in the way pointed out by James Fenton Lalor – for all their plans were essentially the same – this could be very easily established. All they have to do, instead of bothering themselves about the letting of land, was to tax the land. Let the taxes fall on the landowner, and let the requirements of the State thus be met. When there would be a surplus after this was done let it be taken, and used for the common good of the people. Otherwise, one would not have more advantage than another. He would ask them merely to go from there, and apply this system to the city of Dublin or to any country district they knew of. They would see how it would break up monopoly, and force the men who now held large tracts of land, to look for tenants, and throw in their lot with them.

He rode yesterday into the country with two ladies who had just been released from jail – Miss Hannah Reynolds and Mrs. Moore. (Tremendous cheering.)

They rode through a very magnificent park. It was the property of the Earl of Charle . . . Charle . . . Charleville.

Mr. Sexton: Charlemont.

Charlemont, situated down in the Kings County. It was a very fine place surrounded by a big high wall, containing thousands of acres of rich land which was kept for deer, and with a magnificent castle in the center.

How would the system of the tenant buying out the landlord affect him? No, it would not, but put the pressure of taxation on and that demesne must be abandoned.[69] It could not be kept there.

Men talked about their immense population, and what impressed him was what few houses there were. One could go into New York or New Jersey, 1,500 miles west, or up the Plate [?] Country and he would see more houses than could be seen here or in England. And when he came here, he saw too how the people lived – in miserable villages containing some wretched hovels, the most miserable places he ever saw. The people were huddled together, and there was misery, and there was everything repulsive to the eye and sickening to the heart. A garden there was to be seen nowhere; and yet there was land sufficient for a great many more. But what was to be found – subscriptions being raised to take these poor tenants away. (A voice: "They are coming back with a vengeance.")

Mr. George: Yes, but before they come back [ ]. (A voice: "They are coming back now.")

This whole movement was in a large measure, but a reaction of the feeling in America and in Europe. It was the result of the influence of those who had *left*; and he did not believe the movement could cease until they had to settle the Land Question here; as [it] would not only make secure for the future of every Irishman who wanted to remain in his native land, an equal share and participation in his native soil, but would also set an example to the world. (Applause.)

But he wanted them to think of these things. He could not get up upon a platform like that and talk for an hour or an hour and a half, and go over fully and explain in detail such a subject as that. All he could do was to try and excite their thoughts. No one could teach them. They must teach themselves, they must think for themselves. (Loud and continued applause during which the lecturer resumed his seat.)

A vote of thanks to the lecturer was proposed by Mr. Sexton, and seconded by Mr. McInervy [ ]. It was carried with acclamation. Mr. George upon rising to respond was greeted with loud applause.

He said that Mr. McInerny expressed the belief that Michael Davitt derived his knowledge of the great scheme of reform which he advocated from reading

Progress and Poverty. He (Mr. George) did not think this was true. This reform movement was not a new thing; it was the [ ] plan that land should be considered private property [that] was a modern idea. The original perceptions of man always were that land was common property. In England this system existed until very near the time of the Reformation. Private property in land was the result of spoilation, usurpation, and fraud. And in going back to the old plan they would be simply giving effect to the original perceptions concerning land. They had a great deal yet before them to do; what had been done one should spur them forward.

With reference to Peasant Proprietary and Land Nationalization, he did not think these roads led in the same direction. He did not think that Peasant Proprietary would lead to the nationalization of the land. On the contrary, he thought by having recourse to the first they would be going away from the others, and so did the English House of Lords think too, because it would be a bulwark to their power as landlords, and they knew that by this means it would be more difficult for those left outside to get their rights. That was the difficulty they had to deal with in the United States. There would be inequalities still. It was off the fertile land the people had been driven. Supposing there could be such a thing as Peasant Proprietary, under which they made a man the owner of his holding. Would it be justice that one man should have fine land from which the people were driven [off] and another man should have but a little bit of a holding upon which he could not live?

He (Mr. George) did not believe purchasing out the landlords. He differed from Mr. Davitt on that, though perhaps circumstances, or the force of political necessity might give some justification for it; but it would not be a greater injustice to the landlords to buy them out, that it would be to the other people who would be materially injured by that. If the land belonged to the people why should they pay anything to resume what was their right? Was it because they were robbed for the last few years and their fathers before them? Was that the reason why the landlords should be [allowed to] continue until they were bought off? Michael Davitt was right when he said that the landlords had not a claim to their fares from Dublin to Holyhead. He (Mr. George) did not agree with any plan that gave them that much.

The simple plan was to shove the taxes on them and to tax them out. (Loud applause.) In this he did not think there was involved as much Communism as was really involved in the plan of Mr. Gladstone's, if he had his own way. He (the lecturer) would make land common property without giving those who owned it one penny of compensation. . . .[70]

Under the plan Mr. Davitt proposed every man, whether he be a farmer, mechanic, or laborer would have his equal share, but under this other plan the landlords would be turned out. What was more Communistic in that than in taking the land without giving them anything for it. Some landlords . . . might establish a claim to get something through charity; but he claimed as much for the poor ragged children he was speaking about, and for others. This country was rich enough to support its people, and the only plan was that by which *the people* could be well maintained.

Mr. George concluded with much applause and when he was leaving the building accompanied by Dr. Kelly he was followed by the people who cheered him most enthusiastically.

GEORGE AND THE *IRISH TIMES*
Correcting Some Misstatements of the Landlord Organ;
The Old Irish Doctrine; "The Entire Soil of a Country Belongs
of Right to the Entire People of that Country"; From the
Irish Times – Anti-Land League
(*Irish World*, June 24, 1882).[71]

Sir: Will you kindly permit me to correct some of the misapprehensions in your article this morning on Mr. Davitt's Manchester speech?

1. You style me the "new American ally of Mr. Davitt." I do not know that there is anything new about it. I have been an ally of Mr. Davitt ever since I first met him in New York, and found that he believed as I believe, that there can justly be no individual property in land, but that the land of every country belongs rightfully to the whole people of that country.

2. You say that I would find on American platforms and before American audiences much greater reason to proclaim my views than in England or Ireland, and that in my own model Republic I will find bad landlords and poor tenants.

The truth is that I never, either at home or abroad, held up the American Republic as a model Republic, but have, on the contrary, told my fellow countrymen for years that a Republic permitting private property in land could be a Republic only in name; that we had not established the Republic, and never could establish the Republic until we carried out in its entirety the truth enunciated in our Declaration of Independence, and secure to every citizen his equal and inalienable right to the soil.

Nor have I ever pretended that the Irish land system was in essence a whit worse than the American land system.

Permit me to quote a paragraph from the pamphlet entitled "The Irish Land Question" to which you allude:

> Never had the parable of the mote and the beam a better illustration than in the attitude of so many Americans towards this Irish Land Question. *We* denounce the Irish land system!

> *We* express our sympathy with Ireland. *We* tender our advice by congressional and legislative resolution to our British brethren across the sea! Truly our indignation is cheap, and our sympathy is cheap, and our advice is very, very cheap! For what are we doing? Extending over new soil the very institution that to them descended from a ruder, darker time. With what conscience can we lecture them? With all power in the hands of the people, with institutions yet plastic, with millions of virgin acres yet to settle, it should be ours to do more than vent denunciation, and express sympathy, and give advice. It should be ours to show the way.[72]

This is the spirit of everything I have ever written or spoken in regard to the Irish Land Question.

Of course, I recognize the fact that so far as political institutions are concerned, we of the United States are very far in advance of you on this side of the water, and it would give men great pleasure to see you follow our example in this direction, and establish a democratic Republic. But I hold with Michael Davitt [that] the social question includes the political question; and, while I believe that democratic institutions must follow the nationalization of the land, I believe that without the nationalization of the land the democratic institutions cannot avail to prevent the growth of a plutocracy in some respects even worse than an hereditary aristocracy.

There is nothing pharisaical in the feeling or the attitude towards the Irish people of Americans than myself. We desire to do all we can to help our brethren on this side of the sea to assert the equal right of the humblest to the soil of his country, not merely because we wish them well, but because we desire their aid in return. And just as America has acted on Ireland in promoting this land agitation, Ireland is reacting on America. "The gospel of the land for the people is a universal gospel," and its enunciation here is already leading hundreds of thousands of American voters to an acceptance of its truth.

3. The following words which you attribute to Mr. Davitt were merely quoted by him:

> Now, therefore, of every country is the common property of the people of that country, because its real Owner, the Creator who made it, has transferred it as a voluntary gift to them. *Terram autem dedit filis hominum.* [The earth He hath given to the children of men.] Now, as every individual in that country is a creature and child of God, and as all His creatures are equal in his sight, any settlement of the land of this or any country that would exclude the humblest man in this or any country from his share of the common inheritance would be not only an injustice and a wrong to that man, but would, moreover, be an impious resistance to the benevolent intentions of his Creator.

These words are from the pastoral letter addressed last year by the Most Rev. Dr. Nulty, Bishop of Meath, to the clergy and laity of his diocese. Mr. Davitt, of course, thoroughly endorses them, but the honor of having put in such compact and striking form the true doctrine as to the ownership of land belongs not to him, but to one of the most devout and learned of living Irish prelates.

4. You say: "Mr. George found no audience in Dublin for his novelties."

In the first place, this doctrine that the land of every country is of inalienable right the common property of the people of that country is no novelty of mine. It was believed by Michael Davitt and proved by Dr. Nulty before either of them had ever read any of my writings. It was stated by James Finton Lalor, the most logical and thoughtful of the Irish patriots of 1848. "I hold and maintain," said Mr. Lalor in the *Irish Felon*:

> that the entire soil of a country belongs of right to the entire people of that country, and is the rightful property not of any one class, but of the nation at large, in full, effective possession, to let to whom they will, on whatever tenures, terms, rent, services, and conditions they will.

So far from being a novelty, this is the old Irish doctrine, and is, as I am informed by Catholic clergymen who speak the Irish language, still treasured by the peasants of the West.

Nor is it a novelty anywhere else. Herbert Spencer, in his *Social Statics*, has proved that there can be rightfully no individual property in land. John Stuart Mill declared that "the land of every country belongs to the people of that country." Thomas Carlyle believed that "private property in land must be abolished, and would be abolished when Cosmos replaced Chaos," and H. M. Hyndman, President of the English Democratic Federation, has recently reprinted a lecture delivered in Newcastle a century ago by Thomas Spence, in which the same doctrine is advocated in the most striking manner.[73] Further than this, if you will do me the honor to read the chapter in *Progress and Poverty* in which the history of private property in land is considered;[74] or, still better, if you will read the work of Professor Emile de Lavelaye, of the University of Liege, on *Primitive Property*, you will see that private property in land is little better than a modern absurdity.[75] As for this "novelty" finding no audience in Dublin, you are mistaken as to that. There are many men in Dublin, and many men throughout Ireland, who believe, as Dr. Nulty and Michael Davitt and myself believe, and their number is daily increasing.

5. You argue as though the recognition of the equal right of the whole people of a country to the land of their country involved the cutting up of farms into garden patches, and the parceling out of land into equal lots. But it involves nothing of the kind. All that it is necessary to do to make land the common property of the whole people is to appropriate for the benefit of the whole people the rent now wrongfully taken for private uses.

Permit me, in conclusion, Mr. Editor, to make a quotation from the pamphlet to which you allude, which has pertinancy at the present time:

It is only to earnest men capable of feeling the inspiration of a great principle that I care to talk, or that I can hope to convince. To them I wish to point out that caution is not wisdom when it involves the ignoring of a great principle; that it is not every step that involves progression, but only such steps as are in the right line and make easier the next; that there are strong forces that wait but the raising of the true standard to rally on its side.

Let the time servers, the demagogues, the compromisers, to whom nothing is right and nothing is wrong, but who are always seeking to find some halfway house between right and wrong – let them all go their ways. Any cause which can lay hold of a great truth is the stronger without them. If the earnest men among the Irish leaders abandon their present halfhearted illogical position, and take their stand frankly and firmly upon the principle that the youngest child of the poorest peasant has as good a right to tread the soil and breathe the air of Ireland as the eldest son of the proudest duke, they will have put their fight on the right line.

Present defeat will but pave the way for future victory, and each step won makes easier the next. Their position will not only be logically defensible, but will prove the stronger the more it is discussed; for private property in land, which never rises from natural perceptions of men, but springs historically from usurpation and robbery, is something so utterly absurd, so outrageously unjust, so clearly a waste of productive forces and a barrier to the most profitable use of natural opportunities, so thoroughly opposed to all sound maxims of public policy, so glaringly in the way of future progress, that it is only tolerated because the majority of men never think about it or hear it questioned. Once fairly arraign it, and it must be condemned; once call upon its advocates to exhibit its claims, and their cause is lost in advance. . . . Once force the discussion on this line, and the Irish reformers will compel to their side the most active and powerful of the men who mold thought.

And they will not merely close up their own ranks, now in danger of being broken; they will "carry the war into Africa," and make possible the most powerful of political combinations.

It is already beginning to be perceived that the Irish movement, so far as it has yet gone, is merely in the interest of a class; that, so far as it has yet voiced any demand, it promises nothing to the laboring and artisan classes. Its opponents already see this opportunity for division, which, even without their efforts, must soon show itself, and which, now that the first impulse of the movement is over, will the more readily develop. To close up its ranks and hold them firm, so that, even though they be forced to bend, they will not break and scatter, it must cease to be a movement looking merely to the benefit of the tenant-farmer, and become a movement for the benefit of the whole laboring class.

And the moment this is done the Irish land agitation assumes a new and a grander phase. It ceases to be an Irish movement; it becomes the van of a worldwide struggle.

This, to my mind, is the significance of Michael Davitt's Manchester speech.

May 28, 1882.

HENRY GEORGE AND HIS CRITICS:
Davitt's Step in Advance Not the Result of a Sudden Conversion
(*Irish World*, July 15, 1882).[76]

Mr. Henry George writes the following letter to the London *Echo* regarding Michael Davitt's position on the Land Question:

Sir: In the article entitled "Land Nationalization," in your issue of Thursday, there occurs, after reference to some works of mine, the following:

> It is to be presumed that Mr. Davitt has read these works. At any rate, since his liberation from Portland he has been much in company with Mr. George, who is staying in this country. There is no evidence to show that Mr. Davitt had any idea of Land Nationalization before he became acquainted with Mr. George, or with Mr. George's books. Therefore, the so-called explanation of the demand, "the land for the people," must be regarded as an afterthought.

Permit me to say that Mr. Davitt has been very little in my company since he left Portland, while the only times we ever met before were as he was about leaving New York on his last visit to the United States [sic]. Mr. Davitt has, I believe, read my books, and there is, I think, on the Land Question, substantial agreement between us, except as to the matter of compensation to existing landowners; but nothing could be more erroneous than to suppose he had no idea of land nationalization until he had read what I have written, or that his recent explanation of "the land for the people" is an afterthought.

It is well known to those acquainted with the history of the Land League that at the time of its formation (before Mr. Davitt knew anything of me or of what I had written), he, with Mr. Brennan and Mr. Ferguson, wished to make the nationalization of the land the declared program of the League, but was overruled by others who thought this too radical a platform for the time. And it has been well understood, both in America and in Ireland that, whatever others might mean, Mr. Davitt has meant by "the land for the people" precisely what he now says – the land for the *whole* people, and not for any class, be it large or small. That the land of any country is by natural right the common property of the whole people of that country is as obvious a truth that the wonder is not how any man came to perceive it, but how anyone fails to perceive it. Instead of being a new doctrine, it is one of the primary perceptions of humanity. That the land on which and from which all men must live should be made the private property of some of them is as artificial and unjust a system as that of slavery.

As for the criticisms which your correspondent mingles with the praise he bestows upon my book, *Progress and Poverty*, I can well leave them to be answered by the book itself. I would like, however, to correct one mistake. The London publishers of *Progress and Poverty* are not Messrs. Trubner & Co., but Messrs. C. Kegan Paul, Trench & Co.

Yours very respectfully,

INTERVIEW ON IRISH NATIONALIST POLITICS
FOLLOWING THE RELEASE OF PARNELL
(sometime in December, 1882).[77]

Question: What do you think of the vote of the Irish party in Parliament for Mr. Gladstone's Cloture Bill?[78] Does it surprise you?

Mr. George: No. It cannot surprise anyone who has watched the course of the Irish Parliamentary leaders since Mr. Parnell's release from Kilmainhaim Prison. It is the last proof if that were needed that Mr. Parnell and his followers have reconciled their differences with the Gladstone government and have adopted the policy of seeking their needs by falling in with Mr. Gladstone's will rather than opposing it. Of course, a year ago nothing seemed more improbable than this. One could have imagined stranger things as coming to pass within the twelfth month, but I do not think anyone would have imagined this, at least it never entered into my head when thinking of the future nor did I ever hear of a suggestion of its being in the range of possibilities from anyone I talked with. Gladstone has certainly achieved a most remarkable victory.

At the beginning of the year it seemed as though he was plunging deeper and deeper into difficulties from which there could be no escape. His Irish policy was proving a disastrous policy and the moment when he should be forced to abandon it and retire in disgrace seems only a matter of time. But the year draws to a close with the Irish members who were then his most bitter enemies voting in his train. All his difficulties have disappeared and he is now the most powerful minister who ever held sway in England since the days of William Pitt, and, is far more powerful than he, for the crown which when worn by George III was really a power in England is in these days little more than a cipher.[79]

There is something superb in the success of this old man, the imperious dictator of the British Empire. Everything and everybody has seemed to bend to his will or fall before him; and this conversion of the Irish "Irreconcilables," as they were called, into his supporters, this carrying of his pet measure of the Cloture, by the votes of the very men whom it was proposed for the purpose of musseling[80] is indeed a crowning triumph.

Question: This removes all doubt as to the Kilmainham treaty, does it not?

Mr. George: I do not think there has been any doubt about what is called the Kilmainham treaty for a good while. The disposition of Mr. Parnell to act with the government has been apparent since he took his seat again in Parliament after his release from Kilmainham; and the influence of the Parliamentary party has been steadily exerted in Ireland to calm down the agitation and bring the Irish people into such conservative lines as would enable them to work with the government. But, of course, this vote shows with great distinctions that the Irish party has become, for the present, at least, a part of the Liberal Party.[81]

Question: From when does the Kilmainham treaty date? You were in Ireland during the time watching the course of events and ought to know something about it.

Mr. George: Whether there was any formal agreement between Mr. Parnell and the government I do not know, but am inclined to think there was not, but that an understanding was come to whether formal or otherwise, that Mr. Parnell on his part would cease opposition and act as near as possible with the government and that the government on its part would adopt a milder policy towards Ireland, facts show.

The first beginning of this understanding seems to date from the presentation by the Irish party of the bill for the amendment of the land act. This bill was drawn up by Mr. Healy, the brother of the member, a young solicitor who had been for some time in the office of Mr. McGouf, the Land League solicitor assisting him in attending to Land League business. It was revised in Kilmainham and then sent to Mr. Tim Healy for presentation as embodying the amendment which the Irish party thought ought to be made in the land act.

I first heard of the bill in one of the lobbies of the House of Commons where Mr. Healy showed it to me saying, in effect, that it was not an endorsement of the land act, but merely meant to put forward the ideas of the Irish party as to how that act should be amended in case it was amended. I paid no attention to it at the time but I have since learned that it gave a shock to some of Mr. Parnell's most active and intelligent friends and followers in Ireland. The gentleman through whose agency it passed from Kilmainham to Mr. Healy has since told me that from that moment he saw that the fight was really over and that the policy of compromise with the government had been resolved on. And his confidence and hope gave way to a feeling of despair. And so it was with at least some others who had been most active in the Irish movement.

After this bill had been presented, though before anything had been said about it in Parliament, there came the release of Mr. Parnell on parole. That at a time when over five hundred Irishmen were lying in jail, when the ladies of the Land League were going to prison rather than give bail and when it was deemed a point of honor not to make any conditions with the government. For the leader of the Irish people to ask for and obtain release and to stay out of jail, such a time could hardly be explained on any other theory than that some sort of an understanding with the government was being sought or being patched up.

When a general of an army asks for a flag of truce to visit the camp of his enemy the natural supposition is that he is disposed to come to terms. I remember it was my impression at the time that something of this kind was going on and I know since that at least some of the most intelligent and best-informed Irishmen both in Dublin and in London regarded the matter in the same light, though they only spoke of their fears privately. I was in London at this time and an English friend of mine told me that he had dined the night before in company with an Irish member. One of Mr. Parnell's lieutenants, well-known in English literary and social circles had informed him that the trouble between the Irish party and the government was probably over and that they would soon be acting together. I mentioned these things in writing to Mr. Ford but did not [feel] sure enough to state them publicly in my letters to the papers.[82]

I remember going to one gentleman who is in a position to know what is going on in Parliamentary circles and whose cool judgment I had a great deal of faith and saying to him that it looked to me that Mr. Parnell's coming out on parole could only mean some sort of a compromise with the government. He said he did not think so and that in his opinion Parnell, who was a man of weak rather than of determined character, had got very tired of imprisonment and had taken the opportunity of coming out in order to rest himself.

But before Mr. Parnell had got back to prison again there came an unmistakable disposition of a coming together between the government and the Parnellites in the bringing up of the Parnellite bill for the amendment of the land act. When Mr. Redmond and, if my memory serves me, one or two of the other Irish members spoke respectfully and moderately of Mr. Gladstone, and Mr. Gladstone on his part spoke in such moderate terms of the Parnellites that the action was regarded everywhere as most remarkable and significant. It was more the tone of what was said than what was actually said as in conflict and striking contrast of all that had gone before on either side and showed clearly the disposition of both sides to find common ground of agreement.

Question: Was this realized in Ireland?

Mr. George: It was felt in Ireland and much talked of privately, though there was no public expression, but there was a vague feeling of distrust and suspicion. It was very marked in Parnell's release which did not excite the enthusiasm and a little which before would have been confidently looked for when at last he came forth from Kilmainham.

I went up into the room of the Ladies Land League on the night of Mr. Parnell's release and it resembled a [ ]. They were all sitting around a table as though the work was done and had lost their best friend. Enthusiasm broke forth a few days afterwards when it was announced that Davitt was free.

Question: [Do y]ou remember the interview which you had with Mr. Parnell after his release and which was telegraphed to the *Irish World* denying in the most unequivocal terms anything like an understanding with [the] government?

Mr. George: Yes. That interview was printed just as Mr. Parnell wished it to appear, but was about as strong as the English language could make it. The next time I met Mr. Parnell was in London the morning after the assassinations in Phœnix Park. He then stated to me, as though in explanation of what he had said in this interview, that he had written a letter to Mr. Justin McCarthy stating that if the government released the members and in short adopted a more moderate policy, outrages would in his opinion cease and the country become peaceable. He had written this letter, he said, to be shown to Mr. Chamberlain and at the time I saw him in Dublin, did not know that it had gone further, but that it had as he had since learned, been laid by Mr. Chamberlain before the cabinet.[83]

Question: Has this letter ever been published?

Mr. George: No. I think not. The letter to Captain O'Shea[84] was brought up by Mr. Foster shortly afterwards in the debate and although the McCarthy letter was the other document alluded to by Gladstone, it was never brought out.

Question: How did Mr. Parnell's release affect the feeling or rather, as you say, the vague suspicion that there was a treaty with the government?

Mr. George: I do not think it affected it at all. From the tone of the English press it was evident that the coercive policy of the government was falling into utter discredit in England and that the release of Mr. Parnell and his companions was only a matter of weeks, if not days. I do not think that at the time there were more than two or three influential journals in all England that were not favoring such a cause. It was Mr. Parnell's conduct after the release that produced this feeling.

His going off suddenly to England without giving anyone an opportunity to see him, followed by the remarkable speech which signalized his reentrance into Parliament, and instead of assuming the defiant tone which characterized him just before he had been imprisoned he flattered Gladstone in the most impressive way by declaring that he was too great and too strong to be opposed. Then came that dramatic scene in which Foster compelled the production of that letter to O'Shea and brought out that clause which Mr. Parnell in sending the letter had omitted – the clause announcing that thereafter the Irish members would act with the Liberal Party.

Question: Was this a surprise to the Irish members?

Mr. George: I think it was to the majority of them at least. As I understand it even Dillon did not know anything about the visits of Captain O'Shea or any correspondence that had passed. You will remember that on the night in which Gladstone in defending himself against those who had taunted him with weakness in releasing the Irish leaders and demanded to know what grounds he had for [ ] conduct themselves better in the future stated that he had documentary evidence. Dillon, you will remember, got up and denied this most emphatically so far as he was concerned. O'Kelly made a similar denial. Parnell, however, said nothing. This, as I understand, very much surprised and alarmed the others of Mr. Parnell's followers and in the lobby they asked him what there could be in Gladstone's assertions, whereupon he told them of the letter to O'Shea, omitting, however, the suppressed clause. This, I am told, produced a feeling of surprise and dismay among Mr. Parnell's followers, the majority of whom did not [know] anything of the kind, and when Mr. Foster compelled the reading of the full letter the effect, of course, was very marked both there and in Ireland.

Question: But everybody has been since protesting undiminished confidence in Mr. Parnell and he has certainly for a long while seemed to retain his popularity with the masses of the people?

Mr. George: Yes. But the expressions of confidence of which you speak did not really [ ]. I was going to say the feelings of those who made them but at least did not represent the feelings of the intelligent and watchful Irishmen ... [85] [ ] the feeling so far as I could gather it seemed to be of utter dismay but they also felt that anything like discussion would be ruinous and Mr. Parnell had so fully and so thoroughly the confidence of the people that even those who felt they could no longer believe in him hesitated to express their feelings publicly.

And then came a sort of reaction. The Parliamentary party rallied around him and the idea gained ground among the people that Mr. Parnell was playing some deep and astute game with the English government – and that if he pretended to treat with them it was only in his own time to more forcibly strike.

The feeling of the Irish people towards Mr. Parnell has not been simply that of unbounded confidence in his devotion but of unbounded confidence in his strategical capability. They have looked on him as a very "long-headed" man who laid his plans far ahead; who knew always what he was doing and who could beat the government by his superior management. The very qualities in which he most differs from the typical Irish character gave them the greatest idea of his power, and as I have often heard expressed that Parnell was the best leader Ireland could have because he was so unlike an Irishman.

To put it roughly, the feeling among those who would have been first to advance anything that they believed to be a real surrender to the government was that he was only fooling the government. But there was still left on the part of large numbers, though it has not found much public expression, a feeling of distrust and this seemed to me steadily growing up to the time I left Ireland.

Question: Does that account for the lessened energy in the Irish movement?

Mr. George: Certainly the change in the policy in the Land League leaders accounts for it. Revolutions never go backwards because they lose their force as soon as it is attempted to turn them backwards and whether there was or was not anything like the treaty of Kilmainham it is certain that the policy of the Irish leaders has ever since that time been a reactionary one. Their whole influence has been exerted to prevent the spread of radical ideas and to bring the people into lines of thought and of action which agreed with the notions of the English Liberals.

Question: How has this been shown?

Mr. George: In various ways. From the moment of Mr. Parnell's release, when he went off to Kingstown to [ ] in order to avoid any demonstration, up to the recent conference, where by the influence of the Parliamentary party a milk-and-water resolution was adopted as the land platform. And Davitt was assisted by T. P. O'Connor for proposing that the managers of the new league be elected by the people.

Everything that could excite or maintain the spirit of the people has been discouraged and the word passed along to keep quiet. The No Rent Manifesto was ignored, tenants advised to settle and to take the benefit of the land act, Brennan silenced,[86] Davitt attacked, the ladies disbanded, the Labor League got into the hands of the Parliamentary managers, and nationalization, which simply means the land for the people attacked as though it meant treason to the Irish cause.

Question: But it is charged that Davitt's advocacy of nationalization was calculated to produce dissensions and divisions?

Mr. George: Yes, I know. That is the charge with which Davitt has been bulldozed and Brennan silenced but before Mr. Parnell went into Kilmainham and when his rule was opposition to the government in conjunction with it, the advocacy of nationalization would not have been thought productive of dissension and divisions. Every Land League meeting rang for the cry of the land for the people and the most radical sentiments were proclaimed. Even Mr. Parnell said things that could mean nothing less than nationalization. Even his famous theory of "Prairie[?] Value" is a very different thing from the fair rent that is now talked of. Had the policy of the leaders been to carry the movement forward they would have welcomed even if they had not endorsed Davitt's advocacy of the nationalization of the land and Brennan's ringing utterances when he first got out of jail. To allow the movement to thus go on to the actual abolition of landlordism would have broken up any understanding with the government and would have made action with the Gladstone Liberals impossible.

The attacks upon Davitt were, it seems to me, not that he made a new departure, for his utterances on the land question when he came out of prison were consistent with his utterances when he went in, but it was that in the meantime the Parliamentary party had made a new departure and repudiated the gospel according to Dr. Nulty and accepted the land gospel of William Ewart Gladstone and Joseph Chamberlain.

Question: Why then was it that so many of the Irish Nationalists who have always been opposed to anything like a combination with the government or even the Parliamentary representation and leadership join in the outcry against Davitt?

Mr. George: That I do not understand except by supposing that they did not stop to think what they were doing; [they] were carried away for the moment by the misrepresentation made to them. The moment they stopped to think of it, and I am inclined to believe that a good many of them now are stopping to think of it, Irish Nationalists must see that the nationalization of the land which means the complete abolition of landlordism in Ireland is the only settlement of the land question consistent with the Nationalists' aims.

Even those who do not realize the importance of the [ ] of the land as the only basis for political independence must see that the surest way to maintain the nationalistic sentiment is to inspire the people to demand [more] of the land question than the government is prepared to give; to arouse not merely the tenant-farmers but all classes of the people to demand their natural rights in the soil. While they must also surely see that in any such scheme as that proposed by Mr. Parnell to have the government buy out the landlords and sell again to the agricultural tenants would raise up a considerable body of men upon whom the government could confidently rely to oppose agitation and change.

Without any reasoning from general principles this can be seen clearly in the history of Ireland. That the land question is the fundamental question may be seen in the history of Ireland even if it were not seen anywhere else. It was by giving the Irish chieftains the land, which under Irish law was really the common property of the [ ] that they were bribed into extending English dominion and accept English law and the Cromwellians among whom Irish land was presented by donation or sale in the seventeenth century turned out at once from being the most bitter of opponents of prelacy into its most loyal supporters, the moment they found this [they] imagined that monarchy and prelacy would best secure their lands.[87] Strong as England is, it is not by force of British bayonets that Ireland is held in subjection to it so large as it is by the power of those classes in Ireland who are made supporters of the government by their special interests and special privileges.

THE GREAT QUESTION:
How it is Bound to Revolutionize Things in England and Ireland; But One Settlement Only, and that is the Making of the Soil National Property and Opening Up Equal Opportunity to Its Use; A Year's Experience Across the Water Only the More Confirms Henry George in the Opinion that Land Monopoly is the Bottom Evil; How Parliamentarianism Killed the Land League; Peasant Proprietary Does Not and Cannot Satisfy the National Feelings of Ireland; English Rule in Ireland a Brutal Despotism?
(Irish World, January 27, 1883).[88]

I am requested to write an article for the Christmas [issue of the *Sacramento*] *Bee*, a request with which I gladly comply as well as the time at my disposal will permit. For not only is the editor of the *Bee* an old and valued friend, but in writing for his newspaper I feel as though I were again coming into communication with the many friends I have left in California.

I am asked to write upon the subject of the Land Question in the British Islands. This is a subject on which there is so much to say that would be interesting that I can only hope to touch on some parts of it. The first article I wrote for the Christmas *Bee* (and that, by-the-by, was, I think, the first time the special Christmas number of the *Bee* was issued) was upon "The Land Question in Ireland." In that article, which, perhaps, many of the readers of the *Bee* will remember, I contended that the Land Question in Ireland did not essentially differ from the Land Question in any other country, and that the only measures which could settle the Irish Land Question upon a secure and just basis would be measures which ought to be applied to all other countries, including our own – namely, such measures as would make the land virtually the common property of the whole people and secure everyone an equal opportunity for its use, and an equal share in all the benefits arising from it.

There is no need of my going over the same reasoning again. I have only to say that what I saw in Ireland confirmed the opinions I then expressed.

A Brutal Despotism.
I did not fully appreciate before I went to Ireland and mixed with its people (nor do I think that until he has done this an American can fully appreciate) the tyranny and baseness and degrading effect of the Government under which Ireland lies. No man who loves liberty can, when he realizes the character of the Government, fail to feel a most ardent sympathy with those Irish patriots, whose struggle it has been, and yet is, to free their country from this galling yoke – the yoke of a blind and brutal despotism, hardly less hurtful to England

whose strength imposes it, than it is degrading and hateful to the people on whom it rests.

And I think, to fully appreciate the Irish character, one must go to Ireland and realize the terrible disadvantages under which her people struggle. That they have preserved so much manhood and have maintained such a high standard of morals, in the face of conditions calculated to debase and corrupt, is a thing of which Irishmen may be justly proud. So far from being a turbulent people they really seemed to me as a most peaceable people. The wonder is not that there are occasional outrages in Ireland. The real cause of wonder is that there are not more. Over and over again, while I was in Ireland, I found myself wishing that ten or twenty thousand of such Americanized Irishmen, as some I have known in California, could be transported into the Green Isle and settled down among the people. I have a notion that Ireland would become a very much more turbulent country.

The Bottom Question
But of these things I can speak only incidentally. What I have been asked to write about is the Land Question.

Yet, in the Land Question of Ireland is to be found the center and cause of all Irish difficulties and troubles. The bloody wars, relentless exterminations, and ruthless penal codes, that mark Irish history, have all their cause in the effort to get and to keep the land of the Irish people, and the political tyranny exercised in Ireland today springs from the same cause. Ireland is not governed by the English people. It is governed by the landholding class, who use the English power to maintain their dominance and preserve what they call their rights. They are as brutal and unscrupulous as a privileged class always is when it finds a popular revolt beginning to make headway against it, and they are backed by their fellows, the privileged oligarchy that rules England. For though it has become the fashion in the United States of late years to talk of England as a republic in all but name, nothing is more fallacious. The Government of England is not in form or in fact a government of the people, but a Government of a privileged class – an aristocracy of rank and an aristocracy of wealth.

The Question the Same the World Over
And, as I wrote in the *Bee* before I had gone to Ireland the Irish Land Question does not essentially differ from the Land Question in England or Scotland. The Land Question in Ireland is essentially the same as the Land Question in California. The landholder in Ireland has no legal powers that the landholder does not have in California; in fact he has not the same unrestricted ownership. It is the fashion of many Irishmen to talk about the "feudal ownership of

land in Ireland" and "feudal Landlordism." I notice that Mr. Parnell uses this term when he wishes to say anything against Landlordism. And in the lecture which I heard him deliver during his recent visit to this country, Mr. A. M. Sullivan, a gentleman whom I much respect and esteem, constantly used the same term in denouncing Irish Landlordism.[89] But this is a misuse of words. Feudal Landlordism does not exist in Ireland any more than it exists in England or in California. The essence of the feudal system of land tenure was that the holding of land was a trust or peculiar privilege for which a return of some sort had to be made to the whole community.

But the land system which has supplanted it, and which now exists in the United States as well in Great Britain, treats land not as peculiar trust, but as individual property. And all the tyranny, and extortion, and impoverishment which can be justly charged upon Landlordism in Ireland springs from this fact – the fact that the laws which make the land upon which, and from which the whole people must live, the private property of some individuals among them.

Landlordism in Cities

This is just as true of the cities of the country; among the artisans and laborers as among the farmers. If I were asked by a stranger in Ireland to point out to him the worst examples of the misery and degradation caused by Landlordism, I would not take him to the seacoast of Donegal, or Mayo, or Galway, where the people are compelled to pay rent for miserable patches of rock and bog, hardly capable, in the best of seasons, and if rent free, of furnishing them the bare necessaries of life. I would not take him to those poor districts, where the cry of famine is again going up, but I would rather show him, as even worse than the poverty and suffering of these poorest districts, sights that are to be seen in the best of times in such cities as Dublin, Cork, and Belfast.

Nor is Landlordism on the one side of the Irish Sea a whit more merciful than Landlordism on the other. The crofters of Skye and other parts of the Scottish Highlands are, if that be possible, under even worse conditions than their Irish brethren; and the English laborer is a much more degraded being than his fellow of the same class in Ireland. Even the poorer classes of the Irish peasants have at least some recollections of their rights, but this has almost been crushed out of the English laborer, who, working like a beast and living little better than a beast, can look forward only to the workhouse for his declining age.

Absentee Landlords

I well remember a remark which an English gentleman made to me when pointing out in one of the most fertile and beautiful counties of England, how even the footpaths through which he used to roam when a boy had been closed, and how the lord of the manor had fenced in even the little common on which the children used to play, so that now there was literally no place for them except the roads, although on each side lay thousands of acres tenanted only by deer and cattle. "The Irish," he said bitterly, "have had one great advantage over us – their landlords have been absentees!" And there is a sense in which this has been an advantage. The difference of race and religion that has so long existed between the landlords of Ireland and the people, and the fact that many of the former have been and are absentees, have prevented Landlordism in Ireland from losing the character of an alien system, and preserved a better recollection of ancient rights. The English laborer, the descendant of the ancient yeomanry of England, has been so thoroughly robbed he has lost consciousness of the robbery. The Irishman, though he bowed to force that he could not resist, never confounded it with justice.

The Gladstone Land Act

Readers of the *Bee* will recollect that in the Christmas article before alluded to I pointed out that no halfway measures, such as were at that time proposed by the Irish leaders, could settle the Land Question in Ireland upon a just basis, or effect any general improvement in the condition of the Irish people. Since then the measure known as the Gladstone Land Act has been passed and has now been over a year in operation. It has upon the whole effected a considerable reduction of rents to a large number of the agricultural tenants, and had given those who have taken advantage of it security that for fifteen years their rents will not be increased. But while the reductions have on the whole fallen much short of the desire of the tenants, no general gain in which other classes may share has been affected.

Where rent has been reduced, the value of the tenant right has commensurately gone up, so that the man who wants to get a farm must, even where the rent has been reduced, pay as much as before, for the reduced rent which he will have to pay to the landlord is offset, by the increased price which he must pay to the outgoing tenant. Wages have not been increased at all, nor will they be increased by the operation of this law.

A Peasant Proprietary

The same shortcomings, it is plain, would characterize any measure for the creation of a peasant proprietary, while any law which would enable a number of farmers to buy out their holdings, or in any other way to become landowners themselves, would only increase the landlords' strength and would enable the Government to more effectually resist popular demands. Under the operations of the Encumbered Estates Act a number of Irish estates were divided up and sold to the tenants.[90] It is noticeable that in these localities the new proprietors have become intensely conservative and, during the recent agitation, would have nothing to do with the Land League and its works.

And so, while any scheme for the purchase of their holdings by the agricultural tenants would do nothing whatever for those who have no land, while the position of the future tenant, of the laborer or the artisan, would be in nowise improved, the great landlords would get a sufficient bodyguard of smaller landowners, and the English Government would be planting another garrison in the country to support its misrule. And as in nine cases out of ten the proprietors whose estates were sold to their tenants through the intervention of the Government, would take out of the country the purchase money they had received, the drain which absenteeism now makes upon Ireland would for a long term of years be increased rather than diminished.

Course of the Irish Movement

At the beginning of the Irish agitation the popular demand was only for a reduction of rent; but it is in the nature of things that such an agitation should become more and more radical, and that a people who began by asking why the landlords should get extortionate rents, should before long, begin to ask why the landlords should get any rent at all. This has been and yet is the course of the Irish Land agitation, but its development has been for the time checked by the action of the popular leaders.

Davitt, Brennan, Ferguson, and most of the men who with them were the first starters of the Land League had clear and radical ideas upon the Land Question, though in order to secure the cooperation of Mr. Parnell and the more conservative element, they made the mistake of consenting to the enunciation of a halfway platform. Mr. Parnell and his coadjutors, as has been evident to anyone who has watched their movements, have never been thoroughly in earnest on the Land Question. After Davitt was sent to Portland and Brennan to Kilmainham the Parliamentary party gained complete control of the organization of the League, and their control was made all the more powerful from the fact that they were the disbursers of the large funds sent from America.

A Great Opportunity

While Mr. Parnell and the Parliamentary followers never grasped the Land Question, and had in them nothing of the social reformers, they were carried on by the strength of the movement. Davitt's demand for the land for the people was reechoed from every platform, though unfortunately a precise meaning was too seldom attached to the phrase; and the conference held in Dublin in

September of '81 (and which, by-the-by, was forced on the Parliamentary party), showed that on the Land Question the Irish people were far more radical than their leaders. Then came the arrest of Mr. Parnell, the issuance of the No Rent Manifesto, the suppression of the Land League, and the dragonade of Ireland under Forster's administration.[91] All this excited and intensified the popular feeling to a much greater degree than the operation of Gladstone's Land Act served to allay it.

The National feeling was thoroughly roused and no matter how radical a position Mr. Parnell had taken the mass of the Irish people would have followed him. It was but natural to suppose that the effect of imprisonment upon the Parliamentary leaders, backed by the spirit which had been wakened in the country, would have carried them forward to more and more advanced ground.

The Kilmainham Treaty

But this expectation doomed to disappointment by that understanding between Mr. Parnell and the Gladstone Government, which is called the "Treaty of Kilmainham." Instead of taking more advanced ground, it was found upon Mr. Parnell's release that he had fallen back to much lower ground, and, with perhaps a moment's wavering, his Parliamentary Lieutenants, accustomed to obedience, and without any clear ideas of social questions, followed him, and the whole personal influence of the Parliamentary party and the whole force of the organization was turned against the natural course of the popular movement, and exerted in the attempt to turn the rising demand of "the land of Ireland for the whole Irish people" into a miserable program for the purchase of their holdings by the farmers – a program which would not be offensive to the governing oligarchy either in Ireland or in England, and would coincide with Mr. Gladstone's wishes and plans.

The Attack on Davitt

Davitt came out of prison amid the whirl of excitement caused by the Phœnix Park assassinations to find that some sort of a compact had been made for the lowering of the standard which he had raised. In his Manchester and Liverpool speeches he endeavored, without giving any offense to the Parliamentary party, to raise the true standard of the Land League, and to declare unequivocally that the land of Ireland was by natural right the property of the whole people of Ireland. But he was immediately set upon as a disorganizer, and the charge (a most terrible one to him) was rang in his ears, both in Ireland and when he came to America, that he was creating dissension and division, and was striving to overthrow Parnell and to become the popular leader. Davitt is a man of large views, of clear insight, of frank and generous impulse, and of a most ardent and devoted patriotism – patriotism of that highest kind which, while seeking all things good for its own country, recognizes also the common interests of mankind. But he is not made of stern enough stuff to be unaffected by such an onslaught as was made upon him, and instead of simply doing what Dr. McGlynn[92] told him at his reception in New York he should do: "Go ahead and preach the Gospel," he began to apologize and defend himself, and to profess devotion and obedience to those from whom he ought to have held himself aloof.

Radicalism Discouraged

The radicalism of the Parliamentary party, acting upon the proverbial devotion of the Irish people to their leaders; the influence given by the control of the Land League funds and of what was left of its organization, has been, since Mr. Parnell's release from Kilmainham, added to that of the Government and the landowning class to discourage radical expression upon the Land Question and to prevent the growth of radical ideas. Brennan, from whom much was to be expected, made a ringing speech an hour after his release from Kilkenny, in which he asserted the equal right of every Irishman to his native soil, and denounced any attempt to patch up a temporary settlement of the Land Question by increasing the number of landowners and making special concessions to the agricultural tenants.

But he was immediately gagged by the same influences that were brought to bear on Davitt, and has not since opened his mouth, for fear that the charge of creating dissensions will be brought against him. The ladies of the Land League, who were Nationalizers to a woman, were disbanded. Associations that had nothing radical on the Land Question in their program, were started to take the place of the Land League, and were securely placed under the control of the Parliamentary managers, while the *United Ireland*, a paper which was purchased with Land League money, largely devoted itself to the effort to "draw a red herring across the trail" of the Land agitation by getting up other agitations.

But the Movement Still Goes On

I have not space in this article to go over this phase of the history of the Irish land movement. All I wish to point out is this, that the natural course of the movement still goes on, not with the rush and sweep that would have been the case had not its natural development been checked by the new departure of Mr. Parnell, but possibly in a more thorough manner. All over Ireland and among all classes of people the question of peasant proprietary versus nationalization is being discussed, and that the popular verdict will be given in favor of the nationalization, and that in a form which does not involve the compensation of the landlords, is certain.

An educational process is going on which will not, and which cannot stop short of the demand that the whole soil of Ireland shall be recognized as the common property of the whole people of Ireland. Mr. Parnell and his Parliamentarians can no more stop the march of ideas than could Mr. Forster. Even if the Gospel of the Land for the People had not been heard in Ireland, even if Bishop Nulty's denunciation of any settlement of the Land Question which does not secure to the humblest his equal share in the bounty of the Creator as a wrong and an impiety had not rung forth, the agitation which has been, and yet is, going on in Ireland could have no other result. I am inclined to think that the Irish people were in many respects the most conservative people in Europe, and that Isaac Butt was right when he used to tell Irish landlords that Home Rule would enable them to keep possession of their estates for fifty years after the agrarian wave, which he foresaw must sooner or later arise in England, had swept away Landlordism on the other side of the St. George's Channel.

But the breach has been made; the leaven is working. Radical ideas on the Land Question have entered the Irish mind through that channel by which alone they could effectively enter – the passionate desire for national freedom, and they cannot now be cast out. Peasant proprietary might have satisfied the Irish people had it been offered soon enough, but it is now too late. The tide of national thought and feeling is slowly but irresistibly setting in favor of the ideas represented by Davitt and against the ideas represented by Parnell. Peasant proprietary will not satisfy the national feeling; it will not satisfy the aspirations, strong though vague, which have been aroused among other classes than the agricultural tenants, and every day the idea of nationalization is gaining.

In England and Scotland

In the meantime a still more important movement than the Irish land agitation has commenced on the other side of the Channel. The agrarian wave which Isaac Butt foresaw would one day sweep away the English aristocracy has at last begun to swell. In some degree it is a direct result of the Irish agitation; but in the main it is an independent movement; though springing from the same great causes and converging to the same end.

The example of the Irish strike against rent, aided by the missionary work of some of the Scotch Land Leaguers, has infused a new spirit into the crofters of the Scotch Highlands, and the stand made in Skye, by calling popular attention to the depopulation of a great part of Scotland has aroused Scotch feeling; while both the Scotch farmers and the English farmers have been invited by the passage of the Irish Land Act to demand something of the same kind for themselves.

But the strength of the movement that is beginning in Great Britain lies not with the agricultural tenants, but with the people of the cities and towns. And for this reason the land movement in England and Scotland is from the first assuming a more radical form than in Ireland. The Irish land agitation has done good in England by bringing into discussion the subject of land tenures, but no proposition for the reduction of agricultural rents or the establishment of a peasant proprietary could secure much sympathy in England and Scotland, because the farmers are a comparatively small class of capitalists with whom the masses have no special sympathy. Land nationalization on the other hand appeals to the interests and the sympathies of the main body of the people, and on both sides of the Tweed men[93] are beginning to realize that the Land Question involves town rents as well as agricultural rents – that it is really nothing more nor less than the great Labor Question.

The adoption by the recent Congress of the Trade Unions at Manchester of a resolution in favor of the nationalization of the land, was more ominous to English Landlordism than any action of the Farmers' Alliance could be; and it is only one of the many indications of the direction in which the tide of public opinion is setting in England.

Discussion Fatal to Landlordism

Whoever reads the English papers can see how the question of land nationalization has within the year come to the front in England, and is now admittedly a topic that must be discussed where a few months ago it was treated with contemptuous silence; and when discussion begins, the rest may be predicted. As I told the English clergymen, who did me the honor of inviting me to a conference in London, if only half a dozen of the principal landholders would start in to defend the existing order of things and to prove to the people of England the rightfulness of private property in land, we whose aim it is to overthrow it might fold our hands. They would do our work. For so absurd and so unjust is the institution which makes the land, on which the whole people of the country must live, the private property of but a few of their number, that it cannot bear even to be defeated. It can only continue to exist when it is not questioned. Discussion is fatal to it.

And the discussion that is beginning in the English press is merely a faint indication of the amount of discussion going on. There are now scattered through England – and it is one of the significant signs of the times – a great number of clubs and societies who hold meetings for the discussion of public questions, but which are never reported in the press. In these the Land Question has become a favorite topic, and there is no lack of men to take the most radical ground. The enormous sale which the cheap edition of *Progress and Poverty* and *The Irish*

Land Question are having show, in fact, how much attention the subject is exciting. Of my own personal knowledge I know that there are, from the universities down to the ranks of the day laborers, many Englishmen who will not rest until the English Land Question is forced into practical politics. And among those who have fully accepted the truth, that land is rightfully the common property of the whole people are many of the clergy of all denominations. They are yet a minority, it is true, but their numbers are steadily growing. On both sides of the Atlantic the Land Question is the coming question.

AN ARTICLE FROM THE *IRISH TIMES*
(April 10, 1884).[94]

Mr. Henry George, of *Progress and Poverty*, has arrived in London, and contemplates lecturing under the auspices of the Land Reform Union in a number of English and Scotch towns.[95] He wants the State to own everything and do everything – possess every acre, every steam engine, every water pump, every apothecary's mortar and pestle, for it comes to that, and every cobbler's stall. If the State is bound to protect the people in all their interests, why not ensure to them sound boot leather? It has a closer connection with sanitary reform and the prevention of disease than the whitewasher's brush. Mr. Henry George does not propose to come to Dublin. Our city, in whatever else it has demerits, always did try perfumers of the sort. It was too fierce a furnace of criticism for John Bright once, and Mr. Gladstone himself escaped from it with difficulty.[96] Mr. Henry George is invited hereby to come to Dublin and show us how he would nationalize the land and abolish all private bargaining, and we promise to give him a fair hearing.

THE IRISH QUESTION FROM AN AMERICAN STANDPOINT
(sometime in 1886).[97]

Mr. Parnell's article in the last number of the *North American* is notable as a statement of the views and aims of the leader of the Irish agitation, and like all similar statements which have fallen under my notice, it seems to me also notable in its failure to grasp the real cause of Irish distress, or to prepare any adequate remedy.

I say this in no want of sympathy for Mr. Parnell or the movement he represents. While it seems to me that he and his colleagues have opened a question far greater than they appreciate, it also seems to me that they are doing a far greater work than they may yet understand. They have begun an agitation which is not likely to end in Ireland or with the measures they propose – an agitation which is everywhere bringing up the question of land tenure and causing men to think.

But on neither side of the Atlantic does the real significance of the "Irish land question" seem yet to be fully realized. It is talked about and written about as though it grew out of conditions peculiar to Ireland. In truth, however, what is coming to the front in this Irish land question is a question of universal and growing importance.

For, except as to circumstances which give it prominence, there is nothing peculiar about the Irish land question. In essence, it in no wise differs from the English land question or the American land question. The great fact from which it springs is the fact that as in every other civilized country, some of the people of Ireland are permitted to treat as their exclusive property the land on which and from which all the people of Ireland must live. The great issue which it presents is whether this system shall be continued.

Mr. Parnell and the Irish Land Leaguers do not seem to perceive this fact or as yet to fully realize this issue. As a consequence, the position they take is illogical, and the remedies they propose are insufficient. The only logical question they can assume for attacking the present system is that the soil of Ireland belongs of right to the people of Ireland; the only adequate remedy they can propose is one which fully recognizes this principle.

To this position they must ultimately come if the agitation they have commenced is to run its course. It is probable that the really earnest and thoughtful among them already feel this, though ideas of expediency may yet prevent the avowal. But on the other side it is certainly felt. For it can hardly be that the cold indifference to the fact of the famine[98] on the part of the ruling classes of England, of which Mr. Parnell complains, springs, as he intimates, from the desire to keep Ireland poor and miserable, so much as from the fear of encouraging any questioning of the titles to their own estates. The slaveholders of the South were not as a class wanting in human sympathies. But they naturally avoided and resented any giving of prominence to the ugly facts of slavery as a virtual attack upon the system itself.

This fact, that Irish land tenures are substantially the same as the land tenures of other countries, is at once the weakness and the strength of the Irish movement. It is the real secret of an opposition which Mr. Parnell has doubtless discovered is not confined to England. Yet it not only gives to the Irish land question an interest deeper and wider than can attach to any local matter, but it gives to the Irish agitators, whenever they have the frankness and courage to raise the true standard, the power to summon to their aid forces of which they have not availed themselves.

The cruel wrongs of centuries of alien rule, made more hateful and more hated by sectarian animosities and race prejudices, have left in the Irish mind bitter memories, and it seems natural to Irish men to attribute the present miseries of their country to English oppression, and to talk vaguely of national independence as though that of itself would redress all the wrongs of the Irish peasant. But, however much English domination may in the past have affected the production and distribution of wealth, or the restrictions to which Mr. Parnell alludes may have cramped industrial development, there is today in an economic sense no particular oppression of Ireland by England.

That, as Mr. Parnell says, more revenue is raised in Ireland than is spent in Ireland, is true of Great Britain as well, as it must be true of every country that maintains fleets, ambassadors and costly colonies, and indulges in the imperial amusement of foreign wars. Certainly Irish trade is as free as English trade (and a good deal freer than American trade) while Irish land laws only differ from those of England in recognizing certain rights in the tenant. Nor yet do they materially differ from our own land laws.

We have a good deal of Irish rack rents, and evictions, and absentee landlords. But in these there is nothing peculiar to Ireland. For, to say nothing of England, or of other countries, there are rack rents and evictions and absentee landlords in the United States, as well as in Ireland.

What is a rack rent? Simply a rent fixed by competition at short intervals. This is the common rent among us today. In our towns most of us live in houses rented from month-to-month, from year-to-year, for the highest price the landlord thinks he can get; in our agricultural districts land is rented from season-to-season to the highest bidder. This is what in Ireland is called a rack rent.

What is an eviction? Simply a legal ejectment; as well-known here as in Ireland. Only the other day one of our Federal Courts passed sentence of eviction upon 1,800 settlers who had in good faith made homes on what they believed to be government land, and in all our cities evictions for nonpayment of rent is constantly taking place.

Nor is the absentee landlord peculiar to Ireland. Most of the large English and Scotch landholders are absentees. A majority of our large American landholders are also absentees. Much of our farming land in California is held by men who live in San Francisco, in the East or in Europe. The great "bonanza farms" of the new Northwest, as we are told by a writer in the January *Atlantic*, are owned by and worked for professional men and capitalists who live in Minneapolis, Chicago, Boston, or New York – hundreds and thousands of miles away. Corporations located in the Atlantic cities own much greater bodies of land, at much greater distances, than do the London corporations who have landed estates in Ireland, and although landlordism in its grosser forms is only beginning in the United States, there is probably no American, wherever he may live, who cannot, in his immediate vicinity, see some instance of absentee landlordism. The fact is that in a civilization like ours absenteeism is the necessary result of large landed estates. The tendency to concentration shows itself in this way, as in many others, for the great cities have irresistible attractions for those who can live where they please.

And while it is to be remarked that the owner of land in Galway or Kilkenny would be to his tenantry and neighborhood just as much an absentee if he lived in Dublin as if he lived in London, it is also to be remarked that in itself absenteeism is but a minor evil. So far as Irish producers are concerned, the produce drawn away as rent might as well be destroyed or never produced. But if these landlords were compelled to live in Ireland instead of abroad, all the Irish people would gain would be, metaphysically speaking, the "crumbs that fell from their tables." If the butter and eggs, the pigs and the poultry, of the Irish peasant must be taken from him and expected to pay for his landlord's wine and cigars it makes no difference to the peasant whether the wine is drunk and the cigars smoked in Ireland or somewhere else.

But the main fact to which I wish to call attention is, that the Irish land system is simply the general system of modern civilization – the system which we of the United States have adopted and still maintain.

How little we have to boast of – with what ill-grace denunciations of Irish landlordism come from us, may be clearly seen if we imagine Ireland peaceably separated from the British Empire and made a state of the American Union under such a Constitution and laws of those of any of our states.

Sweep away every vestige of British rule, extend over Ireland today the Constitution and laws of an American state and of the American Union, and Irish peasants would gain no shadow of right to their native soil, while Irish landlords could rack-rent, evict, and drain away the produce of the land as remorselessly and rapaciously as now. Our laws would just as fully recognize this right to ask what price they pleased for the use of this property; our Courts would just as quickly issue a writ of ejectment against any tenant who failed to pay his rent, and the officer who went to serve it would have at his back, if need be, not merely the whole power of the country, not merely the whole power of the State, but the army and navy of the United States; and absentee landlords could still spend in London or Paris incomes drawn from the labor of men who in the best of times cannot decently keep their families.

A bigot landlord could still declare that no Catholic who refused to permit his children to be instructed by Protestant Bible readers should live on his estate. A lecherous landlord might still ruin by a raise of rent the farmer whose pretty daughter turned from his advances. Any bailiff could still give notice to quit to the cottagers who neglected to raise his hat as he passed by. Any merciless creature in whom was vested the legal ownership of land could still issue the edict which compels his tenants to deny the warmest claims of charity, the dearest ties of kinship, and leaves evicted families to the shelter of the heavens and the cutting winter blast. Any owner of land who preferred sheep to men could still reenact the scenes which drew from Goldsmith his deathless plaint and convert the smiling village into a dreary solitude.[99]

For these powers do not spring from any special laws; they are not vestiges of anti-Catholic codes or relics of feudalism. They arise wholly from the ownership of land. Every American landowner has them, as fully and completely as any Irish landowner. The only difference between the two is that with us competition for the use of land is not yet as great. But when it becomes as great, and that time is fast approaching, American landlords can and will do precisely the same thing that any set of men clothed with the same power and placed in the same circumstances would naturally do. The Irish landlords are neither any better nor worse than other people. Anyone who looks around him may see that human nature does not alter with longitude.

But it may be asked: If the causes which produce distress and famine in Ireland are general ones, how is it that they do not elsewhere produce the same results?

I answer that they *do* everywhere produce, or tend to produce, the same results. The only difference is that owing to peculiar conditions the relation between cause and effect is more easily seen in Ireland than elsewhere. Ireland is an agricultural country, a well-populated country, and a country where the ownership of the land is in the hands of a clearly defined class, and where the simplicity of the industrial organization brings laborer and landowner into direct connection; and hence the relation between Irish famine and Irish landlordism is clearly seen. But the same relation exists between English pauperism and English landlordism, between American tramps and the American land system.

I fully agree with Mr. Parnell that Ireland is not, and never has been overpopulated; that all the results which English economists so complacently attribute to overpopulation are really the results of bad government and unjust laws. I fully agree with him that there is no reason in the nature of things why Ireland could not maintain a vastly increased population in vastly increased comfort; but I cannot agree with him that the lessening of population by emigration has really made it harder for the lower class of Irish people, or the building up of manufactures would have decreased the competition for land. Had the English clauses for "protection to home industry" not been suffered to secure the strangling of Irish industries, Ireland might now be more of a manufacturing country, with larger population and greater aggregate wealth.

But the tribute which the owners of the land would have levied upon labor would likewise have been greater. Put a Manchester, a Glasgow, or a London in one of the poor Irish counties, and where the landlords now take pounds they would be enabled to take hundreds and thousands of pounds. And the competition for land, which Mr. Parnell seems to think would be less were there more manufacturers, would be as intense as now, though it might, in fact, take the form of competition for employment.

For, competing with each other for the opportunity to make a living, the Irish peasants are obliged to give for this opportunity all the produce of their labor above what in ordinary times is a bare living. Sometimes they promise to give more, but this manifestly is all they can give. Now what, for the same privilege, are corresponding classes of English laborers obliged to give? Precisely the same price, since all that remains to them is a bare living in ordinary times, while with every season of disaster, or as the only recourse in ill-health or old age, they are obliged to resort to the poor rates or to charity, and large numbers are constantly maintained as paupers. And what is true of England is true of other countries where industry is diversified. Wages in their lowest stratum everywhere touch or tend to this minimum – a bare living.

And so, too, out of Ireland as in it, is rent the devourer of the real earnings of labor. The Irish peasant hires his land from a landlord and pays rent directly. The English laborer is hired by an employer, who is generally not a landowner but a capitalist. But that in the one case as in the other it is rent that consumes the earnings of labor is clear, for interest, the return to capital, is no higher in England than in Ireland.

Mr. Parnell mentions, as though it were something peculiar to Ireland, the fact that within the last twenty-five years rents have largely increased while wages, measured in produce, have even fallen. But the same general fact is obvious in every civilized country, and obvious just in proportion as the country is a progressive one. Everywhere rent has largely increased,[100] but everywhere that twenty-five years ago wages [ ] at the minimum of a bare living, they [ ] these still, while in places where the minimum was [ ] that they have sunk or are sinking to it. When I [ ] I got, when a boy, for unskilled manual labor more than many skilled mechanics are now glad to work for. But sands then all but valueless are now worth hundreds and thousands of dollars a front foot, and agricultural land which could then have been had for government price now rents for a third and even a half of the crop.

Why and how material progress must tend to the augmentation of rent; and, so long as land is treated as private property, must crowd down wages, I have recently endeavored to show at length.[101] It is sufficient here to point to the fact. For the fact makes it clear that the cause of the Irish people is in reality the cause of the masses all over the world.

The [ ] distress in Ireland results ... from a succession of bad seasons; but this famine is in truth what an English writer (H.M. Hyndman) styled the recent Indian famine – a financial famine.[102] In any part of Ireland the man who has money in his pocket can get all he wants to eat, and if the masses of the people had money or its equivalent, any shortness of Irish crops would at once determine towards Ireland the flow of abundant harvests, without any call upon charity. It is not that food is so dear in Ireland; it is that the masses of the people are so poor. They get so little in ordinary times that when anything occurs to reduce or interrupt their accustomed income they have nothing to fall back on.

But is not this the case with the working classes all over the civilized world? Though in a country of more diversified industry the failure of a crop or two could not have such wide effects, yet the more highly organized communities are liable to a greater variety of disasters, from which similar though generally narrower results are constantly occurring. A war on another continent produces famine in Lancashire; a change of fashion, and Coventry operatives are only saved from starvation by public alms; a bank breaks in New York and in all the large American cities soup houses must be opened.

The fact, the terrible fact, is, that famine, just such [a] famine as this Irish famine, constantly exists in the most highly civilized lands. It persists even through good times when trade is "booming;" it spreads and rages whenever from any cause industrial depression comes. It is kept under, or at least kept from showing its worst phases, by poor rates and almshouses, by vast organized charities, by private benevolence; but it still exists. In the very centers of civilization, where the machinery of production and exchange exhibits the latest improvements, where bank vaults hold millions and millions, where show windows flash with more than a prince's ransom, and markets are piled up with all things succulent and toothsome, never the sun goes down but on hungry men and women and little children; never the sun rises but on human beings who must prowl like starving beasts for food; and in good seasons as in bad seasons famine steadily claims its victims. Most striking of all the wondrous sights of London to some savages who were taken there was the plenty of meat they had never dreamed of. But in that very London the mortuary reports have a standing column for deaths by starvation! And even in our great American cities such deaths are not unknown.

Yet even in widespread famines it is only in rare cases that the famine stricken dies of starvation, for the human machine must be perfect in all its parts to run on till the last bit of available tissue is drawn on to feed its fires. It is under the guise of diseases for which the doctors have other names than famine – especially the chronic famine of civilization – kills. But the statistics of mortality, and especially of infant mortality, among the poorer classes in the richest communities show that famine is constantly at its work. Insufficient nourishment, inadequate clothing and warmth and unwholesome surroundings steadily swell the death rates in the very centers of plenty. And mad to escape the torture of want, men, every day, rush out of life unbidden, and, daily, women sink! Alas!, that it should be so; but let us not [ ] the [ ], famine is not peculiar to Ireland.

Most terrible of satires, a New York paper, *Puck*, recently published a cartoon representing James Gordon Bennett[103] sailing away to Ireland in a boat loaded down with contributions, while a sad-eyed hungry-looking tattered group gazes wistfully from the pier. The bite and bitterness of it, the humiliating sting of it, is in its truth.

Our population is yet sparse; our country is yet new; our great public domain is not yet gone; our crops have been bountiful; we are the greatest food exporters in the world; yet in the midst of abundance there is want and misery and destitution – beggars, tramps and paupers, men who do not know where to turn to get for their families the necessaries of life; children growing up amid such soul-destroying poverty and squalor that only a miracle can keep them pure. Go into the lower wards of New York and look at the little children; stand on Broadway at night and see the stream of girls who might have been happy wives and mothers; sit by the roadsides of our new states and talk to the tramps!

If this be, in bud and blossom, what will be the fruit? If this be the green ear, what of the dry?[104]

Mr. Parnell and his fellows do not appreciate the importance of the agitation they have begun. The Irish Land Question is but the local phase of a question of worldwide importance. Nor yet do the measures they propose show any adequate grasp of the subject. They are illogical and insufficient to the "last degree." They neither disclose any clear principles, nor do they aim at any result worth the struggle. For it is illogical to urge men not to pay excessive rents. Either the land is rightfully the property of the landlords or it is not. If it is, the landlord has the right to say what rent should be paid. If it is not, why should he be paid any rent at all?

And it is inefficient, for though temporarily it may have served a good purpose, as Mr. Parnell shows, nothing permanent can be effected in this way. Rent is fixed by general laws. The landlord always wants to get as much as he possibly can; the tenant always wants to pay as little as he possibly can. What really, on the average, will be paid depends upon an equation over which either, individually, has little control. And if rents are temporarily under the influence of agitation somewhat reduced, they will soon, under the influence of confiscation, spring up again.

So, too, is the scheme for creating a peasant proprietary illogical. Either the land of Ireland rightfully belongs to the Irish landlords or it rightfully belongs to the Irish people. If it rightfully belongs to the landlords, then is the whole agitation wrong, and every scheme [ ] in any way interfering with the landlords is condemned. If the land rightfully belongs to the landlords, then it is nobody else's business what they do with it, or where or how they spend the rent they draw from it, and whoever does not want to live upon it on the landlord's terms is at perfect liberty to emigrate or to starve. If the land is rightfully the landlords then is any agitation to make them part with it on terms they would otherwise refuse, a bad and dangerous precedent.

For if a man may be made to part with one species of rightful property by agitation, why not with another? If a man's title to land is as rightful as that to the possession of his watch, then it is only a difference of kind and degree between agitation by monster meetings and parliamentary proceedings to make him give up the one, and agitation with a cocked pistol to make him give up the other. But if, on the contrary, the land rightfully belongs to the Irish people, then why in the name of Justice should the landlords be paid an enormous sum for it? Why should all this be done merely to substitute one privileged class for another privileged class? The Irish agricultural tenants constitute a larger class than the Irish landlords, but no more than the landlords do they constitute the Irish people. Does injustice become any the less unjust by enlarging the number who profit by it?

And so, too, this scheme is as inefficient as it is illogical. Waiving all practical difficulties, and supposing the landlords of Ireland to be bought out by the government and their estates resold to the tenants to be paid for in a term of years, what would be accomplished? Nothing real and permanent. For no sooner were the lands thus divided than a process of concentration would infallibly set in which would be all the more rapid from the fact that the new landholders would be so heavily mortgaged.

The tendency to concentration which has so steadily operated in Great Britain, and is so plainly showing itself in the United States, must operate in Ireland to weld together again the little patches of peasant proprietors. For this tendency springs from influences constantly becoming stronger and more penetrating – the same influences which are concentrating population into large cities, business into the hands of great houses and for the blacksmith making his own nails or the weaver working his own loom substitutes the great corporation with its millions of capital and thousands of employees.

Free trade in land, from which a certain class of English and Irish publicists seem to expect so much, is in itself good, but it would simply operate to permit landed property to assume the most profitable forms. And under the conditions which already exist in Ireland, and toward which modern progress more and more tends, that is not the form of five, nor two nor forty-acre holdings; but farms so large as to permit the use of machinery and of considerable amount of capital.

If the Irish reformers would keep their peasant proprietors after they got them, they must take some lessons [from] the Chinese mandarins and prohibit machinery and improvements. For it is not merely the steamplow and threshing machine. Even in such things as butter, cheese, and poultry, the steam dairy and steam hatching will soon be able to discount the cottages. And so, too, does it seem to me, they must banish the schoolmaster and the newspaper.

But even if the agricultural tenants of Ireland could be changed into proprietors, what about the mere laborers, the mechanics, the poorer classes of the cities? Have they not, also, some right to the soil of their native land? How under such a scheme, is this to be acknowledged?

Even if the hundreds of Irish landowners were multiplied into thousands, land monopoly would still continue to bear its bitter fruit. If a majority of the people of Ireland owned the whole land, the minority would be no better off than now. Rent would still levy upon labor, and wages tend to the minimum which in good times means but a bare living and in bad times means starvation.

Even if it were possible to permanently cut up the soil of Ireland into the little tracts into which the soil of France and Belgium is cut in those districts where the *morcellement*[105] prevails (which, clearly, under the habits and conditions of the Irish people it is not) this would not be the attainment of a healthy and just social state. For in those districts the condition of the mere laborer is, if possible, even worse than in Great Britain and the tenant farmers – for even there tenancy largely prevails – are, according to such authority as M. de Laveleye, rack-rented with a mercilessness unknown in Ireland, and are forced to vote just as their petty landlords dictate.

What, then is the true solution of the Irish land question? At what should [sic] Irish patriots aim as the adequate remedy for the cruel injustices of the present state. The answer must apply not merely to Ireland, but to all other countries, for the problem to be solved is the same in all. The answer must be logical – that is to say it must conform to principles. And, conforming to principle – that is to say, to Justice – it will, if there be a true harmony between the moral perceptions of man and the laws of the universe in which he finds himself, be at the same time practicable and efficient.

There is no difficulty in obtaining an answer if we look solely to principle. What does Justice decree? That must be the answer.

To whom rightfully does the soil of Ireland belong? That is to say, who are justly entitled to its use and enjoyment? – for Nature laughs at our ideas of a fee simple in land,[106] of our passing by deed and grant exclusive title to the surface of a globe that existed before man was, and will continue to exist long after he had gone; on which our succeeding generations are but the tenants of a day.

There can be but one answer. Unless we suppose that all but a few hundred or few thousand out of all the millions of the Irish people have any natural right to live in Ireland; unless we suppose that the Creator has intended one set of men to be masters and another set of men to be serfs; unless we suppose that one human being, could he concentrate in himself the legal title to the whole soil of Ireland, would have the moral, as under existing laws he would have the legal right to drive all the millions of Irish people across the sea or out of existence, and give back again the whole island to the wolves and foxes, as William the Conqueror gave the New Forest, there can be but one answer.

> The soil of Ireland rightfully belongs to the people of Ireland! Not to landlords; not to the tenants; not to some people, be they hundreds or thousands; but to all the Irish people.

This is the great principle upon which the Irish reformers, if they would be logical, must take their stand – the principle that the youngest child of the poorest peasant has as good a right to live on the soil and breathe the air of Ireland as the eldest son of the proudest duke. And this assumed, it evidently follows that the landlords have no right either to rent or compensation, and that to take the land from one set of men and give it to another set of men would only be to change the parties while continuing the wrong.

Now if the Irish agitators abandon their present halfhearted, illogical attitude, and take this stand frankly and firmly on this principle, it is probable that they will arouse at first even a bitterer opposition than now, and that for a time their following may largely fall away. But they will have put their fight on the right line, so that every step won will be a real advance and will make easier the next. Their position will not only be logically defensible, but one on which they must ultimately win.

> For private property in land is something so utterly absurd, so outrageously unjust, so clearly a waste of productive forces and a barrier to the most profitable use of natural opportunities, so glaringly in the way of real progress, so thoroughly opposed to all sound maxims of public policy, that it is only tolerated because the majority of men never think about or question it. Any general discussion can have but one result. Once arraign it, and it is condemned. Once put its advocates on the defensive, and their cause is lost in advance.[107]

If the Irish leaders take the ground it will at once be seen that they are not demagogues, playing for votes and power, but earnest men contending for a great principle. And they will gain the strength that comes from an honest avowal of principle. They will not only strengthen themselves by the moral support of earnest men all over the world, but they will disarm opposition. They must now encounter and call to their aid political forces that are more [ ] to them.

For the moment they do this they "carry the war into Africa." They assert a clear and tangible principle which will not only bring to their support the working classes of England and which has only to be discussed to enlist in its favor the intellectuality and the true conservation of the nation, but which makes possible such a combination between the laboring classes and the capitalistic classes as that which against the landed interests [won] the battle of free trade in 1846.[108]

Consider the practical measures which alone are necessary to the assertion of this great principle – this self-evident truth – of the equal right of all the people to the soil of their common country. They involve no harsh proceedings, no forcible dispossession even of the largest landholding, no shock to public confidence or violence to the rightful sense of property, no retrogression to a lower industrial organization, no loaning of public money or the establishment of commissions. On the contrary, the only measures necessary are such as will recommend themselves to the judgment and the interests of the masses of the British people.

The dictates of Justice cannot be fulfilled by carving out of the estate of the landlord an estate for the tenant, by buying up the land from one set of men and selling it to another; but they can be fulfilled by simply appropriating rent to the public service; and the way to do this is the easy way, abolishing one tax after another *until the whole weight of taxation falls upon the value of land*. In this way all objections are avoided save the objection of those who now reap when they have not sown, and against the selfish interests of the landlords can be united the interests of all other classes and even the interests of the landholding class insofar as it is also a capitalistic or laboring class.

To put all taxes upon the value of land and thus appropriate rent (the creation not of the individual but of the community for common purposes and the common benefit) would be not a novel and unheard of thing, but the return, in a form adopted to modern social life and the modern industrial organization, to the principle involved in the primitive tenure, and which the first perceptions of men have everywhere recognized. It would be a return to the first principle of the feudal system, which imposed upon the holders of the common property burdens which have been gradually shifted to the shoulders of the masses, and from which substitution the enormous debts and widespread pauperism of modern times proceeds. While it would fulfill the aspirations of the Irish people to gain back the rights in the land of which they have been robbed, it would also, as soon as properly understood, commend itself to the conservative instincts of the English, and their strong [ ] for the rights of property – for discussion would soon bring out the truth

that private property in land becomes, as development goes on, is inconsistent with the right of property in other things;[109]

and it would have, in all the steps necessary to its practical appreciation, the support of great interests and strong sentiments.

No political economist will deny that the tax upon land values or rent is of all taxes that which combines the maximum of certainty with the minimum of cost; that unlike taxes upon capital, or exchange, or improvements, it does not check production, or enhance prices, or fall ultimately upon the consumer; and any discussion of this proposition will only make its advantages clearer, while the practical steps towards it will bring the *argumentum ad hominem* to bear on those who might never comprehend an abstract principle. England has been educated up to a belief in free trade. Let the Irish leaders take advantage of this by preparing to sweep away all customs duties and give absolute free trade. The income tax is onerous and unpopular with those who pay it. Let them take advantage of this by demanding the abolition of the income tax. The manufacturers, the national corporations, the capitalistic classes generally will readily see the advantages of abolishing all taxes upon capital or improvements. And so on, until nothing remains but a tax upon land values.

Yet all the expenses of the United Kingdom, local and general, would nothing like consume the whole of rent – still less would they do so when reduced as they might be reduced by such a simple system and by the general prosperity and content which would ensure from the greater production and fairer distribution of wealth. But when discussion had brought home to popular apprehension the essential difference between property in land and property in things which are the produce of human labor, there would be no danger of the landholders being suffered to appropriate the remainder of rent. For it would be clearly seen that rent, which is the creation of the whole community, belongs of right to the whole community, and that this result of natural laws is not a creative sanction of the horrible inequalities which material progress is now producing, but a beautiful adaptation, which when rightly viewed and rightly treated makes progress in civilization progress towards equality and brotherhood.

And perhaps, as the discussion goes on, we of the United States – we, who [ ] imagine we can build up a democratic republic on a foundation which as development goes on must become more and more inconsistent with democracy, before it is too late, be led to see that equal political rights become but a mockery and a danger when the equality of national rights is denied.

FROM THE *IRISH TIMES*
(July 15, 1889).[110]

On Saturday evening a Dublin audience in the Rotunda had the advantage of hearing Mr. Henry George expounding his single-tax principle, or land nationalization. Mr. Davitt occupied the chair, and was careful to explain that there was no connection between the lecture, though it was taking place under the auspices of the National Club, and the intended Tenants' Defence League. The visit of the American notable had been for some months in contemplation. There was a hope expressed, however, by Mr. Davitt, that the program of the new combination would "leave open" the question of adopting Mr. Henry George's solution – "would leave open the question of what the final settlement of the agrarian problem was to be."

Mr. Davitt indicated that those demanding an Irish parliament should go for land nationalization, to be ... immediately brought about, and should, we presume, require from Mr. Gladstone now nothing less than sanction for that idea. Any trades union association in the interest of agriculturists, he thinks, should have the extinction of landlordism for its ultimate aim – and not only, as we understand, of landlords, but houselords also in cities. There should be no rent at all, either for acres cultivated or sites for warehouses or factories. For rent there should be a universal tax paid into the national chest, out of which all public charges should be met. There was an amusing illustration in connection with the occupation of land in cities, for shops or residences, in the story which Mr. Davitt told over again of the Pembroke estate, but as we do not suppose he suggest that all other estates built over on skirts of towns were obtained in like easy ways, there is no argument that can be rested on the particular case.

When Mr. Davitt talks of the idlers who live on ground rents, what he is doing is denying to the community the opportunity of dealing in land as a commodity or investment. If a man buys with his money fields in a suburb, and lets them out for building, and gets rents as representing the interest of his capital, and lives upon these, it may be as some would describe it in idleness, what has been wrongly done, and who has a right to complain? If such an owner cannot be removed from his property except by giving him his money back, and an Irish parliament should give the money to him, where is it to get it for him except by levying the same as interest that was paid as rent, in which case nationalization would mean merely change of owner.

Land is being set free from rent by the only process that is possible, and in the interest of a class – but it is the tenant and not the landlord class – in the operation of the Land Purchase Act.[111] By the most rapid and easiest method the rent is made the means of extinguishing rent, and to proclaim any fresh or further design of nationalization of land is put forward what is palpably unpractical as a scheme, and calculated at this moment greatly to alarm statesmen of both parties.

Eight years have elapsed since Mr. Henry George was here before, and if his views were sound they would have made some way during that interval. But who in Ireland accepts them? They have not prospered in America. Mr. George substitutes the relation of man to the planet that he inhabits – his own words – for his relation to the capitalist who built the house that shelters him, and to whom he owes rent as a payment for the service done.

During the lecture when he was describing the land as the property of all and of none, a Voice asked: "What about the Dublin Corporation?" The Corporation is a landlord who would never dream of countenancing the common fund project of Mr. Henry George. In allusion to the Purchase Act, he predicted that "after a little while by the operation of causes that could not be prevented great estates would concentrate again." That will be to some extent so, but even the single-tax principle would not make any real difference, since when any individual ceased to pay his tax – Mr. George's name for rent – and fell into arrear, the State would evict him and give his holding to another, until by uniting farm to farm some one industrious man would secure an extended space, and become a landlord, and on the largest scale.

We do not imagine that the leader of the Irish party will in this matter take Mr. Davitt's advice, and associate the land nationalization of Mr. Henry George with the new departure. The public are patiently awaiting the disclosure of the better Plan with which Mr. Parnell will extinguish the Plan of Campaign. Until the principles and methods of this third League in the succession of movements are defined, and no one can do more than conjecture about it upon such hints as its friends have presented us with. But we do not expect that it will "leave open" the question of acceptance of the Henry George platform. Alliance with the land nationalists would bring no strength to the Irish party, and it would put Mr. Gladstone more deeply in the very difficulty that he is probably struggling to escape from. He wants to get the whole party, English and Irish, down to a more rational position, but in doing this he can have little help from the author of *Progress and Poverty*.

IN BELFAST: HENRY GEORGE'S MEETING
IN ULSTER HALL:
A Splendid Address and a Brilliant Speech; Much Enthusiasm and a Raking Fire of Questions; Short Addresses by Silas M. Burroughs of London, England, and Charles L. Garland, M.P., of New South Wales, Australia

(The Standard, July 27, 1889).[112]

The following interesting account of the meeting which greeted Henry George at Belfast, Ireland, is taken from the *Belfast Morning News*:

Mr. Henry George, whose fame as a land and labor advocate is worldwide, and whose works on the social problems of the day have done more, perhaps, than any other intellectual agency to agitate Europe, arrived in Belfast yesterday morning, according to appointment, having traveled from London via Strauraer and Larne. He was received at the Northern Counties Railway by a number of his sympathizers, and, after the customary greetings, the distinguished visitor drove to the residence of Mr. Charles Hunt, 28 University Street, whose guest he remains during his stay in the city. After breakfast Mr. George visited the Queen's Island Shipbuilding Yard and inspected the White Star liner *Majestic*, sister ship of the *Teutonic*, and therefore one of the two largest ships afloat. Mr. George was greatly interested with all that he saw, and expressed himself pleased that he had had an opportunity of seeing so remarkable a center of industry. He then drove through some of the principal thoroughfares.

The Ulster Hall had been specially engaged for a lecture by Mr. George on the land and labor question, and the building was fairly well-filled with his admirers by eight o'clock. The attendance numbered about 1,000, and included several ladies. Contrary to expectation, there was only one dissentient – a young gentleman who sat in front of one [of] the balconies, and interjected an occasional "No" as Mr. George developed his theory of the single tax. The reception accorded was enthusiastic in the extreme, and he was warmly applauded at frequent intervals. The friends accompanying him – Mr. Garland, a member of the New South Wales legislature, and president of the Single Tax League in that colony; and Mr. Burroughs, an American gentleman employing labor both in New York State and at Deptford in England – were listened to with much attention.

Among those present were: The Hon. Judge Savage, United States consul....[113] One or two Episcopal and Presbyterian clergymen were also present, but it was found impossible to obtain their names. The arrangements for the conduct of the meeting were excellent, and the stewards rendered every assistance and courtesy. Mr. George's published works, by the way, were on sale in the vestibule, and copies of *Brotherhood* were placed on each seat in the hall.

Mr. George was escorted to the platform by the leading members of the reception committee, and was loudly cheered.

On the motion of Dr. Hyndman, seconded by Mr. Shone, the chair was taken by Mr. J. Bruce Wallace, M.A.

The chairman, in introducing the lecturer, said there was a movement gaining strength the world over that might be called the movement of the new conscience. It had sprung from a quickened realization in the minds and hearts of earnest men of the Fatherhood of God, and of human brotherhood. The abolition of chattel slavery, with its manacles and whips and hideous immoralities, was a step in that movement – a stride forward in the march of humanity. A subtler slavery, no less cruel, disguised under the forms of freedom, came next to be dealt with. At the basis of their social fabric today they saw a stratum of workers who were but little better off than the Negroes that used to cower under the lash of slave drivers in American plantations. The freedom of contract they enjoyed was found in practice to be little more than freedom to starve.

A Crushed, Stunted, Cheerless Life was the lot of multitudes among them. As it came home to the conscience of Moses, in the midst of all the luxurious ease of Pharaoh's palace, that the Hebrews toiling in the brickfields and at the building of the pyramids and treasure cities were his own kindred, and there arose in him the ambition and the hope to be their deliverer, so was it coming home to the conscience of many who were enjoying opportunities of culture and of comfort not available for all, that the toiling masses, burdened and crushed, were their brothers and sisters, children likewise of the All-Father, and that these must, at all cost, be emancipated from all degrading conditions and lifted up into the enjoyment of equal opportunities with the most favored of them for healthy and happy physical, mental, and moral development. The old superstition that ascribed to an inscrutable Providence the contrasts between rich and poor was giving way to a strong conviction that wherever industrious people were badly off, wherever life was for them a dreary bondage and subsistence precarious, underlying such a condition of things there was some grave injustice. The earth was amply stored with the means of yielding all its inhabitants abundant sustenance and comforts in response to a moderate amount of work. With the growth of population, making of course enlarged demands on the earth's resources, there had come also a wonderfully enlarged knowledge as to how to turn those resources to fruitful account.

All Might Be Wealthy. Under a deep sense of responsibility for their downtrodden fellows, earnest thinking men had been setting themselves to discover wherein precisely the injustice lies – what was the fundamental injustice of their sad social order. ("Hear, hear.") No longer content with ministering mere relief to want, they were determined to sweep away its causes and to build a new order on foundations of righteousness. (Applause.) In this great worldwide movement of thought and feeling and action – this new crusade – Mr. Henry George is one of the most prominent leaders. His great work, *Progress and Poverty*, had not only found its way on its mission of light wherever the English tongue was spoken, but had been translated into well-nigh all the languages of Europe, and had everywhere carried conviction and kindled enthusiasm. (Applause.) The party that had gathered round him in the United States of America was indeed still in a comparatively small minority, but it was growing with extraordinary rapidity. It was coming to be recognized as the successor of the abolition party, and was attracting not a few of the best men of America to the standard. In England and in Scotland Mr. George has been received with open arms, and his lectures had given a tremendous impulse to the cause of social progress. ("Hear, hear.") At the international conference on the land question held last month at Paris, composed largely of men who in many points differed considerably from Mr. George, he was unanimously elected honorary president, in recognition of his immense services.[114] (Applause.) He (the chairman) did not think the land nationalizers in Ireland are prepared to adopt the method – the "single tax" – which Mr. George advocates, but they accepted the principle and claimed "the land for the people" (applause), and they honored Mr. George as a great teacher. (Applause.)

Henry George's Address

Mr. Henry George, on coming forward, was greeted with prolonged applause. He said if the people who were hostile to them, as the chairman had suggested, were away, he was all the more sorry, for it was to them he would rather talk than to the sympathetic people. It was, in fact, at the hostile people he wanted to get. ("Hear, hear.") There was nothing in the doctrine that they stood for, in the proposals that they made, that ought to excite the hostility of any man, and if there was hostility it was through ignorance. (A voice: "No.") Well, he, at any rate, thought so, and he had a right to show how it was and how it came that they in all they did and said stood for everything that was right and everything that was wise. They wanted to show that in every class of the community their theories were calculated to bring about a state of things in which there should be no undeserved poverty, and in which everyone should be wealthy, not in the sense that they should have more than their neighbors, for that would be impossible for all, but in the sense of having abundance of the necessaries of life and a reasonable amount of the luxuries of life as well – in the sense of having a full and free development of every faculty of mind and body, and enjoying the advantages that advancing civilization brought ("Hear, hear.") He had been to see that day a most suggestive and a most glorious thing. He went to the works of their great shipbuilding firm of Harland & Wolff and saw that great ship, the White Star liner *Majestic*, the largest ship, he believed, that had been built with the exception of the *Great Eastern*, and he expressed a hope that the reputation of the great firm by which she was built would be maintained in her. (Applause.) That great vessel typified to his imagination that quality in man which made him really

The Child Of God, that quality which proved to them that man was indeed in the likeness of the Creator. Such a ship as that clearing her way against wind and sea across the Atlantic typified to him the earnestness and power of man. It led them to think of the insignificance of man. If they thought of man in the water, how slowly he went! Yet by his power of adjusting things it was possible for him to lift himself above the animals that lived in the sea and to build a ship which could cross the wide expanse of the Atlantic in less than six days – by this great ship the journey might be made in five days, a distance of 3,000 miles! Let them look back at the ships of Nelson's time,[115] or, going further back, at the rude canoes of early war, and think of the development that man had made, his growing power of adapting means to ends, and asserting his will. It was to him (Mr. George) significant that

These Great Ships Were Built In Ireland, here in Belfast. Thirty years ago he fancied they had not any shipbuilding in Belfast. The firm of Harland & Wolff was not then in existence. The firm was now known over all the seas of the world. In the time of which he spoke, thirty years ago, his people, the Americans, were the great shipbuilders, and American ships were sent to every part of the civilized world. Where were they now! Why, the great army of Americans who crossed the ocean every summer not one of them crossed in an American vessel. They came in vessels built in Belfast or on the Clyde. What was the reason? It was not that the American mechanics had lost their cunning or American seaman their skill. It was merely that they in America had applied a baby act, and that a protective tariff had so taxed everything in connection with ships that they could not build them. Especially significant to him was it that the best ships in the world were built in Ireland. That was a significant answer to those who said that Irish industry could not be developed without a protective tariff. (Cheers.) It was an answer, too, to those in the United States who said that a protective tariff could not be done without. From what he saw in Belfast he held that it was

Not Restriction But Freedom, they wanted in the development of industry. (Cheers.) The law of liberty was the law of life and the law of development.

Let them think of that marvelous product of human skill traveling 500 miles per day across the wastes of the Atlantic. Could they think that a human creature of the very same kind – a man like to the man whose ingenuity constructed and contrived that ship – could suffer for want of the primary necessities of life? Could they think that men who could produce such works would pass their lives in a struggle for mere existence, in bitter want and poverty? Yet it was so – there were men so situated. He asked them to suppose that if the *Majestic* was crossing the Atlantic, and that every man, woman, and child on board were counting the days, hours, and minutes to the time of her arrival beyond, and a woman or child was seen clinging to a raft in midocean, would it not be expected – nay, would there not be an outcry of indignation if it were not done – that the engines would be reversed, and that every effort would be made to save the woman and child! ("Hear, hear.") Well, here in their so-called Christian cities, under the very shadows of their churches,

Women And Children Died Weekly, died nightly, the worst deaths – deaths not only of the body, but of the soul. Was it not their duty to rescue these? And how were they to do it? How did man construct such mighty engines as the *Majestic*? It was by seeking out the laws of matter and of mind and exercising them by the highest quality of adapting means to ends. Let them think of the advance of civilization in our times – only yesterday he was speaking to an old friend of his, a man who rode on the first railway train that ever ran. Let them think of the advance in human industry, and invention, and discovery, and of the wonderful progress that had been made, and yet there was vice and crime born of poverty, and there was a great majority who by the hardest toil only got a bare living, while all the advantages of progress were enjoyed by a few. Was it not time that they gave thought to the social questions that surrounded them, and that they sought out the laws that governed the production and distribution of wealth, just as the laws of matter and mind – the physical laws – had been sought out. (Cheers.)

What Was The Cause of Poverty? Their chairman had told them it was not the fault of the Creator. If there were any of their fellowmen who did not get wealth enough it was not because of any scarcity in the elements of wealth. What were the factors of wealth? He answered, labor and land. It was only by a metaphorical phrase they could speak of man as a creator. He was only the producer, and his productive work consisted merely in the union of his labor with the raw material of the land. Man made the ships by taking the iron and the wood and shaping them together, and he made the sails by growing the cotton and changing its form into the required shape. Without land, as they saw, he would be absolutely helpless. Without land he could not exist. His very body was drawn from the soil, and to the soil it returned again. He was, therefore, not the creator but the producer, and all his production consisted in modifying land by labor. (Cheers.) Capital was but a secondary factor. It could only be that, for at the beginning there was simply man, and the union of labor with land led to production.

The Creator Supplied The Land, and gave to man the power to labor on the land and to modify the raw material to his own purposes. Everywhere there was plenty of land to spare; yet they had men crying out that they wanted work. They saw men going about their streets earning a living by holding boards, one in front, and one behind; and they built ships in which their men were sent across the ocean in search of work. He called their attention to this question, and put it to them whether when they examined it, they did not find a great injustice, a great wrong, a monstrous violation of the laws of the Creator. (Cheers.) They were all land animals. To every human being land was absolutely indispensable; to his life it was absolutely necessary. They could not live without the use of land. They found themselves here in the world with land adapted to their needs and hands for the adaptation of that land to their needs. Had they not clearly and plainly equal rights to the use of the earth? Was not the title of each human being to use the earth as his workshop [and] reservoir – was it not proved by the very fact of his existence, was it not the natural right conferred on him by the power, was it not an inalienable right that could not be sold, and that he could not be deprived of by the edict of any king or by the act of any parliament?

Look At Ireland, at the monstrous injustice involved in the fact – that of the people of Ireland only some had a right to the use of Ireland. (A voice: "Quite right.") They saw the same in England and Scotland and elsewhere, and on that day saw the cause of the social evils of today. That was the reason why the laboring classes were always the poorer classes – because labor without land was absolutely hopeless. Here was the great fundamental wrong, here was the primary injustice they had to deal with. They proposed to take from no one that which was originally his. So far from denying the rights of property he and his friends were sticklers for the rights of property. They believed there was a sacred right of property – that whatever a man produced from the reservoir placed at his disposal, and whatever he adapted to human use was and ought to be his – his, to use himself, to sell if he pleased or to give by gift or devise to anyone else. (Cheers.)

There Was A Sacred Right Of Property – the right framed by the command "Thou shalt not steal." No one had a right to claim the ocean or the air, and no one could assert a right to them without denying the sacred right of property. (Cheers.) If the Western ocean between Ireland and America had been parceled out by kings and held as private property, a ship like the *Majestic* could not cross without paying toll to the owners of the ocean. (Laughter.) Would it not amount to a robbery of labor that the owners of the *Majestic* should have to pay for the right of crossing the ocean? Fortunately, no one had laid claim to the ocean. (Laughter.) When the salmon that lived in the sea came near certain coasts they became the property of private individuals. (Laughter.) It was at one time tried to put herring in the same category, but that failed. (Laughter.) Did not the same principle apply to land? The *Majestic* must burn coal, and the royalty paid upon coal for her, he was told, would amount to more than the wages of the seamen and the men in the engineer's department. What was the difference between royalty for coal and the

Royalty For The Ocean? The men who took up the coal from the earth had a right to demand payment for their exertion; the men who furnished capital had a right to a return; but the men who owned the coal – what had they to do with it? (Laughter.) Coal he described as merely the sun's heat stored in the earth for the use of man, not for the use or profit of an individual man – the Duke of Newcastle or the Marquis of Bute, or, as in his own country, the Lehigh Coal and Navigation Company, or Messrs. Pardee and Cox. Consider the grants made by kings – what right had any man to give away this world or any portion of it? He might as well give away the sun. (Laughter.) What right had a dead man to bind generations of people? What right had a dead man to this world, anyhow? The world was for living men. When a man died they hoped that he went to a better world; this was intended for living men. Now, he and his sympathizers proposed to end the wrong which was perpetrated by the system of which he had been speaking to them. They did not propose charity, or benevolence, or help for the poor, or to do anything for the working classes in particular. What they proposed was

Simply To Have Fair Play, to put all on an equality, and to leave individual power to do the rest. They wanted the land that was not in use put to proper use. They wanted a man who made improvements to be secured the value of his improvements. They wished to take over all land for the use of the community, and to impose a single tax on the land for the purpose of raising revenue – that, they were convinced was the easiest way of dispensing with taxes, on industry, or labor, or the production of labor. They called it the single tax, as they would abolish all other taxes, and they would raise all revenue by its means, irrespective of improvements. (Applause.) He explained at some length the advantages which would arise from this system in the abolition of monopoly of land, of speculation in land, of which they had had some experience in Ireland. They could not do away with injustice in Ireland by increasing the number of owners of land; so long as one Irishman was denied the right to his native soil, so long would there be injustice. If they took the national credit or the national money, it did not matter which, and gave it to one special class, the landlords, as the price of their unjust privileges, and then turned a greater number of people into owners,

What Did That Do For Artisans And Laborers? (Cheers.) Why, if they bought farms for the farmers, should they not buy boats for the fishermen, printing presses for the compositors, donkeys for the costermongers,[116] goods and chattels for the clerks? (Laughter.) Why not buy every man a house? (Laughter and cheers.) Mr. George elaborated this argument to some extent, afterward concluded by impressing on his audience the benefits which should result from the adoption of the single tax. He had intended to address them at greater length, but he was desirous that other gentlemen should be enabled to state their views. He hoped afterward to answer any questions that might be written and sent up to the chairman (Cheers.). . . .[117]

Answering Questions

The first was: Suppose a man had bought an estate with capital that he had saved from his earnings, would he think it just to tax what that man had bought any more than a ship, a watch, or a horse, that another man might buy with his money?

Mr. George: Oh!, certainly, as a horse, an estate, or a watch were the production of labor, they were the production of individual exertion to which the right of individual ownership properly attached. Land was not the production of exertion, and its value was not produced by individual exertion, but by the growth and improvement of the whole community, and in taking it for the whole community, they were merely taking for the community what rightfully belonged to them. The fact that a man had bought that which belonged to the community had nothing to do with the question. (Applause.)

The second question was: Would not the landlord make the tenant pay the taxes after?

Mr. George: No; and whoever might think that, the landlords would not think so. (Laughter.) If they did they would not object to the taxes. It was a settled principle of political economy that taxes on land value or rent must be paid by the rent owner, and could not be thrown upon the raiser of rent. Taxes upon capital must fall ultimately upon the consumer, because the prices of such things were determined by the equation between supply and demand.

The third question was: Would not Christianity, if properly understood, especially in its social aspect, solve the social and industrial problem?

Mr. George: Unquestionably, because the first principle of Christianity was Justice (Cheers.) That was the first principle of true Christianity, not the Christianity that prayed to God to relieve poverty, but did nothing itself to relieve it. (Cheers.) That blamed every class of poverty upon the inscrutable decrees of Providence. The Christianity of those who said when a poor man died that he had gone to a place where there was more room for him, because that was a place where no one was allowed to take up too much. (Laughter.)

The fourth question was: Did the land tenure of the Jews – the only system of land tenure that had ever received the divine sanction – not stand in the way of the single-tax method of land tenure, as it seemed based on the same principle as the various systems of peasant proprietary?

Mr. George said the laws of the Almighty remained always the same, and the Bible never said men were to take lands because they conquered them or got them by grant, but always: "Take thou the land the Lord thy God giveth thee." (Applause.)

The fifth question was: Would the abolition of Irish landlordism benefit the working classes of Belfast?

Mr. George: It would certainly benefit the working classes of Belfast. What was the rate of wages paid to the artisans of Belfast fixed by? It was fixed by the number of workmen who offered their services to the employer of labor. He (Mr. George) had been in Donegal some years ago, and he had followed one of those eviction armies, and he had seen them throwing out on the roadside the people who had lived there from time immemorial – old men and women and little children, and this at the instigation of a man who had never set his foot in Ireland (shame) – a Scotchman, living on the other side of the Channel, and some months afterwards, in one of their Belfast slums, he had been taken by a good priest to these very families who had been driven from their homes into this city to sell their labor at what price they could get in their factories. (Cheers.) Why were wages higher in New South Wales than they were in Ireland? Simply because the access to the land there was less difficult than here. How was it that when gold had been discovered abroad the rate of wages increased so enormously? If gold had been discovered in this country when the land was private property wages would not have gone up at all. But wages had gone up in that country, because then the land was let to people so long as they worked it, and only so long as they did work it. (Cheers.)

The [sixth] question was: What would you pay the purchasers under Lord Ashbourne's act of 1887, or how would you deal with them?[118]

Mr. George said he would deal with them as he would deal with everybody else. He thought that the people purchasing under this act should understand perfectly what they were doing. A friend had suggested to him that these people were not purchasing their farms with their own money. He did not think that that made any difference. It was value all the same.

The [seventh] question was: Under the Parnell scheme of buying land at seven years' purchase, how would he deal with the purchasers in case his scheme should come into law in this country?

Mr. George said he would not make any distinction between the man who purchased land under the Land Act and the man who got it by inheritance, or the man who just took it. (Laughter.) The mere purchase of anything did not give possession. That was the argument used in the slavery days. It was said it would be all right if they could get the men who had brought over the niggers, but what were they to do with the men who had purchased their niggers? Were they going to rob them of what they had purchased? They did not propose to take anything from anybody, but they were going to take from landowners that which belonged to the community.

The [eighth] question was: Do you deny the law enunciated by Charles Darwin – that in all animated Nature there is a constant struggle for existence? If not, why claim for your single-tax system that it is in itself the cure for our social evils? Do you not recognize the fact that even if your system obtained here today the operation of this law would involve us in a generation or two in the same evils, which are, in reality, not due to landlordism but to the non-intelligent use of man's reproductive powers?

Mr. George said that his idea was that man had the means of increasing his subsistence, and that Providence had not created more creatures than there was food to maintain.

The [ninth] question was: Do you respect vested interests?

Mr. George said that he did not recognize vested interests.

In reply to the question: Would you sacrifice the natural beauties of the earth for factories, works, etc.?

Mr. George said the natural beauties of the earth would be preserved by his theory, and instead of having parks closed for the special benefit of individuals they would have people's parks, and in a great many instances gardens would surround the houses of the owners of the land.

[He was then] asked: How many members of the United States Senate supported his views?

Mr. George said he could not exactly tell, but he was aware that many of them did, and also that many members of the English and other parliaments agreed with him. It was not improbable that the whole question would be put to the test at the next election, and if it should be, they would find that it had thousands of supporters more than they thought; at all events that he predicted it was the burning question of the immediate future.

In reply to further questions, he said that his views in regard to trades unionism had not changed since he had written his chapter on that subject.[119]

As to arbitration on trades disputes it had only the advantage of bringing about an amicable settlement where a difference existed. He would not interfere with the moneyed people who had no land in connection with his theory if it were put in practice, money being a created property. . . .[120]

NOTES

1. These two articles come from the *Irish Times* (Nov. 15, 1881).

2. The Great Bonanza was but one example of a large-scale industrial farm ("bonanza farm") with an enormous amount of acreage that overshadowed European landholdings.

3. The Land Act of 1881 established judicial fixing of rents, and its scope was extended by the Land Acts of 1882 and 1887.

4. This passage comes from *The Deserted Village*, published in 1770 by Oliver Goldsmith (ca. 1730–1774).

5. The editor has deleted the following line: "The chair was occupied by Mr. Alfred Webb. Among others present were . . ."; then a list of forty-eight names, the line: "The Chairman said he had received letters of apology for non-attendance from . . .," and four additional names were also deleted.

6. Helen Taylor (1831–1907) was an outspoken English advocate of woman's rights and step-daughter of J. S. Mill. (See footnote no. 73.) Michael Davitt (1846–1906) was an Irish agitator. Although a Fenian (see footnote no. 50) and incarcerated from to time to time, he was elected to Parliament in 1880. Charles Stewart Parnell (1846–1891) was an Irish nationalist leader elected to Parliament in 1875, who united different Irish factions. His agitation on behalf of the Irish land question helped pass the Land Act and the first Home Rule Bill. (See footnote no. 38.)

7. A note on British monetary symbols: £ = pound; s. = shilling (20s. = 1 £); and d. [denarius] = penny (12 d. = 1s.).

8. Robert A. T. Gascoyne-Cecil, third marquess of Salisbury (1830–1903), was a Conservative foreign minister under Disraeli. He was also prime minister three times and noted for maintaining peace with foreign powers. William E. Gladstone (1809–1898) was a Liberal British statesman who was prime minister four times and promoted a range of important reforms, among them relating to Ireland, the civil service, and the military.

9. The Irish National Land League was founded in 1879 in Dublin with Parnell as president and Davitt as a secretary. It was created to ameliorate the unjust land relations in the countryside. Some members sought to end rack rents (see footnote no. 4, page 15) and create a peasant proprietary; others desired a broader-base appeal for a return to the land to the entire people.

10. During much of the eighteenth century, the conservative Tories were staunch supporters of the landed gentry. Falling from power, they were revived by William Pitt (see footnote no. 79), then fell again in 1832, and finally evolved into the Conservative Party. Presently, the word "Tory" is used describe anyone with conservative political leanings. British loyalists during the American War of Independence were also dubbed "Tories."

11. The following sentence was deleted by the editor: "The proceedings soon afterwards terminated."

12. These handwritten notes by Henry George were probably written sometime in 1882; # 9, HGP.

13. Repetitive phrase struck out by George: "as it . . . Galway."

14. The words "adding gold to the green and light to" were deleted and replaced by "lighting up."

15. The words "it is" were struck out after "landscape."

16. The original word was "beautiful."

17. The words "A donkey cart in the most pretentious vehicle" were struck out.

18. These words were deleted: "Shortly after leaving Galway we pass . . . fine gentleman's place but" and "Nothing else was visible on the road save. . . ."

19. The following words were struck out: "stopped near a little cabin and jumping from the car . . . back and entered. Coming up in out in a minute from . . . our shadow had alighted and gone up to the door of our car, and was asking when he was going."

20. The words "stop at" replaced "go back to."

21. A number of indecipherable crossed-out words had originally appeared in this part of the sentence.

22. The words "was blind" were struck out.

23. The word "massive" replaced "great."

24. Two undecipherable words were crossed out.

25. The word "good" and one undecipherable word were crossed out.

26. Henry George, "England and Ireland: An American View," *The Fortnightly Review*, Vol. XXXI (June, 1882) Periodical Division, McKeldin Library, College Park campus, University of Maryland. Hereafter abbreviated to PDM, pp. 780–794.

27. The phrase "white terror" usually suggests punitive action taken by the authorities, and "red terror" repression by radicals.

28. The English conquest of Ireland began in 1169. Ireland and England were united in 1800, and Ireland achieved independence in 1921.

29. The Norman William I, (ca. 1027–1087), known as the Conqueror, reigned in England from 1066 after the Battle of Hastings.

30. Religious schism in sixteenth-century Western Europe that led to diminution of power of the Roman Catholic Church and the rise of Protestanism.

31. The Coercion Act of 1881, passed by the Gladstone ministry, sought to curb the increasing violence in Ireland through the suspension of the writ of habeas corupus.

32. See footnote no. 3.

33. The Poor Laws were various pieces of legislation, from the sixteenth century onwards, that imposed on parishes the duty to make provision for paupers.

34. The British government for Ireland was seated in Dublin Castle.

35. References to the Swiss Guards who have guarded the pope since 1505 and the Hessians who were hired by the English to fight the colonists in the American War of Independence.

36. J.P. is an abbreviation for justice of the peace.

37. The four Inns of Court, Gray's Inn, The Inner Temple, Lincoln's Inn, and The Middle Temple, are the unincorporated legal societies for the regulation of professional standards and means of admittance for barristers to law practice.

38. Isaac Butt (1813–1879) was a Protestant Dublin attorney who coined the phrase "Home Rule." He desired to have his fellow countrymen control internal affairs, preferring that London handle foreign policy and defense.

39. Edward III (1312–1377) reigned from 1327.

40. A yeoman was a farmer who owned a small landholding.

41. Lord Frederick Cavendish, the chief secretary for Ireland, and his undersecretary Mr. Burke were murdered in 1882.

42. The Magna Carta (1215) issued by King John after his loss at the Battle of Runnymede insured feudal rights for the aristocracy and protection from royal encroachment on their privileges. Other provisions included assessment without consent, the rights of churches, towns, and other subjects. It has been considered a starting point of the supremacy of the British constitution over the monarch with due process of law, and hence, while in no sense democratic, a remote ancestor of democratic rule.

43. For the Magna Carta see the previous footnote. John Hampden (1594–1643) was part of the antiroyalist faction in the House of Commons who opposed the raising of loans by Charles I without parliamentary sanction and monarchical usurpation of this body's prerogatives. The king's attempt to arrest him and others precipitated the Great Rebellion of 1642. Hampden died in battle against the Royalists.

44. Herbert J. Gladstone (1854–1930), the son of William E. Gladstone (see footnote no. 8), was a member of the Liberal Party and served in a number of capacities for the English government.

45. Americans can well understand these exaggerations of Irish outrage. After the [Civil] War, while "carpetbaggers" and "scalawags" were maintaining themselves in the South by Northern power, and political purposes were to be served by appeals to Northern passion, our Northern papers had their regular columns of Southern outrages, just as English papers have now their columns of Irish outrages. Worse stories were told of Kuklux Klan [sic] than are told of "Captain Moonlight;" especially as elections were coming on, what we got to know as "the outrage mill" was worked with redoubled energy. But now that the "carpetbaggers" have gone – now that South Carolina and Louisiana are as truly sovereign members of the Union as New York and Massachusetts – we hear no more of Southern outrages and the masked horsemen of the Kuklux Klan. Not that all the stories of Southern outrages were exaggerations or fabrications. On the contrary, many of them were horribly real, just as there are many horribly real Irish outrages. But reality and exaggeration were both products of the same state of things (H.G.).

46. The Great Famine, which began in 1845 and lasted for three years, immeasurably worsened life in an already poverty-stricken Ireland: 750,000 people died and over a million emigrated. A recurrence of agricultural distress in the 1870s inflamed Irish nationalism and violence.

47. George III (1738–1820) reigned from 1760 until his death, but bouts of insanity forced establishment of a regency under his son in 1811.

48. A bummer in this sense is a soldier who temporarily deserts his ranks and goes foraging.

49. A rough is someone belonging to the lower classes and inclined to violence or rowdiness.

50. Fenianism was a nationalist revolutionary movement that sought independence for Ireland during the nineteenth century. There were also branches in England and the United States. Ribbonism was an Irish movement that began around 1808. It was composed of secret associations of Catholics, or Ribbon societies, who fought the Irish Protestants.

51. Dogberry was a foolish constable in Shakespeare's *Much Ado About Nothing*: hence an ignorant but consequential official.

52. Henry George, "Lecture by Henry George: The Irish Land Question; What 'The Land for the People' Means." This speech of June 10, 1882 comes from transcribed notes; # 9, HGP. The phrase "Hearty recognition of his views" followed the title.

53. Nineteen names have been omitted.

54. Notation by transcriber; material not available.

55. Patrick Egan was an Irish Land Leaguer who lived in exile in France.

56. In the nineteenth century France and Britain vied for the control of Egypt. The completion of the Suez Canal in 1869 increased Egypt's importance. A nationalist uprising in 1881 brought a British conquest and subsequent control by London.

57. The British gained hegemony over India in the eighteenth century after ousting the French and taking advantage of internal rivalries.

58. Mexico achieved independence from Spain in 1821. The French, however, occupied this country from 1864 to 1867 under the rule of Maximilian and in 1876 Porfiro Diaz assumed dictatorial control that lasted until 1910.

59. Advocates of peasant proprietarship sought to secure private tracts of land for people, and advocates of land nationalization desired to compensate landowners before returning the land to the community as a whole. Henry George rejected the former and accepted the latter but without compensation.

60. See Genesis 3:19 in the Hebrew Scriptures.

61. A line is missing since the bottom of the microfilmed page was torn.

62. Probably a reference to large-scale Irish Land League agitation on behalf of evicted tenants.

63. Repetitive phrase struck out by the editor.

64. James Finton Lalor (1808–1849), a militant Irish radical of the 1840s who advocated tenant and private property rights but rejected absolute ownership in land since it belonged to the community; he also rejected rent payment to landlords until economic security might be obtained.

65. Bracketed notation by transcriber; a blank page follows; and quotations are not available.

66. Thomas Nulty, the Bishop of Meath in Ireland, irritated the Vatican by his pastoral letter affirming common rights in land.

67. Bracketed notation by transcriber ("Here insert extract from Dr. Nulty's letter") was placed here; a blank page follows; and the extract from this letter is not available.

68. Bracketed notation by transcriber.

69. A demesne is the manorial land held by the lord; tenants neither hold nor use it.

70. Bracketed notation by transcriber: "Explain how under land courts the landlords would get fourteen percent instead of eleven percent; I was knocked about in taking your reply; amendments are necessary" was placed here.

71. Henry George, "George and the *Irish Times*," *Irish World and American Industrial Liberator* (June 24, 1882): Interlibrary Loan Division, McKeldin Library, College Park campus, University of Maryland. Hereinafter abbreviated as ILD.

72. George, "The Land Question," in *The Land Question*, pp. 73–74.

73. John Stuart Mill (1806–1873), a noted English philospher, was a proponent of utilitarianism and liberalism. The Scotsman Thomas Carlyle (1795–1881) wrote prolifically on history and philosophy in a German metaphysical vein. Henry M. Hyndman (1842–1921) was a founder of British socialism. In 1881, he established the Democratic Federation, later renamed the Social Democratic Federation. (See pages 281–293, 334–348 for his debate with George.) In 1775, Thomas Spence (1750–1814),

an English promoter of the single tax, published *The Meridian Sun of Liberty, or the Whole Rights of Man Displayed and Most Accurately Defined*. For more information about Spence, see chapter 3 ("Thomas Spence") by Roy Douglas in Wenzer, *An Anthology of Single Land Tax Thought*, pp. 41–72.

74. Consult chapter four ("Property in Land Historically Considered') of Book VII ("Justice of the Remedy") in any edition of *Progress and Poverty*, pp. 368–384.

75. Emile de Laveleye (1822–1892) was a noted Belgian political economist and publicist.

76. Henry George, "Henry George and His Critics," *Irish World and American Industrial Liberator* (July 15, 1882): ILD.

77. Henry George, "Interview on Irish Nationalist Politics Following the Release of Parnell," Dec. 1882; #9, HGP.

78. A cloture is used in a legislative body for the limitation or closing of debate by a vote.

79. William Pitt (1759–1806), the younger, was prime minister from 1784 until 1801 and from 1804 to 1806. For George III see footnote no. 47.

80. Mussel could be an archaic form of muzzle.

81. The Liberal Party, which grew out of the Whigs around 1830, advocated domestic reforms, free trade, extension of the franchise, and greater liberties.

82. Patrick Ford, editor of the *Irish World*, was an ardent supporter of Irish agitation in the United States who had hired George as a correspondent. Ford, once an admirer of George, broke with him later on.

83. Joseph Chamberlain (1836–1914) was an English statesman and colonial secretary from 1895 to 1902 who advocated domestic social reforms and imperial expansion. A member of Parliament, a representative of the Radical wing of the Liberal Party, he entered Gladstone's ministry.

84. The wife of Captain William H. O'Shea (1840–1905) was the mistress of Parnell.

85. A line is missing since the bottom of the microfilmed page was creased.

86. The No Rent Manifesto was issued in 1881 by Irish nationalist leaders who were imprisoned. Thomas Brennan, along with Michael Davitt, was a founder of the Irish Land League movement in 1879, which lasted until 1882.

87. Oliver Cromwell's forces first landed in Ireland in 1649 in an attempt to conquer Ireland for Protestantism.

88. Henry George, "The Great Question," *Irish World and American Industrial Liberator* (Jan. 27, 1883): ILD. The next caption, which has been deleted by the editor, reads: "Written for the Christmas Number of the *Sacramento Bee* by Henry George."

89. A. M. Sullivan was an Irish member of Parliament.

90. The Encumbered Estates Act of 1849 attempted to facilitate the sale of bankrupt large estates for the creation of smaller holdings.

91. William E. Forster (1818–1886) became the chief secretary in Ireland in 1880. He opposed Home Rule and resigned in 1882 when Parnell was released from prison.

92. Reverend Edward McGlynn (1837–1900), an eloquent orator, was the president of the Anti-Poverty Society and one of the most ardent disciples of George. Political disagreement led to a break in their relations, which was healed later on. The excommunication of McGlynn became a *cause célèbre*. The struggle between this priest and Archbishop Corrigan of New York revealed fissures in the Roman Catholic Church in the United States and even the world between advocates of political liberties and defenders of hierarchal control. Roman Catholic enemies of George may have seen his

economics as threatening the landed order on which much of Catholic culture was based. Scholars (and George) regard Leo XIII's *Rerum Novarum* of 1891 as perhaps, in part, a defensive attempt to present an alternative to George's political economy.

93. The Tweed River forms part of the boundary between England and Scotland.

94. This article comes from the *Irish Times* (April 10, 1884): ILD.

95. The Land Reform Union, established in 1883, promoted George's ideas in the United Kingdom.

96. John Bright (1811–1889) was the vociferous laissez-faire champion of free trade and repeal of the Corn Laws.

97. Henry George, "The Irish Question from an American Standpoint," 1886; # 9, HGP.

98. For the famine see footnote no. 46.

99. For Goldsmith see footnote no. 4.

100. I do not, of course, mean to say that the rent of all lands has increased, since rent being determined by relative advantages rent generally may advance while in particular places it falls. And since it is so often forgotten, I may be permitted to remark that rent does not relate exclusively to agricultural land (H.G.).

101. [See] *Progress and Poverty. An Inquiry Into the Cause of Industrial Depressions and of Increase of Want with Increase of Wealth. The Remedy.* New York: D. Appleton & Co. (H.G.).

102. Possibly a reference to the Orissa famine of 1866, which carried away 1,500,000 people. Frequent famines, however, ravaged India at this time.

103. James Gordon Bennett (1841–1918) was an American journalist known for well-popularized stunts.

104. Notation placed under here reads: "[blank line or dark line]."

105. *Morcellement* is the French noun form of the verb *morceler*, which means to cut up or parcel out.

106. For fee simple see footnote no. 30, page 177.

107. Formatting and italicization added by the editor.

108. The free-trade battle refers to the repeal of the Corn Laws by Robert Peel in 1846. The laws had blocked the importation of grain products unless prices were high.

109. Formatting and italicization added by the editor.

110. This article comes from the *Irish Times* (July 15, 1889): ILD.

111. The Land Purchase Act of 1885, known as the Ashbourne Act, gave financial support to Irish tenants towards the entire purchase price of their holdings. It was followed by the Land Purchase Act of 1888, which extended more funds.

112. Henry George, "In Belfast: Henry George's Meeting in Ulster Hall," *The Standard* (July 27, 1889): General Research Division of The New York Public Library; Astor, Lenox, and Tilden Foundation. Hereinafter abbreviated as GRD.

113. A list of fifty-two names follows which was deleted by the editor.

114. For the Paris conference see footnote no. 71, page 253.

115. The English naval hero Horatio Nelson, Viscount Nelson, was born in 1758 and died at Trafalgar in 1805.

116. A costermonger hawks vegetables or fruits.

117. Addresses by Mr. Garland and Mr. S. M. Burroughs and the following sentence have been deleted by the editor: "Several written questions addressed to Mr. George were then taken up and read by the chairman.

118. See footnote no. 111.

119. The chapter on unions possibly refers to *Progress and Poverty*, pp. 310–316.

120. Deleted lines by the editor read: "The different questions having been answered, a vote of thanks was passed to Mr. George on the motion of Mr. George Hunt, seconded by Mr. Shipperbottom. Mr. George acknowledged the compliment, and moved a similar vote to the chairman which having been acknowledged, the meeting concluded with cheers for Mr. George and Mr. Wallace. Mr. George lectures this (Thursday) evening in the Temple of Liberty, Toomebridge, on 'The Land and the People.'"

SECTION II: SCOTLAND

LAND AND PEOPLE
(1881).[1]

Whose Land is It?

To whom rightfully does the soil of Scotland belong? Who are justly entitled to its use and to all the benefits that flow from its use?

Let me go to the heart of this question by asking another question: Has or has not every child born in Scotland a right to live? There can be but one answer, for no one would contend that it is right to drown Scottish babies, or that any human law could make it right. Well, then, if every human being born in Scotland has a right to live in Scotland, these rights must be equal. If each one has a right to live, then no one can have any better right to live than any other one. There can be no dispute about this. No one will contend that it would be any less a crime to drown the baby of a Scottish cottar[2] than it would be any less a crime to drown the baby of the proudest duchess, or that a law commanding the one would be any more justifiable than a law commanding the other.

Since, then, all the Scottish people have the same equal right to live, it follows that they must all have the same equal right to the land of Scotland. If they are all in Scotland by the same equal permission of Nature, so that no one of them can justly set up a superior claim to life than any other one of them; so that all the rest of them could not justly say to any one of them: "You have not the same right to live as we have; therefore we will pitch you out of Scotland into the sea!;" then they must all have the same equal rights to the elements which Nature has provided for the sustaining of life – to air, to water, and to land. For to deny the equal right to the elements necessary to the maintaining of life is to deny the equal right to life. Any law that said: "Certain babies have no right to the soil of Scotland; therefore they shall be thrown off the soil of Scotland;" would be precisely equivalent to a law that said: "Certain babies have no right to live; therefore they shall be thrown into the sea." And as no law or custom or agreement can justify the denial of the equal right to life, so no law or custom or agreement can justify the denial of the equal right to land.

It therefore follows, from the very fact of their existence, that the right of each one of the Scottish people to an equal share in the land of Scotland is equal and inalienable: That is to say, that the use and benefit of the land of Scotland belong rightfully to the whole people of Scotland, to each one as much as to every other; to no one more than to any other – not to some individuals, to the exclusion of other individuals; not to one class, to the exclusion of other classes; not to landlords, not to tenants, not to cultivators, but to the whole people.

107

This right is irrefutable and indefeasible. It pertains to and springs from the fact of existence, the right to live. No law, no covenant, no agreement, can bar it. One generation cannot stipulate away the rights of another generation. If the whole people of Scotland were to unite in bargaining away their rights in the land, how could they justly bargain away the right of the child who the next moment is born? No one can bargain away what is not his; no one can stipulate away the rights of another. And if the newborn infant has an equal right to life, then has it an equal right to land[?] Its warrant, which comes direct from Nature, and which sets aside all human laws or title deeds, is the fact that it is born.

Here we have a firm, self-apparent principle from which we may safely proceed. The land of Scotland does not belong to one individual more than to another individual, to one class more than to another class; to one generation more than to the generations that come after. *It belongs to the whole people who at that time exist upon it.*

Landlords' Right is Labor's Wrong

To say that the land of a country belongs to the whole people, and then merely to ask that rents shall be reduced, or that the tenant right be extended, or that the State shall buy the land from one class and sell it to another class, is utterly illogical and absurd.

Either the land of Scotland rightfully belongs to the Scottish landlords, or it rightfully belongs to the Scottish people; there can be no middle ground. If it rightfully belongs to the landlords, then every scheme for interfering in any way with the landlords is condemned; it is nobody else's business what they do with it, or what rent they charge for it, or where or how they spend the money they draw from it, and whoever does not want to live upon it on the landlords' terms is at perfect liberty to starve or emigrate. But if, on the contrary, the land of Scotland rightfully belongs to the Scottish people, then the only logical demand is, not that it be bought from a smaller class and sold to a larger class, but that it be resumed by the whole people. To propose to pay the landlords for it is to deny the right of the people to do it. It is to admit that the Scottish people have no more right to the soil of Scotland than any outsider. For, any outsider can go to Scotland and buy land, if he will give its market value; and to propose to buy out the landlords is to propose to continue the present injustice in another form. They would get in interest on the debt created what they now get in rent. They would still have a lien upon Scottish labor.

And why should the landlords be paid? If the land of Scotland belongs of natural right to the Scottish people, what valid claim for payment can be set up by the landlords? No one will contend that the land is theirs of natural right, for the day has gone by when men could be told that the Creator of the universe intended His bounty for the exclusive use and benefit of a privileged class of His creatures – that He intended a few to roll in luxury while their fellows toiled and starved for them. The claim of the landlords to the land rests not on natural right but merely on municipal law – on municipal law which contravenes natural right. And, whenever the sovereign power changes municipal law so as to conform to natural right, what claim can they assert to compensation? Some of them bought their lands, it is true; but they got no better title than the seller had to give. And what are these titles? Titles based on murder and robbery, on blood and rapine – titles which rest on the most atrocious and wholesale crimes. Created by force and maintained by force, they have not behind them the first shadow of right. That men, now dead, have had the power to give and grant Scottish lands is true; but will anyone contend they had the right? Will anyone contend that in all the past generations there has existed on the British Isles or anywhere else any human being, or any number of human beings, who had the right to say that in the year 1889 the great mass of Scotsmen should be compelled to pay – in many cases to residents of England, France, or to the United States – for the privilege of living in their native country and making a living from their native soil? Even if it be said that might makes right; even if it be contended that in the twelfth, or seventeenth, or eighteenth century lived men who, having the power, had therefore the right, to give away the soil of Scotland, it cannot be contended that their right went further than their power, or that their gifts and grants are binding on the men of the present generation. No one can urge such a preposterous doctrine. And, if might makes right, then the moment the people get power to take the land the rights of the present landholders utterly cease, and any proposal to compensate them is a proposal to do a fresh wrong.

Should it be urged that, no matter on what they originally rest, the lapse of time has given to the legal owners of Scottish land a title of which they cannot now be justly deprived without compensation, it is sufficient to ask, with Herbert Spencer: At what rate per annum wrong becomes right?[3]

And, even supposing that in their ignorance the masses have acquiesced in the iniquitous system which makes the common birthright of all the exclusive property of some. What then? Does such acquiescence turn wrong into right? If the sleeping traveler wake to find a robber with his hand in his pocket, is he bound to buy the robber off – bound not merely to let him keep what he has previously taken, but pay him the full value of all he expected the sleep of his victim to permit him to get? If the stockholders of a bank find that for a long term of years their cashier has been appropriating the lion's share of the profits, are they to be told that they cannot discharge him without paying him for what he might have got, had his peculations not been discovered?

The Great-Great-Grandson of Captain Kidd

I apologize to landlords for likening them to thieves and robbers. I trust they will understand that I do not consider them as personally worse than other men, but that I am obliged to use such illustrations because no others will fit the case. I am concerned not with individuals but with the system. What I want to do is to point out a distinction that in the plea for the vested rights of landowners is ignored – a distinction which arises from the essential difference between land and things that are the produce of human labor, and which is obscured by our habit classing them altogether as property.

The galleys that carried Cæsar to Britain, the accoutrements of his legionaries, the baggage that they carried, the arms that they bore, the buildings that they erected; the scythed chariots of the ancient Britons, the horses that drew them, their wicker boats and wattled houses – where are they now?[4] But the land for which Roman and Bri[ton] fought, there it is still. That British soil is yet as fresh and as new as it was in the days of the Romans. Generation after generation has lived on it since, and generation after generation will live on it yet. Now, here is a very great difference. The right to possess and to pass on the ownership of things that in their nature decay and soon cease to be is a very different thing from the right to possess and to pass on the ownership of that which does not decay, but from which each successive generation must live.

To show how this difference between land and such other species of property styled wealth bears upon the argument for the vested rights of landholders, let me illustrate again.

Captain Kidd was a pirate.[5] He made a business of sailing the seas, capturing merchantmen, making their crews walk the plank, and appropriating their cargoes. In this way he accumulated much wealth, which he is thought to have buried. But let us suppose, for the sake of the illustration, that he did not bury his wealth, but left it to his legal heirs, and they to their heirs, and so on, until at the present day this wealth or a part of it has come to a great-great-grandson of Captain Kidd. Now, let us suppose that someone – say a great-great-grandson of one of the shipmasters whom Captain Kidd plundered, makes complaint, and says: "This man's great-great-grandfather plundered my great-great-grandfather of certain things or certain sums, which have been transmitted to him, whereas but for this wrongful act they would have been transmitted to me; therefore, I demand that he be made to restore them." What would society answer?

Society, speaking by its proper tribunals, and in accordance with principles recognized among all civilized nations, would say: "We cannot entertain such a demand. It may be true that Mr. Kidd's great-great-grandfather robbed your great-great-grandfather, and that as the result of this wrong he has got things that otherwise might have come to you. But we cannot inquire into occurrences that happened so long ago. Each generation has enough to do to attend to its own affairs. If we go to righting the wrongs and reopening the controversies of our great-great-grandfathers, there will be endless disputes and pretexts for dispute. What you say may be true, but somewhere we must draw the line, and have an end to strife. Though this man's great-great-grandfather may have robbed your great-great-grandfather, *he* has not robbed *you*. He came into possession of these things peacefully, and has held them peacefully, and we must take this peaceful possession, when it has been continued for a certain time, as absolute evidence of just title; for, were we not to do that, there would be no end to dispute and no secure possession of anything."

Now, it is this commonsense principle that is expressed in the statute of limitations – in the doctrine of vested rights. This is the reason why it is held – and as to most things held justly – that peaceable possession for a certain time cures all defects of title.

But let us pursue the illustration a little further:

Let us suppose that Captain Kidd, having established a large and profitable piratical business, left it to his son, and he to his son, and so on, until the great-great-grandson, who now pursues it, has come to consider it the most natural thing in the world that his ships should roam the sea, capturing peaceful merchantmen, making their crews walk the plank, and bringing home to him much plunder, whereby he is enabled, though he does no work at all, to live in very great luxury, and look down with contempt upon people who have to work. But at last, let us suppose, the merchants get tired of having their ships sunk and their goods taken, and sailors get tired of trembling for their lives every time a sail lifts above the horizon, and they demand of society that piracy be stopped.

Now, what should society say if Mr. Kidd got indignant, appealed to the doctrine of vested rights, and asserted that society was bound to prevent any interference with the business that he had inherited, and that, if it wanted him to stop, it must buy him out, paying him all that his business was worth – that is to say, at least as much as he could make in twenty years' successful pirating, so that if he stopped pirating he could still continue to live in luxury off the profits of the merchants and the earnings of the sailors.

What ought society to say to such a claim as this? There will be but one answer. We will all say that society should tell Mr. Kidd that his was a business to which the statute of limitations and the doctrine of vested rights did not apply; that because his father, and his grandfather, and his great and great-great-grandfather pursued the business of capturing ships and making their crews walk the plank, was no reason why he should be permitted to pursue it. Society, we will all agree, ought to say he would have to stop piracy, and stop it at once, and that without getting a cent for stopping.

Or supposing it had happened that Mr. Kidd had sold out his piratical business to Smith, Jones, or Robinson, we will all agree that society ought to say that their purchase of the business gave them no greater right than Mr. Kidd had.

We will all agree that this is what society *ought* to say. Observe, I do not ask what society *would* say.

For, ridiculous and preposterous as it may appear, I am satisfied that, under the circumstances I have supposed, society would not for a long time say what we have agreed it *ought* to say. Not only would all the Kidds loudly claim that to make them give up their business without full recompense would be a wicked interference with vested rights, but the justice of this claim would at first be assumed as a matter of course by all or nearly all the influential classes – the great lawyers, the able journalists, the writers for the magazines, the eloquent clergymen, and the principal professors in the principal universities. Nay, even the merchants and sailors, when they first began to complain, would be so tyrannized and browbeaten by this public opinion that they would hardly think of more than of buying out the Kidds, and, wherever here and there anyone dared to raise his voice in favor of stopping piracy at once and without compensation, he would only do so under penalty of being stigmatized as a reckless disturber and wicked foe of social order.

If anyone denies this, if anyone says mankind are not such fools, then I appeal to universal history to bear me witness. I appeal to the facts of today.

Show me a wrong, no matter how monstrous, that ever yet, among any people, became engrafted in the social system, and I will prove to you the truth of what I say.

The majority of men do not think; the majority of men have to expend so much energy in the struggle to make a living that they do not have time to think. The majority of men accept, as a matter of course, whatever is. This is what makes the task of the social reformer so difficult, his path so hard. This is what brings upon those who first raise their voices in behalf of a great truth the sneers of the powerful, and the curses of the rabble, ostracism and martyrdom, the robe of derision and the crown of thorns.

Am I not right? Have there been states of society in which piracy has been considered the most respectable and honorable of pursuits? Did the Roman populace see anything more reprehensible in a gladiatorial show than we do in a horserace? Does public opinion in Dahomey see anything reprehensible in the custom of sacrificing one or two thousand human beings by way of signalizing grand occasions? Are there not states of society in which, in spite of the natural proportions of the sexes, polygamy is considered a matter of course? Are there not states of society in which it would be considered the most ridiculous thing in the world to say that a man's son was more closely related to him than his nephew? Are there not states of society in which it would be considered disreputable for a man to carry a burden while a woman who could stagger under it was around? – states of society in which the husband who did not occasionally beat his wife would be deemed by both sexes a weak-minded, low-spirited fellow? What would Chinese fashionable society consider more outrageous than to be told that mothers should not be permitted to squeeze their daughters' feet, or Flathead women being restrained from tying a board on their infants' skulls?[6] How long has it been since the monstrous doctrine of the divine right of kings was taught through all Christendom?

What is the slave trade but piracy of the worst kind? Yet it is not long since the slave trade was looked upon as a perfectly respectable business, affording as legitimate an opening for the investment of capital and the display of enterprise as any other. The proposition to prohibit it was first looked upon as ridiculous, then as fanatical, then as wicked. It was only slowly and by hard fighting that the truth in regard to it gained ground. Does not the American Constitution bear witness to what I say? Does not the fundamental law of that Republic, adopted twelve years after the enunciation of the Declaration of Independence, declare that for twenty years the slave trade shall not be prohibited nor restricted?[7] Such dominion had the idea of vested interests over the minds of those who had already proclaimed the inalienable right of man to life, liberty, and the pursuit of happiness!

Is it not but yesterday that in the freest and greatest Republic on earth, among the people who boast that they lead the very van of civilization, this doctrine of vested rights was deemed a sufficient justification for all the cruel wrongs of human slavery? Is it not but yesterday, when whoever dared to say that the rights of property did not justly attach to human beings; when whoever dared to deny that human beings could be rightfully bought and sold like cattle – the husband torn from the wife, and the child from the mother; when whoever denied the right of whoever had paid his money for him to work or whip his own nigger was looked upon as a wicked assailant of the rights of property? Is it not but yesterday when in the South whoever whispered such a thought took his life in his hands; when in the North the abolitionist was held by the churches as worse than as an infidel, was denounced by the politicians, and rotten-egged by the mob?[8] I was born in a Northern state;[9] I have never lived in the South; I am not yet grey; but I well remember, as every American of middle age must remember, how over and over again I have heard all questionings of slavery silenced by the declaration that the Negroes were the *property* of their masters, and that to take away a man's slave without payment was as much a crime as to take away his horse without payment. And whoever does not remember that far back, let him look over American literature previous to the war, and say whether, if the business of piracy had been a flourishing business, it would have lacked defenders? Let him say whether any proposal to stop the business of piracy without compensating the pirates would not have been denounced at first as a proposal to set aside vested rights?

But I am appealing to other states of society, and to times that are past, merely to get my readers, if I can, out of their accustomed ruts of thought. The proof of what I assert about the Kidds and their business is in the thought and speech of today.

Here is a system which robs the producers of wealth as remorselessly and far more regularly and systematically than the pirate robbed the merchantman. Here is a system that steadily condemns thousands to far more lingering and horrible deaths than that of walking the plank – to death of the mind and death of the soul, as well as death of the body. These things are undisputed. No one who will examine the subject can deny that the chronic pauperism and chronic famine which everywhere mark our civilization are the results of this system. Yet we are told – nay, it seems to be taken for granted – that this system cannot be abolished without buying off those who profit by it. Was there ever more degrading abasement of the human mind before a fetish? Can we wonder, as we see it, at any perversion of ideas?

Consider: Is not the parallel I have drawn a true one? Is it not just as much a perversion of ideas to apply the doctrine of vested rights to property in land, when these are its admitted fruits, as it was to apply it to property in human flesh and blood; as it would be to apply it to the business of piracy? In what does the claim of the Scottish landholders differ from that of the hereditary pirate, or the man who has bought out a piratical business? "Because I have inherited or purchased the business of robbing merchantmen," says the pirate, "therefore respect for the rights of property must compel you to let me go on robbing ships and making sailors walk the plank until you buy me *out*." "Because we have inherited or purchased the privilege of appropriating to ourselves the lion's share of the produce of labor," says the landlord, "therefore you must continue to let us do it, even though, in their poverty and misery, they are reduced to wallow with the pigs." What is the difference?

This is the point I want to make clear and distinct, for it shows a distinction that in current thought is overlooked. Property in land, like property in slaves, is essentially different from property in things that are the result of labor. Rob a man or a people of money, or goods, or cattle, and the robbery is finished there and then. The lapse of time does not, indeed, change wrong into right, but it obliterates the effects of the deed. That is done; it is over; and, unless it be very soon righted, it glides away into the past, with the men who were parties to it, so swiftly that nothing save omniscience can trace its effects; and in attempting to right it we would be in danger of doing fresh wrong. The past is forever beyond us. We can neither punish nor recompense the dead. But rob a people of the land on which they must live, and the robbery is continuous. It is a fresh robbery of every succeeding generation – a new robbery every year and every day; it is like the robbery which condemns to slavery the children of the slave. To apply it to the statute of limitations, to acknowledge for it the title of prescription, is not to condone the past; it is to legalize robbery in the present, to justify it in the future. The indictment which really lies against the Scottish landlords is not that their ancestors, or the ancestors of their grantors, robbed the ancestors of the Scottish people. That makes no difference. "Let the dead bury their dead." The indictment that truly lies is that here, now, in the year 1889, they rob the Scottish people. And shall we be told that there can be a vested right to continue such robbery?

How to Restore the Land to the People[10]

I have dwelt so long upon this question of compensating landowners, not merely because it is of great practical importance, but because its discussion brings clearly into view the principles upon which the Land Question, in any country, can alone be justly and finally settled. In the light of these principles we see that landowners have no rightful claim either to the land or to compensation for its resumption by the people; and, further than that, we see that no such rightful claim can ever be created. It would be wrong to pay the present landowners for "their" land at the expense of the people; it would likewise be wrong to sell it again to smaller holders. It would be wrong to abolish the payment of rent, and to give the land to its present cultivators. In the very nature of things, land cannot rightfully be made individual property. This principle is absolute. The title of a peasant proprietor deserves no more respect than the title of a great territorial noble. No sovereign political power, no compact or agreement, even though consented to by the whole population of the globe, can give to an individual a valid title to the exclusive ownership of a square inch of soil. The earth is an entailed estate – entailed upon all the generations of the children of men, by a deed written in the constitution of Nature; a deed that no human proceedings can bar, and no prescription determine. Each succeeding generation has but a tenancy for life. Admitting that any set of men may barter away their own natural rights (and this logically involves an admission of the right of suicide), they can no more barter away the rights of their successors than they can barter away the rights of the inhabitants of other worlds.

What should be aimed at is thus very clear. The "three F's" are three frauds;[11] and the proposition to create peasant proprietorship is no better. It will not do merely to carve out of the estates of the landlords minor estates for the tenants; it will not do merely to substitute a larger for a smaller class of proprietors; it will not do to confine the settlement to agricultural land, leaving to its present possessors the land of the towns and villages. None of these lame and impotent conclusions will satisfy the demands of Justice, or cure the bitter evils now so apparent. The only true and just solution of the problem, the only end worth aiming at, is:

to make ALL THE LAND *the Common Property of* ALL THE PEOPLE.[12]

This principle conceded, the question of method arises. How shall this be done? Nothing is easier. It is merely necessary to divert the rent which now flows into the pockets of the landlords into the common treasury of the whole people. It is not possible to so divide up land as to give each family, still less each individual, an equal share. And, even if that were possible, it would not be possible to maintain equality, for old people are constantly dying, and new people constantly being born, while the relative value of land is constantly changing. But it is possible to equally divide the rent, or, what amounts to the same thing, to apply it to purposes of common benefit. This is the way, and this is the only way, in which absolute Justice can be done. This is the way, and this is the only way, in which the equal right of every man, woman, and child can be acknowledged and secured. As Herbert Spencer says of it:

> Such a doctrine is consistent with the highest state of civilization; may be carried out without involving a community of goods, and need cause no very serious revolution in existing arrangements. The change required would simply be a change of landlords. Separate ownership would merge into the joint-stock ownership of the public. Instead of being in the possession of individuals, the country would be held by the great corporate body – Society. Instead of leasing his acres from an isolated proprietor, the farmer would lease them from the nation. Instead of paying his rent to the agent of Sir John or His Grace, he would pay it to an agent or deputy agent of the community. Stewards would be public officials instead of private ones, and tenancy the only land tenure.
>
> A state of things so ordered would be in perfect harmony with the moral law. Under it all men would be equally landlords; all men would be alike free to become tenants.... Clearly, therefore, on such a system, the earth might be enclosed, occupied, and cultivated in entire subordination to the law of equal freedom.[13]

Now, it is a very easy thing to thus sweep away all private ownership of land, and convert all occupiers into tenants of the State, by appropriating rent. No complicated laws or cumbersome machinery is necessary. It is only necessary to tax land up to its full value. Do that, and without any infringement of the just rights of property, the land would become virtually the people's. What under this system was paid as rent by the tenant would be taken by the State. The occupiers of land would come to be nominally the owners, though, in reality, they would be the tenants of the whole people.

How beautifully this simple method would satisfy every economic requirement; how, freeing labor and capital from the fetters that now oppress them (for all other taxes could be easily remitted), it would enormously increase the production of wealth; how it would make distribution conform to the law of Justice, dry up the springs of want and misery, elevate society from its lowest stratum, and give all their fair share in the blessings of advancing civilization, can perhaps only be fully shown by such a detailed examination as I have made in my books, *Social Problems* and *Progress and Poverty*.[14] Nevertheless, anyone can see that to tax land up to its full rental value would amount to precisely the same thing as to formally take possession of it, and then let it out to the highest bidders.

If it be denied that land justly is, or can be, private property, if the equal rights of the whole people to the use of the elements gratuitously furnished by Nature be asserted without drawback or compromise, then the essential difference between property in land and property in things of human production is at once brought out. Then will it clearly appear not only that the denial of the right of individual property in land does not involve any menace to legitimate property rights, but that the maintenance of private property in land necessarily involves a denial of the right to all other property, and that the recognition of the claims of the landlords means a continuous robbery of capital as well as of labor.

All this will appear more and more clearly as the practical measures necessary to make land common property are proposed and discussed. These simple measures involve no harsh proceedings, no forcible dispossession, no shock to public confidence, no retrogression to a lower industrial organization, no loaning of public money, or establishment of cumbrous commissions. Instead of doing violence to the rightful sense of property, they assert and vindicate it. The way to make land common property is simply to take rent for the common benefit. And to do this, the easy way is to abolish one tax after another, until the whole weight of taxation falls upon the value of land. When that point is reached, the battle is won. The hare is caught, killed, and skinned, to cook him will be a very easy matter. The real fight will come on the proposition to consolidate existing taxation upon land values. When that is once won, the landholders will not merely have been decisively defeated, they will have been routed; and the nature of land values will be so generally understood that to raise taxation so as to take the whole rent for common purposes will be a mere matter of course.

To put the public burdens upon the landholders is not a new proposition. On the contrary, it is the ancient British practice. It would be but a return, in a form adapted to modern times, to the system under which British land was originally parceled out to the predecessors of the present holders – the just system, recognized for centuries, that those who enjoy the common property should bear the common burdens. The putting of property in land in the same category as property in things produced by labor is comparatively modern. In Scotland, as in England and Ireland, as in fact among every people of whom we know anything, the land was originally treated as common property, and this recognition ran all through the feudal system. The essence of the feudal system was in treating the landholder not as an owner but as a lessee. To every grant of land was annexed a condition which amounted to a heavy perpetual tax or rent. National debt, pauperism, and the grinding poverty of the poorer classes, came in as the landholders gradually shook off the obligations on which they had received their land, an operation culminating in the abolition of the feudal tenures, for which were substituted indirect taxes that still weigh upon the whole people. To now reverse this process, to abolish the taxes which are borne by labor and capital, and to substitute for them a tax on rent, would not be the adoption of anything new, but a simple going back to the old plan. In Great Britain, as in Ireland; in the Highlands, as in the Lowlands, the movement would appeal to the popular imagination as a demand for the reassertion of ancient rights.

There are other most important respects in which this measure will commend itself. The tax upon land values or rent is in all economic respects the most perfect of taxes. No political economist will deny that it combines the maximum of certainty with the minimum of loss and cost; that, unlike taxes upon capital or exchange or improvement, it does not check production or enhance prices or fall ultimately upon the consumer. And, in proposing to abolish all other taxes in favor of this theoretically perfect tax, reformers will have on their side the advantage of ideas already current, while they can bring the *argumentum ad hominem* to bear on those who might never comprehend an abstract principle. Britons of all classes have happily been educated up to a belief in Free Trade to its fullest extent.[15] If a revenue tariff is better than a protective tariff, then no tariff at all is better than a revenue tariff. Let them propose to abolish the customs duties entirely, and to abolish as well harbor dues and lighthouse dues and dock charges, and in their place to add to the tax on rent, or the value of land exclusive of improvements. Let them in the same way propose to get rid of the excise, the various license taxes, the tax upon buildings, the onerous and unpopular income tax, etc., and to saddle all public expenses on the landlords.

This would bring home the land question to thousands and thousands who have never thought of it before; to thousands and thousands who heretofore looked upon the land question as something that related exclusively to agriculture and to farmers, and have never seen how, in various direct and indirect ways, they have to contribute to the immense sums received by the landlords as rent. It would be putting the argument in a shape in which even the most stupid could understand it. The British landowners are in numbers but an insignificant minority. And the more they protested against having to pay all the taxes, the quicker would the public mind realize the essential injustice of private property in land, the quicker would the majority of the people come to see that the landowners ought not only to pay all the taxes but a good deal more besides. Once put the question in such a way that the working man will realize that he pays two prices for his ale, and half a dozen prices for his tobacco, because a landowners' Parliament in the time of Charles II shook off their ancient dues to the State,[16] and imposed them in indirect taxation on him; once bring to the attention of the man who grumbles as he pays his income tax, the question as to whether the landowner who draws his income from property that of natural right belongs to the whole people ought not to pay it instead of him, and it will not be long before the absurd injustice of allowing rent to be appropriated by individuals will be thoroughly understood. This is a very different thing from asking the taxpayer to buy out the landlord for the sake of the peasant.

I have been speaking as though all landholders would resist the change which would sacrifice their special interests to the larger interests of society. But I am satisfied that to think this is to do landholders injustice [sic]. For landholders as a class are not more stupid nor more selfish than any other class. And there is that in a great truth which can raise a human soul above the mists of selfishness.

A Little Island or a Little World

Imagine an island girt with ocean; imagine a little world swimming in space. Put on it, in imagination, human beings. Let them divide the land, share and share alike, as individual property. At first, while population is sparse and industrial processes rude and primitive, this will work well enough.

Turn away the eyes of the mind for a moment, let time pass, and look again. Some families will have died out, some have greatly multiplied; on the whole, population will have largely increased. And even supposing there have been no important inventions or improvements in the productive arts, the increase in population, by causing the division of labor, will have made industry more complex. During this time some of these people will have been careless, generous, improvident; some will have been thrifty and grasping. Some of them will have devoted much of their powers to thinking of how they themselves and the things they see around them came to be; to inquiries and speculations as to what there is in the universe beyond their little island or their little world; to making poems, painting pictures, or writing books; to noting the differences in rocks and trees and shrubs and grasses; to classifying beasts and birds and fishes and insects – to the doing, in short, of all the many things which add so largely to the sum of human knowledge and human happiness, without much or any gain of wealth to the doer. Others again will have devoted all their energies to the extending of their possessions. What, then shall we see, land having been all this time treated as private property? Clearly, we shall see that the primitive equality has given way to inequality. Some will have very much more than one of the original shares into which the land was divided; very many will have no land at all. Suppose that, in all things save this, our little island or our little world is Utopia – that there are no wars or robberies; that the government is absolutely pure and taxes nominal; suppose, if you want to, any sort of a currency; imagine, if you can imagine such a world or island, that interest is utterly abolished; yet inequality in the ownership of land will have produced poverty and virtual slavery.

For the people we have supposed are human beings – that is to say, in their physical natures at least, they are animals who can live only on land and by the aid of the products of land. They may make machines which will enable them to float on the sea, or perhaps to fly in the air, but to build and equip these machines they must have land and the products of land, and must constantly come back to land. Therefore, those who own the land must be the masters of the rest. Thus, if one man had come to own all the land, he is their absolute master even to life or death. If they can live on the land on his terms only, then they can live only on his terms, for without land they cannot live. They are his absolute slaves, and so long as his ownership is acknowledged, if they want to live, they must do ... everything as he wills.

If, however, the concentration of landownership has not gone so far as to make one man or a very few men the owners of all the land – if there are still so many landowners that there is competition between them as well as between those who have only their labor – then the terms on which these nonlandholders can live will seem more like free contract. But it will not be free contract. *Land can yield no wealth without the application of labor; labor can produce no wealth without land.* These are the two equally necessary factors of production. Yet, to say that they are equally necessary factors of production is not to say that, in the making of contracts as to how the results of production are divided, the possessors of these two meet on equal terms. For the nature of these two factors is very different. Land is a natural element; the human being must have his stomach filled every few hours. Land can exist without labor, but labor cannot exist without land. If I own a piece of land, I can let it lie idle for a year or years, and it will eat nothing. But the laborer must eat every day, and his family must eat. And so, in the making of terms between them, the landowner has an immense advantage over the laborer. It is on the side of the laborer that the intense pressure of competition comes, for in his case it is competition urged by hunger. And, further than this: As population increases, as the competition for the use of the land becomes more and more intense, so are the owners of land enabled to get for the use of their land a larger and larger part of the wealth which labor exerted upon it produces. That it to say, the value of land steadily rises. Now, this steady rise in the value of land brings about a confident expectation of future increase of value, which produces among landowners all the effects of a combination to hold for higher prices. Thus, there is a constant tendency to force mere laborers to take less and less or to give more and more (put it which way you please, it amounts to the same thing) of the products of their work for the opportunity to work. And thus, in the very nature of things, we should see on our little island or our little world that, after a time had passed, some of the people would be able to take and enjoy a superabundance of all the fruits of labor without doing any labor at all, while others would be forced to work the livelong day for a pitiful living.

But let us introduce another element into the supposition. Let us suppose great discoveries and inventions – such as the steam engine, the power loom, the Bessemer process, the reaping machine, and the thousand-and-one labor-saving devices that are such a marked feature of our era.[17] What would be the result?

Manifestly, the effect of all such discoveries and inventions is to increase the power of labor in producing wealth – to enable the same amount with the same labor. But none of them lessen, or can lessen, the necessity for land. Until we can discover some way of making something out of nothing – and that is so far beyond our powers as to be absolutely unthinkable – there is no possible discovery or invention which can lessen the dependence of labor upon land. And, this being the case, the effect of these laborsaving devices, land being the private property of some, would simply be to increase the proportion of the wealth produced that landowners could demand for the use of the land. The ultimate effect of these discoveries and inventions would be not to benefit the laborer, but to make him more dependent.

And, since we are imagining conditions, imagine laborsaving inventions to go to the farthest imaginable point, that is to say, to perfection. What then? Why then, the necessity for labor being done away with, all the wealth that the land could produce would go entirely to the landowners. None of it whatever could be claimed by anyone else. For the laborers there would be no use at all. If they continued to exist, it would be merely as paupers on the bounty of the landowners!

The Civilization that is Possible

In the effects upon the distribution of wealth, of making land private property, we may thus see an explanation of that paradox presented by modern progress. The perplexing phenomena of deepening want with increasing wealth, of labor rendered more dependent and helpless by the very introduction of laborsaving machinery, are the inevitable result of natural laws as fixed and certain as the law of gravitation.

> *Private property in land is the primary cause of the monstrous inequalities which are developing in modern society.*[18]

It is this, and not any miscalculation of Nature in bringing into the world more mouths than she can feed, that gives rise to that tendency of wages to a minimum – that, in spite of all advances in productive power, compels the laboring classes to the least return on which they will consent to live. It is this that produces all those phenomena that are so often attributed to the conflict of labor and capital. It is this that condemns Highland crofters[19] to rags and hunger, that produces the pauperism of Great Britain and Ireland, and the tramps of America. It is this that makes the almshouse and the penitentiary the marks of what we call high civilization; that in the midst of schools and churches degrades and brutalizes men, crushes the sweetness out of womanhood and the joy out of childhood. It is this that makes lives that might be a blessing a pain and a curse, and every year drives more and more to seek unbidden refuge in the gates of death. For, a permanent tendency to equality once set up, all the forces of progress tend to greater and greater inequality.

All this is contrary to Nature. The poverty and misery, the vice and degradation, that spring from the unequal distribution of wealth, are not the results of natural law; they spring from our defiance of natural law. They are fruits of our refusal to obey the supreme law of Justice. It is because we rob the child of his birthright; because we make the bounty which the Creator intended for all the exclusive property of some, that these things come upon us, and, though advancing and advancing, we chase but the mirage.

When, by lightning flash or friction amid dry grasses, the consuming flames of the fire flung their first glow into the face of man, how must he have started back in affright! When he first stood by the shores of the sea, how must its waves have said to him: "Thus far shalt thou go, but no farther!" Yet, as he learned to use them, fire became his most useful servant, the sea his easiest highway. The most destructive element of which we know – that which for ages and ages seemed the very thunderbolt of the angry gods – is, as we are now beginning to learn, fraught for us with untold powers of usefulness. Already it enables us to annihilate space in our messages, to illuminate the night with new suns; and its uses are only beginning. And throughout all Nature, as far as we can see, whatever is potent for evil is potent for good. "Dirt," said Lord Brougham, "is matter in the wrong place."[20] And so the squalor and vice and misery that abound in the very heart of our civilization are but results of the misapplication of forces in their nature most elevating.

I doubt not that whichever way a man may turn to inquire of Nature, he will come upon adjustments which will arouse not merely his wonder, but his gratitude. Yet what has most impressed me with the feeling that the laws of Nature are the laws of beneficent intelligence is what I see of the social possibilities involved in the law of rent. Rent springs from natural causes.[21] It arises, as society develops, from the differences in natural opportunities and the differences in the distribution of population. It increases with the division of labor, with the advance of the arts, with the progress of invention. And thus, by virtue of a law impressed upon the very nature of things, has the Creator provided that the natural advance of mankind shall be an advance toward equality, and advance toward cooperation, and advance toward a social state in which not even the weakest need be crowded to the wall, in which even for the unfortunate and the cripple there may be ample provision. For this revenue, which arises from the common property, which represents not the creation of value by the individual, but the creation by the community as a whole, which increases just as society develops, affords a common fund, which, properly used, tends constantly to equalize conditions, to open the largest opportunities for all, and to utterly banish want or the fear of want.

The squalid poverty that festers in the heart of civilization, the vice and crime and degradation and ravening greed that flow from it, are the results of a treatment of land that ignores the simple law of Justice, a law so clear and plain that it is universally recognized by the veriest savages. What is by nature the common birthright of all, we have made the exclusive fund, from which common wants should be met, we give to a few that they may lord it over their fellows. And so some are gorged while some go hungry, and more is wasted than would suffice to keep all in luxury.

In this nineteenth century, among any people who have begun to utilize the forces and methods of modern production, there is no necessity for want. There is no good reason why even the poorest should not have all the comforts, all the luxuries, all the opportunities for culture, all the gratifications of refined taste that only the richest now enjoy. There is no reason why anyone should be compelled to long and monotonous labor. Did invention and discovery stop today, the forces of production are ample for this. What hampers production is the unnatural inequality in distribution. And, with just distribution, invention and discovery would only have begun.

Appropriate rent in the way I propose, and speculative rent would be at once destroyed. The "dogs in the manger" who are now holding so much land they have no use for, in order to extract a high price from those who do want to use it, would be at once choked off, and land from which labor and capital are now debarred under penalty of a heavy fine would be thrown open to improvement and use. The incentive to land monopoly would be gone. Population would spread where it is now too dense, and become denser where it is now too sparse.

Appropriate rent in this way, and the present expenses of Government would be at once very much reduced – reduced directly by the saving in the present cumbrous and expensive schemes of taxation, reduced indirectly by the diminution in pauperism and in crime. This simplification in governmental machinery, this elevation of moral tone which would result, would make it possible for Government to assume the running of railroads, telegraphs, and other businesses which, being in their nature monopolies, cannot, as experience is showing, be safely left in the hands of private individuals and corporations. In short, losing its character as a repressive agency, Government could thus gradually pass into an administrative agency of the great cooperative association – Society.

For, appropriate rent in this way, and there would be at once a large surplus over and above what are now considered the legitimate expenses of Government. We could divide this, if we wanted to, among the whole community, share and share alike. Or we could give every boy a small capital for a start when he came of age, every girl a dower, every widow an annuity, every aged person a pension, out of this common estate. Or we could do with our great common fund many, many things that would be for the common benefit; many, many things that would give to the poorest what even the richest cannot now enjoy. We could establish free libraries, lectures, museums, art galleries, observatories, gymnasiums, baths, parks, theaters; we could line our roads with fruit trees, and make our cities clean and wholesome and beautiful; we could conduct experiments, and offer rewards for inventions, and throw them open to public use.

Think of the enormous wastes that now go on: The waste of false revenue systems, which hamper production and bar exchange, which fine a man for erecting a building where none stood before, or for making two blades of grass grow where there was but one. The waste of unemployed labor, of idle machinery, of those periodical depressions of industry almost as destructive as war. The waste entailed by poverty, and the vice and crime and thriftlessness and drunkenness that spring from it; the waste entailed by that greed of gain that is its shadow, and which makes business in large part but a masked war; the waste entailed by the fret and worry about the mere physical necessities of existence, to which so many of us are condemned; the waste entailed by ignorance, by cramped and undeveloped faculties, by the turning of human beings into mere machines!

Think of these enormous wastes, and of the others which, like these, are due to the fundamental wrong which produces an unjust distribution of wealth and distorts the natural development of society, and you will begin to see what a higher, purer, richer civilization would be made possible by the simple measure that will assert natural rights. You will begin to see how, even if no one but the present landholders were to be considered, this would be the greatest boon that could be vouchsafed them by society, and that, for them to fight it, would be as if the dog with a tin-kettle tied to his tail should snap at the hand that offered to free him. Even the greatest landholder! As for such landholders as our working farmers and homestead owners, the slightest discussion would show them that they had everything to gain by the change. But even such landholders as the Duke of Westminster and the Astors would be gainers.[22]

For it is of the very nature of injustice that it really profits no one. When and where was slavery good for slaveholders? Did her cruelties in America, her expulsions of Moors and Jews, her burnings of heretics, profit Spain? Did not the curse of an unjust social system rest on Louis XIV and Louis XV as well as on the poorest peasant whom it condemned to rags and starvation – as well as on that Louis whom it sent to the block? Is the Czar of Russia to be envied?[23]

This we may know certainly, this we may hold to confidently; that which is unjust can really profit no one; that which is just can really harm no one. Though all other lights move and circle, this is the polestar by which we may safely steer.

SCOTLAND AND SCOTSMEN
(February 18, 1884).[24]

On Monday, February 18, 1884, Mr. Henry George, author of *Poverty and Progress*, [sic] delivered the following address in the City Hall, Glasgow; Councillor Crawford presided.

Mr. George said: This is the second time I have had the privilege of standing in this hall. I visited Scotland before, but only again by night in a Pullman car, and I saw nothing of the country.[25] The audience that I then addressed was an Irish audience – it was on St. Patrick's night. This audience is a general audience; I presume a Scottish audience. Now, I have been pretty well abused.

I read in the papers all sorts of things about myself, and if I did not know Henry George pretty well, I had thought he was a cross between a thief and a fool.

These charges I have never noticed; nevertheless, there is one charge that has been made against me since I came to Scotland which I would like to say a word about. I have been accused of flattering Scotchmen. The first place where I spoke in Scotland was in Dundee, and I was glad to get before a Scottish audience. It so happens that in my own country I know very many Scotsmen, and among the men who stand with me are very many Scotsmen. These Scotsmen have always been telling me: "Ah, a Scottish audience is the thing; wait till the Scottish people take hold of this question, and they will go to the logical end." I was glad to get before a Scottish audience, and I told them about my Scottish friends, and I told them about the letter I had received from a good "canny" Scotsman, who said to me: "Don't waste your time on these English people. They are a beery set. Beer confuses and dulls their under-standings. (Laughter.) You can do far more good in Scotland, where they are a logical, clearheaded people; and if they drink anything at all, it is only whisky, which does not have such a confusing effect on the intellect." Well, I told them that in the frankness of my nature, and next morning the papers, in their usual denunciation, said I took an advantage by flattering a Scottish audience. Now, I may have been accused of many things, but I don't think those who know me would accuse me of such a thing as attempting to flatter Scotsmen about Scotland. I doubt if that is possible. (Applause.)

When I came from New York to California, a Scottish banker sought me out and said: "I had a wager about you, and I want to ask you a personal ques-tion. You are an American by birth?" And I said: "I am." "Have you not Scottish blood in your veins?" "Well," I said, "my mother's father was a Glasgow body." Says he: "I have won my bet; it's through your mother that you get your talent." (Laughter and applause.) That man had, and still has, a theory that every great man is a Scotsman, with two or three exceptions, and in these cases a mistake was made. Now, joking aside, I do not want to flatter anybody; and if Scotsmen don't like to be flattered, will you let me tell you tonight some home truths – some things that are not complimentary? (Cries of "Yes, yes," and applause.)

I draw my blood from these islands. But it so happens this is the only place to which I can trace my ancestry with any certainty. I do not know but that some of my own kindred perhaps today live in Glasgow, and it is from Glasgow men some of my blood, at least, is drawn. I am not proud of it. If I were a Glasgow man today I would not be proud of it. Here you have a great and rich city, and here you have poverty and destitution that would appal a heathen. Right on these streets of yours the very stranger can see sights that he could not see in any tribe of savages in anything like normal conditions.

"Let Glasgow Flourish by the Preaching of the Word" – that is the motto of this great, proud city. What sort of a Word is it that here has been preached? Or, let your preaching have been what it may, what is your practice? Are these the fruits of the Word – this poverty, this destitution, this vice and degradation? To call this a Christian community is a slander on Christianity. Low wages, want, vice, degradation – these are not the fruits of Christianity. They come from the ignoring and denial of the vital principles of Christianity. Yet you people in Glasgow not merely erect church after church, you have the cheek to subscribe money to send missionaries to the heathen. I wish the heathen were a little richer, that they might subscribe money and send missionaries to such so-called Christian communities as this – to point to the luxury, the very ostentation of wealth, on the one hand, and to the barefooted, ill-clad women on the other; to you men and women with bodies stunted and minds distorted; to your little children growing up in such conditions that only a miracle can keep them pure!

Excuse me for calling your attention to these unpleasant truths; they are something that every man with a heart in his breast ought to think of. John Bright,[26] in his installation speech to the Glasgow University in 1883, made a statement, taken from the census of Scotland, in which he declared that forty-one families out of every one hundred in Glasgow lived in houses having only one room. He further said that 37% beyond this 41% dwelt in houses with only two rooms; that 78%, or nearly four-fifths, dwelt in houses of one or two rooms; and he went on to say further, that in Scotland nearly one-third of the people dwelt in houses of only one room, and that more than two-thirds, or 70%, dwelt in houses of not more than two rooms. Is not that an appalling statement; in the full blaze of the nineteenth century, in the year of grace 1884, here in this metropolis of Scotland – Christian Scotland!

Now, consider what it implies – this crowding of men, women, and children together. People do not herd that way unless driven by dire want and necessity. These figures imply want and suffering and brutish degradation, of which every citizen of Glasgow, every Scotsman, should be ashamed. Here I take at random from one of your papers of this evening a story, a mere item of an inquest held at Peterborough. The deceased was a married woman, the house had no furniture, and the four children were half-starved. There was no food in the house, and the only protection against the chills of night were three guano bags – a basket of litter for the whole family.[27] The dead body of the mother was found to be a mass of sores, and the left arm was shrivelled up. The daughter stated that when they got food the father would bite first, and pass it round in turn. The dying woman craved a bun, but they could not give her even that. In their verdict of death from natural causes, paralysis, deep-seated sores, and exhaustion, the jury stated that the husband had been guilty of gross and unpardonable neglect to his wife and family. But this seems to be based upon the fact that he had not taken his wife to the almshouse, though, as he stated, he had tried to get her into the almshouse, but had been refused, unless he would go too. There is nothing to show that he was idle or drunken. He was but a laborer, and seems to have tried his best to get what work he could, and came home every night to lie beside that poor woman on the rotting straw.

But take the bare facts. Among what tribe of savages in the whole world, in anything like a time of peace, would such a thing as that be possible? I have seen, I believe, the lowest races on the face of the earth – the Terra del Fuegans, who are spoken of as the very lowest of mankind; the black fellows of Australia; the Digger Indians of California.[28] I would rather take my chances, were I on the threshold of life tonight, among those people, than come into the world in this highly civilized Christian community in the condition in which thousands are compelled to live.

The fault of the husband, the verdict says! I know of this case only what the papers say; but this I do know, from the testimony of men of position and veracity, from officials and ministers of the Gospel, that such things as that are happening every day in this country, not to drunken men, but to the families of men honest, sober, and industrious. Why, in this great, rich city of yours, as a gentleman was telling me, there are today numbers of men who cannot get employment. Here the wages of your engineers were reduced a little while ago, and they had to submit. The engineers of Belfast had also to submit to a reduction of wages, because there were so many unemployed shipwrights and engineers in Glasgow that they feared they could not maintain a strike.

Am I not right in saying that such a state of things is but typical of that which exists everywhere throughout the civilized world? And I am bound to say that it is a state of things you ought to be ashamed of. I speak, not because they do not exist in my own country, for in their degree there is just the same state of things in America. But is not the spirit that, ignoring this, gives thanks and praise to the Almighty Father, cant of the worst kind? Can we separate duty towards God from duty towards our neighbors? Yet here are men who preach and pray, yet look on such things as matters of course, laying the blame upon natural laws, upon human nature, and upon the ordinances of the Creator.[29] Is it not cant and blasphemy of the worst kind? How can a man love a God whom he believes responsible for these things? Is God the Creator a "botch," that He should have made a world in which only a few of His creatures could live comfortably – that He should have made a world in which the great masses have to strain and strive all their lives away to keep above starvation point?

It is not the fault of God! It is due merely to the selfishness and ignorance of men. And when you come to ask the reason of this state of things, if you seek it out, you will come at last, I believe, to the great fact, that the land on which you live has been made the private property of a few of their number. This is the only adequate explanation. Man is a land animal. All his substance must be drawn from the land. He cannot even take the birds of the air or fish in the sea without the use of the land or the materials drawn from the land. His very body is drawn from the land. Take from a man all that belongs to land, and you would have but a disembodied spirit. And as land is absolutely necessary to the life of man, and as land is the source from which all wealth is drawn, the man who commands the land, on which and from which other men live, commands those men. (Applause.)

Take the opposite course; trace up the facts. Why is it that men are crowded together so in Glasgow? Because you let "dogs in the manger" hold the land on which these people ought to live. Here is one fact that I happened to see in a communication in one of your papers recently. There is a field in Glasgow called Burnbank, comprising fourteen acres, worth £90,000 – it is surrounded by houses – and ought to be used for buildings. But the owner is holding it till he can get a higher price from the necessities of the community. You let him hold it. You don't charge any taxes for it. The taxation you put upon the houses. The same article says if that field were feued and covered with houses, these houses would pay not less than £7,000 a year in taxation.[30] You charge and fine a man who puts up a house that would give accommodation to the people, and the man who holds land without making any use of it you do not charge a penny for the privilege. How can there be any doubt as to the reason why you are so crowded together? Or, take the fact that wages are so low; that men are competing with one another so eagerly for employment that wages are brought down to starvation rates.

What is the reason? Simply that men are denied natural opportunities of employment. This city of Glasgow has been crowded with people driven from Ireland and your Highlands, where they were living.[31] When I was over in Ireland two years ago I saw the process. I followed some of those redcoated evicting armies, and saw how, at the behest of men who had never set foot in Ireland, the military forces of the empire were being used to turn out poor people from the cabins and the land on which their fathers had lived from time immemorial. (Applause.)

Where were they forced to go? Into cities to obtain work at any price there. That great man who has stood on this platform, Michael Davitt (applause) is one of that class. His mother, forced from her home, carried him around begging, rather than go to the almshouse, and coming over here, he had, at an early age, when he ought to have been at play and at school, and not at work, to enter one of your factories, and that empty sleeve on his right side is a memento of that tyranny. (Applause.) Thus is your labor market crowded with people who must get work or starve, who can't employ themselves, who are forced into competition for anything they can get.

So with your own people – the people of Scotland. They have been crowded here in the same way. There is the explanation. This is the explanation of the fact that, although during this century, by reason of invention and improvement, the productive power of labor has increased so wonderfully, wages have not increased at all save where trades unions have been formed and have been able to force them up a little. (Applause.) I have now seen something of Scotland, and let me tell you frankly that what I have seen does not raise my estimate of the Scottish character. (Applause.)

Let me tell you frankly – seeing I have been accused of flattering you, and you say you can stand unpleasant truths – let me tell you frankly, I have a good deal more respect for the Irish. The Irish have done some kicking against this infernal system, and you men in Scotland have got it yet to do. The Scots are a logical people, as my friend says. I won't gainsay that; but their major premise must be a very curious one. I have really been wondering, since I have been in Scotland, whether you have not got things mixed a little.

There is a story I heard in Ireland about a little crossroads innkeeper. A woman kept an inn there, and a lord came along and stopped there one night. Oh, she was all in a flutter at attending upon a lord, and so she carefully instructed the boots – a rude boy – as to how in the morning he must go and knock at Lord So-and-So's door, and when his lordship asked who was there, he was to reply: "The boy, my lord." Well, the poor fellow was awfully flustered, and he gave a thundering rap at the door, when Lord So-and-So cried out: "Who's there?" and the boy shouted out: "The Lord, my boy!" (Applause and laughter.) He had got things mixed.

Now, since I have been in Scotland, I have been wondering whether you in Scotland haven't got things mixed a little. The Scots are a Bible-reading people. I have sometimes wondered whether, instead of reading that: "In the beginning the Lord created the heavens and the earth," they haven't got it that: "In the beginning the lairds created the heavens and the earth."[32] Certainly the lairds have it all their own way through Scotland. Their's is the land and all upon it; their's is all that is beneath the land; their's are the fishes in the rivers and in the lochs; their's are the birds of the air; their's are the salmon in the sea, even the seaweed that is thrown ashore, even the whales over a certain length, even the driftwood! Their's are even the water and the air.

Why, in Dundee, do you know, the people there, in order to get water, had to pay £25,000 to the Earl of Airlie for the privilege of drawing water for their use out of a certain loch. The water alone; he retains the right to the fish. The very rain as it descends from heaven is the property of the Laird of Airlie! Why, just think of it! You know how that the chosen people were passing through the wilderness and they thirsted, and Moses struck the rock and the water gushed forth. What good would it have done if that rock had been private property, and some Earl of Airlie had been there who would say: "You cannot take a cupful until you pay me £25,000?" And this Earl of Airlie does not live in Scotland at all – at any rate, he does not live in Dundee! He never drinks a cupful of that water; why – just think of it; and here, when you have dry weather, the preachers pray for rain, and then when the good Lord listens to their prayer, and sends it down, it belongs to the Earl of Airlie! (Laughter and applause.)

But the people of Scotland have the air – that is, what they can get in the streets and the roads! There is at Dundee a hill they call Balgay. It was never cultivated, and the only thing about it is that there is good air to be obtained there, and fine views had. That hill belongs to a non-resident. I think the man's name is Scott, and he lives in Edinburgh. The people of Dundee want to take walks on that hill. How do they get that privilege? By paying him a rent of £14 per acre.

Talk about the taboo. Do you remember these superstitious South Sea Islanders to whom we sent missionaries, and now they are all dying out from rum and disease. Do you know these people had a custom that they called the taboo? Their high chiefs, whom they venerated as gods on earth almost, could say of a certain thing, that is tabooed, and one of the common sort dare not touch it or use it; he would have to go around for miles rather than set his foot on a tabooed path, go thirsty rather than drink at a tabooed spring, and go hungry though fruit on a tabooed tree was rotting before his eyes. You have just precisely the same thing here. There are miles and miles of this Scotland of yours – that is, the Scotland that you common Scotsmen call your country – that is, the Scotland for which you are told you ought to lay down your lives if necessary – there are miles and miles of it in a state on Nature which one of you common Scotsmen dare not set his foot on. ("Hear, hear.")

There is one of my countrymen – an American named Winans[33] – who made a great deal of money in Russia; he comes over here and has a playground stretching from sea to sea, in a state of Nature, tenanted by wild beasts, and from which everyone of you Scotsmen are rigorously excluded. And that is only an example of the country all over. If you were heathens, if you were savages, many of you would be far better off.

People would not have to live on oatmeal and potatoes while the streams were flashing with fish and the moors were alive with game. All the fish are preserved. I got hold of a book the other day, *The Streams and Lochs of Scotland*, and I had the curiosity to look over it. Why, every bit of water in which you can paddle a tub is preserved: it belongs to Lord This, or Lady That, or Mr. Somebody Else. And the quail! Why, to go back to what I was talking about. You remember how, to feed the hungry the Israelites, quail were sent from heaven. If they had been sent into Scotland, you common Scotsmen would not have dared to touch them. Here the quail are preserved. Why, through the country that I have been, the common, ordinary working Scotsmen live on potatoes, and are well off when they get salted herrings or a little herrings or a little oatmeal. If the potato rot were to come, you would have just such famines as occurred in Ireland in 1848.

In point of fact, this year there is on the Island of Skye a crop of potatoes only by the charity of the people who subscribed to the destitution fund, and so furnished those people with seed. Full-fed, comfortable people, who eat hearty dinners every day, professors of universities with good salaries, gentlemen with nice steady incomes and pensions, say: "Oh, everything is going right; the working classes are getting better off;" and they deny most bitterly the assertion that poverty is keeping pace with progress, and they give you long tables of statistics to prove it.

Everywhere that I have been I have asked the working people themselves what they thought, and I found everywhere that the very reverse was their opinion. Certainly, after going through this country, there can be no question that all this progress and civilization has only ground this people lower down, that they were better off hundreds of years ago when they were half-heathen savages. They have now been driven from the good land they used to cultivate, and have been forced upon poor land. Their little holdings have been curtailed, so that they cannot keep enough stock to pay their rent. The rent has been increased and increased, and their only way of paying it is to trench[34] upon their revenue and sell off their stock. There are places where they used to fish, where they have become so impoverished that they have now no fishing boats. There are places where they used to have horses where now they have none, and where women – Scottish women – have to do the work of beasts of burden! You can see them today carrying manure and everything else on their backs ("Hear, hear," and applause.)

Go to the Highlands and you will see a state of society – of industrial society – that belongs to past centuries. You will find people cultivating the ground with a hook, and beating out their little harvest of corn with a flail.[35] Civilization has done nothing for them save to make life harder. Those men, large numbers of them, have to pay rents which they cannot possibly get out of the ground. They are forced to go fishing, or to come down to the Lowlands to seek for work in order to get money to pay for their rents. It is not merely for the ground they are charged, not merely for the virtues of the soil; they are charged for a mere breathing space, a mere living place.

Yet those people who live in that way are called lazy! Lazy! I would like to have some of those well-fed people who talk about their laziness go up and take a week of that sort of work. Let these men go up and dig a little with the "crookit spade,"[36] and then go out and face the rough sea in one of those fishing boats, and let those fine ladies go to the Highlands and carry turf on their back as the women do there. As far as I learned when there, it takes, on the average, about one person's labor to keep up those miserable peat fires in the center of the hut. As for flowers; since I have been in Scotland I have never seen a single flower around one of those miserable cabins, where most of the people live. I asked one crofter in Glendale if they had ever any fruit. "Well," he said, "they use to have some kail."[37] (Laughter and applause.)

I went, as Americans would say, to the jumping-off place – to John o'Groat's – and saw two very bright fellows bringing up stones from the seashore.[38] One of them stooped down upon his knees to help me to hunt for groatie buckies, and we had a talk.[39] He said he was going to build a house. The gentleman who was with me asked if he had any surety in building it except the word of his landlord? He said he was a good landlord. I asked: "How much have you to pay?" I think he said £5. His father lived there, and there were two other sons. I asked: "What do you make out of it?" One of them said: "We generally get the meal." I said: "Do you get enough to pay your rent?" "No, we have got to make it up. I go off to the fishing, and my brother goes off to work. Sometimes we get enough to pay the rent, but generally we don't." I said: "The goodness of this good, kind landlord of yours amounts to this, that he lets you live there, and takes from you all that you make save just enough to live." He said: "That is just about so." But then he said: "He is really better than many other landlords."

Well, so he is; some of those landlords are there skinning the people alive. It is not the crofters who have the worst lot – it is the cottars, who come under the tacksmen.[40] The crofter can only be put out once a year; the cottar can be put out at forty-eight hours' notice. The cottars are the absolute slaves of the tacksmen. There is just as much slavery as there existed in any land where human flesh was bought and sold. Why, there was the testimony before the Royal Commission.[41] By-the-by, that Royal Commission, to a man who does not know anything about it, looks like a committee of wolves to investigate the condition of the sheep. I would like to see laboring people represented on some of these commissions. Anyhow, a very intelligent Gaelic witness said all the land he had was a cabin and grass for a cow. Lord Napier asked how much rent he paid. He replied £5. The Commission did not believe it, it seemed so incredible. They said: "How do you pay it?" He replied: "I work a hundred days in the year at 1s. a day."

Is it any wonder that wages are low in your city when that is the state of labor in the outskirts? Poverty and destitution! There is enough to make you sick at heart if you listen to it. Why, a banker in the Highlands told me that only last week a young fellow had come to him, whom he knew was an honest, sober, industrious, hardworking man, and a cottar. He asked him for the loan of a couple of pounds. "Well," the banker said, "I can't lend you that as a matter of business. What is the matter?" The man replied: "I don't know where to get anything to eat; myself, my wife, and four children have had nothing but potatoes since last November, and not enough of them; and now there is not a particle of food in the house. All I have in the world is a cow and a stirk.[42] If I sell them now, I can get nothing for them. If you lend me this money, I will sell the stirk at the term time and give it back to you." My friendly informant said: "I will give you so much meal, enough to keep you" – I forget how much, so many stones you call it[43] – "to last you up to the time, and bring me the money when you sell the stirk;" and he said the man dropped down and burst into a flood of tears; and my informant said: "I never felt so humiliated in my life as to see a human creature, a fellow man driven to such a pinch." And then he said: "The man told me: 'You don't know what anguish I have suffered. Morning after morning I have seen my little children going to school fearing they would fall down from sheer weakness on the road.'" ("Hear, hear," and applause.)

And the treatment of the poor – the poor broken creatures who have nothing of their own – is something outrageous – this endeavor to keep down the poor rates! Do you know that in some of these parishes there are poor decrepit creatures who get an allowance of 2s. a month, and on other places 14 pounds of meal for two weeks? Well, I asked, over and over again: "How do they live? They can't live on that." What they live on is the charity of the poor people. The landlords, the rich farmers, shove this burden of providing for the poor that their rapacity creates upon the hardworking people, who themselves can hardly keep from starvation.

One of the London papers said, jeering at me, that I proposed to take all the property from the landowners, and they supposed, however, I was very kind – I would send them to the almshouse. Well, now, I wish – I have no ill-will towards them – but I heartily wish that a lot of your ruling classes could be sent to the almshouse. I think if some dukes and duchesses and earls and countesses were treated as these poor people are treated, that the wickedness of it, the sheer cold-blooded barbarity of it, would become apparent to our so-called Christian people. Utter slavery!

Why, as one man said to me: "We have feared the landlords more than we have feared Almighty God, and we have feared the factor as much as the land-lord – perhaps even more – and the ground officer as much as the factor."[44] Why, they are right in their power. There is a case, I am told of, where the factor was a fish merchant, and compelled the people to sell him the fish, and fined them £1 if they sold the fish to anybody else. Why, a gentleman was telling me – a professional man – how he had ridden, just a week or two ago, round with the factor on the estate of one of your Liberal members of Parliament (applause); one of your great Liberals. (Cries of "name.") Sir Kenneth Mackenzie. They came up to a man, and the factor said to him: "Look here, why was not your children at school yesterday?" "Well," the man sheepishly replied, and the factor said: "Look here, don't you allow that to happen again. See that they are at school." "Yes, your honor," the man replied. "Heavens and earth, how can you talk to a man like that?" said the professional man, and the factor said: "I can make him toe the mark; I have plenty of power." Why, take the Island of Skye, the factor there is everything except the parish minister. (Laughter and applause.)

I spoke at Portree the other evening. I went up to Portree, and some of the inhabitants came to me, like Nicodemus,[45] at night, and said: "You must not leave Portree without speaking here." I said that I did not want to thrust myself upon them, but if they secured a hall I would speak. They went away, and by-and-by they came back and said: "There is not one of us who has the courage to ask for a hall." They were afraid, and I said: "I will take the whole respon-sibility, and offer myself, if need be, a vote of thanks." I wrote a letter to the factor. I suppose you have heard of that factor – Mr. MacDonald, I think his name is. He is Justice of the Peace and everything else, and he has charge of the only hall there. I wrote him a polite note, stating that some of the people wanted me to speak on the land question. He wrote back to me to say that he could not let the hall for a lecture, and could not take the responsibility without consulting all the proprietors. Anyway, we got a schoolhouse. A clergyman at the head of the School Board was good enough to grant the use of a school-house, although there were threats of interdicts and other terrible things made against him. (Applause.)

I remember reading in an English book, written some years ago, about an aristocratic Polander[46] in the old times, who took an English traveler over some of his ground, and pointed at some miserable-looking objects. He told the trav-eler he could kick any of them he wanted to. It was much like that in Scotland today. (Applause.) Your aristocracy take a pride in all that sort of thing. They like to keep up those Highland romantic notions, the feather bonnet and the kilt, and all that sort of thing. Well, now, really when you come to think of it, those Scottish Highlanders have been an ideal people with the aristocracy. They fight like lions abroad, and they have been taken abroad at the dictate of the very power which has oppressed them, to rob and plunder, and kill other people; but they are as tame as sheep at home. (Applause and laughter.) Don't you think that alongside of the Scottish lion you ought to put a Scottish sheep? (Laughter.)

There is one other thing that has disgusted me. The most disgusting thing I saw in Ireland was that police force – the Royal Irish Constabulary. Well, now, you are keeping up here in Scotland an institution very much the same. When I was in Skye I saw policemen loafing around just as the Irish Constabulary loaf about. In a little bit of a village named Dunvegan, where I don't think there are more than six or seven houses, there are two policemen, all in uniform. (Laughter.) The police of the county of Inverness have been increased by fifty, at a cost of £3,000 to the ratepayers, and £3,000 more to the whole country, on account of the fears of the landlords.[47] (Applause.)

I have been pointing out the evil. How can it be cured?

Well, it cannot be cured by any halfway measures; it cannot be cured by any measures that will be agreeable to your aristocracy. You know that at the beginning of big sheepfarming in the Highlands, and the eviction of their brethren by chiefs who had become landowners under this infamous English law, there was a good deal of misery, and one the earliest measures to relieve that misery was to get up those Highland regiments. They were got up about the time of the American War, and a lot of them were sent over there to cut the throats of our people.[48] You can't relieve poverty by any such measures as that. (Applause.)

In the beginning of the century, when the Duke of Sutherland and other men of that kind were evicting their people with a barbarity that will hardly find a parallel in the annals of savage warfare, there was another measure got up to relieve the destitution – that was the making of the roads. Some £267,000 of public money, in addition to £5,000 a year from the public funds were, for many years, spent on making roads through the Highlands; but this grant was finally abandoned, on the ground that all it had done was to improve the rents of the Highland landlords. No such measures as that will relieve poverty. (Applause.)

You cannot get rid of it in such measures as you Glasgow people adopted in your City Improvement Trust. You have taxed the masses of people to foster corruption; to put large sums into the pockets of speculators and landlords, to improve the property of other landowners; and you have not a whit relieved overcrowding or destitution. You have simply changed the place of the disease. It is like putting a plaster on a cancer and driving it somewhere else. You cannot cure this deep-seated disease by any such measure as that; you must go to the root, boldly and firmly.

Take no stock of those people who preach moderation. Moderation is not what is needed; it is religious indignation. Grasp your thistle.[49] Take this wild beast by the throat. Proclaim the grand truth that every human being born in Scotland has an inalienable and equal right to the soil of Scotland – a right that no law can do away with; a right that comes direct from the Creator, who made earth for man, and placed him upon the earth. (Loud applause.) You cannot divide land and secure equality. It could be secured among a primitive people, such as the children of Israel, who, under the Mosaic Law, divided the land;[50] but in our complex civilization that cannot be done. It is not necessary to divide the land, when you can divide the income drawn from the land. You can easily take the revenue that comes from the land for public purposes. There is nothing very radical in this; it is a highly Conservative proposition.

Why, I had the pleasure of reading a speech delivered in this hall by your member, Dr. Cameron, proposing substantially the same thing. Dr. Cameron and myself, I am glad to say, stand upon the same platform in this respect. He wants to reestablish the old, ancient tax upon land that the landowners have thrown upon the masses of the people. That is what I want to do; and when we have done that, I want to go a little further. But I have no doubt that Dr. Cameron, when he had got so far, would be quite willing to go a little further. The real fight will come on some such proposition as that made by Dr. Cameron, and I have not the shadow of a doubt that, if the people do their duty, the landlords will be routed – horse, foot, and dragoons. (Applause.)

Now, see the absurdity of the present system, even as a great economic measure. Here, in Glasgow, take that field of Burnbank. The owner allows it to be vacant, and pays nothing; but if he puts houses upon it you will then get £7,000 a year in taxation. Have you got enough of houses in Glasgow? Why should you tax houses and not land? The man is a public benefactor who puts up houses. The more you tax houses, the less houses you have. But you may tax the value of land twenty shillings to the pound and you won't have an inch less land. (Applause.) A good part of this city used to belong to our people. It was purchased by a Lord Provost named Campbell. I don't know how he got it. It reminds me of the story I heard in Cardiff, how an ancestor of the Marquis of Bute got a great part of the common of that town – now [a] most valuable property. A predecessor of Lord Bute gave the freemen a dinner every year. In a fit of generosity they voted the common to him; but he did not continue the dinner. (Laughter.) I don't know how the Lord Provost got this property. But I am informed he paid £1,500 for it. Now, his successor, Sir Archibald Campbell, draws £30,000 in feu duties, and he does not pay a penny of the rates of the town. ("Shame.") Would it not be better to take that £30,000 in taxation, and remit your taxes on some other things? (Applause.)

And I want to call your attention to what an enormous fund you would get for public purposes in this way. The chief advantage of putting taxes upon land is that you would choke off those "dogs in the manger," who are now holding the land without using it, or making deer forests of what ought to be homes of men; who, that they may compel a larger blackmail, are withholding land around your towns from building uses, while whole families are crowded in four-storeyed houses, a family to each room. (Applause.)

A great stimulus would be given to industry, to the investment of capital, to production of all kinds, by the removal of the taxes that weigh and press them down. And by taking that which goes to the landowner and using it for public uses, instead of making poor people pay for the education of their children, as you barbarously do now, you could have all your schools free, and the best possible kind of education given to the children of the poor, as well as to the children of the rich; you could establish libraries and museums, and public parks and gardens, and baths and theaters, if you chose, in every town; you could all around this coast build harbors for your fishermen; you could give a pension of enough to live comfortably on to every widow or helpless one, to every decrepit man; you could dower every girl, and give every young man a start in life. (Laughter and applause.)

Preposterous does it seem? Well, it does – this thing of doing anything for the masses of the people. It is highly demoralizing, we are told, to give the people something for nothing. It would destroy their independence if the poor people didn't have to pay for the education of their children! (Applause.) You don't hear anything about that when the pensions get to thousands and five thousands of pounds. (Laughter.) Your Parliament votes £25,000 a year to a young prince, as though it were nothing at all. (Applause.) Judges, officers, and that sort of thing, get most handsome retiring pensions. It don't hurt them, it don't demoralize them. (Applause.)

And see how enormously your other expenses would be reduced. Why, I saw in an office today a chart showing the expenses of this nation diagrammed, and, according to that chart, it was nearly all for war, and the cost of war, and preparation for war. You have been going round the world robbing and murdering and cutting the throats of other people, and out of the present taxes, according to that chart, you pay 16s. 9d., I think, a year for war, the expense of war, and the costs of war, and 3s. 3d. for other expenses. (Applause.) Why is that expense placed upon you? Because you are governed by a landowning aristocracy. The army is a good place for younger sons. You have been governed by the class that likes to make war, and that finds a profit in making war. With the rule of the people that would cease. (Applause.)

There's enough here for all of us. There's no natural reason for poverty, or even for hard work. The inventions and discoveries that have been already made give man such a command over material conditions, that we all could live in ease and luxury if we did not scramble and tread each other underfoot. (Applause.) Once give the people an opportunity, give [the] mind a chance to develop, and the forces of production would increase at a rate never dreamed of. Where wages are highest, there is labor always most productive, there is invention most active.

And certainly it is time that something were done. Why, think if one of us, having a family of children, were to go away from home, and come back and find the big ones leaving the little ones out in the cold, keeping them in ignorance, in squalor and misery, and disease – what would we say? (Applause.) Do you believe that the laws of Justice can be outraged with impunity? Not so. The whole history of the world shows that, though, on the narrow scale of individual life and individual action, injustice sometimes seemed to succeed, yet on the great scale of national life, the punishment of national crimes always comes sure and certain. And, so sure as God lives, that punishment must overtake such nations as this. The cry of the oppressed cannot go up for ever and ever without bringing down punishment. (Applause.)

Look back at the greatest nation that ever played its part on this world stage – Imperial Rome. What was its fate? That very fate may be seen coming over this nation today. Italy, when the Roman power went forth to conquer the world, was the home of hardy husbandmen, independent and self-reliant. As fortunes grew, these men were drained off to the wars, evicted, driven out, and Italy was given up to sheep and cattle and great estates.

That very same thing is going on in these islands today. What was Scotland made for? What is this earth made for? Was it not for man? Was not man given the dominion over the birds of the air and the beasts of the field? Was it not made his duty to subdue the earth? Is not man the highest thing that [the] earth can produce? And yet here, in this Scotland, you are driving off men and putting on beasts, and the vengeance is coming.

We know something of the laws of the universe. We don't yet know them all. But there is a strange thing that has been noticed in new countries, and that is the influence that man seems to have by his mere presence upon Nature. The bee follows the pioneer across the American continent; where settlements are made more rain seems to fall, new flowers without planting seem to spring up, and the earth to bring forth more abundantly; and, where man retires, Nature becomes more savage. See how in Italy fertile districts, when depopulated, became the haunts of fever. Look to the arid wastes of North Africa, once such a teeming hive of population.

The very same thing can be seen in Scotland today. Upon this land the curse that follows the expulsion of men is coming. Men have been driven off the richest and best land, and the sites of their little homes and their little culti-vated fields given up to sheep, and the sheep fattened. It was good grass where the men had been. That everywhere, I can learn, is giving way. I am told by capable authorities that where a thousand sheep twenty or thirty years ago could be kept in places men had been driven off, not 700 can be kept now. There is a fungus moss creeping over the ground; Scotland is relapsing into barbarism again; even sheep are giving way to the solitude of the deer forest and the grouse moor. Will you, men who love Scotland, let it go on? (Loud applause and cries of "no.")

Questions

The Chairman intimated that any gentleman present would now have an oppor-tunity of putting questions to Mr. George. He said that four questions had been handed in by a journeyman tailor.

The first was: Why does Mr. George address meetings in large cities instead of among the farmers and farm laborers, the large cities being centers of commerce, and their inhabitants having no interest in the question?

Mr. George: Because I think it is in the large cities that the evils of land monopoly are best seen, and that it is to the large cities that I look for the force that is to reform these evils. (Applause.) Those poor cowed people in the Highlands, trembling under the eyes of their factors, what can they do for them-selves? It is to you men of the cities that I mainly and principally look. The towns must carry the standard of advancement, as they always do. (Applause.)

The Chairman [said that] the second question is: How would nationalization of the land tend to raise wages or shorten the hours of labor of the city artisan?

Mr. George: Because it would open the primary sources of all employment. Why are wages, generally speaking, in new countries higher than in old countries? Adam Smith, a hundred years ago, stated the reason, when he said it was because there land was cheap – because a man can there work for himself, and therefore will not work for anybody for less than he can earn for himself. When you open up the land, you relieve the pressure on every industry. It is the pioneers in a new country who furnish the foundation and market for all the others. First you have the herdsmen and farmers, and afterwards you have the operatives.

It is sometimes said we all cannot be farmers; but that is the only thing we all can be. We all might be farmers; because communities have existed in which everybody was a farmer; but you never heard of a community where everybody was a tailor. (Laughter and applause.) It is not necessary, however, for us all to be farmers. But if we break up the monopoly of land, so that in the primary occupations there will be easy employment and high wages, then there will be a brisk demand for labor and high wages in all employments. (Applause.)

The Chairman [said that] the next question is: If Mr. George would not tax labor products, and if the rent of the agricultural and grassland is only about sixty-six millions [pounds] per year – fifty-six millions of this amount being a rent imposed upon the labor of the farmer – would he explain to us how he proposes to abolish the presently existing poverty by the paltry sum of ten millions which remain?

Mr. George: The landlords are very anxious to show how little they get. Mr. Mallock has made a colored diagram in which he pictures it as only £100,000,000. If it is so little, what is the use of making a fuss about it? The fact is, that it is an enormous sum. The agricultural rent is put at £60,000,000; but that is the smallest part of the rent. The rents of towns and cities and mineral lands ought to be at least twice as much. Nor in these estimates is everything given. It is merely rent received by the landlords. There may be feued ground that pays 20s., and which the growth of the city has made worth £10 or £20. All that is rent. The Duke of Westminster gets, besides the rent, all the buildings upon his estates in London at the expiration of the leases. The rent of these kingdoms is at least two hundred millions – enough to pay all your extravagant expenditure in some directions, and a great deal more, and at the same time giving labor a chance. (Applause.)

The Chairman [said that] the next question is: If it be unjust to hold private property in land, is it not equally unjust to build a private house upon land, seeing that to build a house upon land is putting a portion of the people's earth to private uses, and excluding everyone except the owner of the house from the use of that portion of the earth? (Laughter.)

Mr. George: That is just as sensible as you will find in the reviews of your best newspapers. That question must come from the editor of one of your leading dailies. If a man takes a fish out of the sea, the fish is properly his private property; but that fact does not necessitate giving him the sea as private property. (Applause.)

The Chairman [said that] the next question is: Would not the abolishing of taxes benefit the large merchants of a city rather than the artisans or laboring classes?

Mr. George: No, I don't think so. The greatest benefit would be to the laboring classes. The incidence of taxation, as now laid, benefits the capitalist, or the man who has most money. The making of liquor has been concentrated, and distillers have built up great fortunes over in Ireland. The distillers are the men who renovate and build churches. It is the same with all sorts of business. We have in our country, more than in yours, a protective tariff. The duties are paid primarily by the importers. Do you think you can get them to work for free trade? On the contrary, they profit by the duties, as their effect in increasing the amount of capital required for the business keeps competitors out. The effect of all these taxes is to concentrate business in the hands of capitalists.

Now, it is said, why attack the landlord alone; why not go for the capitalist? The capitalist, as a capitalist, is doing nobody any harm. What harm is done by the capitalist is as a monopolist. It is the monopoly that you want to destroy. Now, we find when a man has a great sum of money, this power is, in the phrase of the Socialists, used in exploiting labor. Where does this power come from? Suppose I take a million pounds and go into a country where men can earn for themselves £1 a day and put up my big factory, can I get anybody to work for less than £1 a day? Not much. It is because these men are impoverished that they are forced to compete with each other for starvation wages. Suppose every family had, as it well might have, its own house and garden, enough to live on, would you find people working for a few shillings a week? There's where the pressure comes from. One millstone can't grind. It requires two, the nether millstone as well as the upper millstone. (Applause.)

A Gentleman in the area of the hall asked Mr. George a question: in effect, whether, supposing the rent of the land was paid to the State, instead of to the private owner, would it make any difference?

Mr. George: It might not make any difference to the rent or in the rent, but it would make a great difference to the people who paid the rent. That question was well answered in a London newspaper by my friend Mr. Joynes. A man wrote and said: "What difference was it to the farmer whether he paid his rent to the State, or whether he paid it to the landowners? He said this was the difference: "That the State was not likely to go to the Continent, or go off in its yacht and spend it." It would not be just to the rest of the people to make rents low. Every rent ought to be a proper rent, as much as the land is worth, because that is the only way of securing equality.

There's the mistake our friends in Ireland have made. They have gone and turned that great agitation into a miserable little thing for the tenant-farmers. Now, the tenant-farmers are not entitled to a whit more favor than any other class in the community. The class to look to, the class to strive for, is the very lowest class – not the farmer, but the laborer. He is the man. Improve the condition of the man who has nothing but his hands, and you improve the condition of the whole community. (Applause.)

A Gentleman in the gallery asked: Does Mr. George propose to confiscate the interest on bonds held by widows and orphans, which absorbs a large part of the income of the land?

Mr. George: I would propose to confiscate the whole value of the land.

The Gentleman: Well, what I refer to belongs to widows and orphans.

Mr. George: Do not be deluded by this widow and orphan business. That is a matter that is always put to the front. When men talked about abolishing slavery in my country, the cry was raised about the widow and the orphan. It was said: "Here is a poor widow woman who has only two or three slaves to live upon; would you take them away?" It reminds me of the story of the little girl who was taken to see a picture of Daniel in the lion's den. She began to cry very bitterly, and her mother said: "Do not cry, do not cry; God will take care that no harm will befall him." To which she replied: "I ain't crying for him, but for the poor little lion in the back – he is so little I am afraid he won't get any." I propose to take care of the widows and the orphans. As I told those people in London whom I addressed recently, every widow, from the highest to the lowest, could be cared for. There need be no charity or degradation; everyone of them could have an equal pension. It will only take twenty million pounds to give every widow in the three kingdoms a pension of £100.

And in the state of society which would ensue from breaking up land monopoly, no one need fear that the helpless ones he left behind would come to want. This is not the case now. Take your Duke of Argyll or Duke of Sutherland[51] – nothing is more certain than that their descendants will be yet tenanting your almshouses. John o'Groat was sent by one of your kings up to Caithness, and made a rich laird. But the lot of the o'Groats now existing there is just as poor and miserable as any people there. The best blood of England, as it is called, runs in the almshouses. How much better it would be for the richest man to know that he left his widow and children in a state of society where they could not possibly want, where all the influences around them were healthy, than in such a state of society as this! Why, look at its moral aspects. The vice and disease that are bred of poverty, do they rest merely with the poor people? No; they climb up through the ranks of the rich to the highest. (Loud applause.)

The Chairman read the next question, as follows: If Mr. George would abolish ownership of land, what compensation would he give to those owners of land who have acquired it by purchase, sanctioned by existing law?

Mr. George: I would not give them a penny. I don't think this matter of compensation comes into practical politics. Why should you make any discrimination between a man who purchased his land, and a man who did not purchase it? Does it make much difference whether I am the robber or I bought the thing of the robber? Supposing I was big enough to steal one of you, and run you off to a country where I could hold you as my slave, you would have a moral right to get away from me as soon as you could; but would that moral right cease the moment I had sold you to somebody else? If you were to say we will recompense anybody who can show that they bought their land, what would be the result? Why, by the time you came to take the land, everybody would have sold it to somebody else.

A gentleman said to me tonight: "Oh, Scotsmen will not hear of anything else but compensation." I don't believe that. (Applause.) I have a very much higher notion of Scotsmen than that. I believe that the Scots are too logical a people to tolerate the idea of compensation. I will tell you a story I heard about this matter of compensation. There was one of your Highland lairds – a Gordon something or other – in a railway train with a gentleman, and he was talking about these wicked ideas that were floating about – this theft and Communism. The gentleman said to him: "How did you get your land?" He said: "We got our land by bringing our men into the field to fight for the country." The gentleman said to him: "What did the men get?" Well, he had to admit that the men had not got anything. But he said: "We have had the land for a long time, and sanctioned by law. It would be robbery to deprive us of it." The other gentleman said: "How long have you had it?" "We have had it for 800 years." Well, the gentleman said: "If you have had it for 800 years, don't you think you have had it long enough?" (Applause.)

Compensation is preposterous. Why, all titles to land are nothing but robbers' titles, and the titles to a large part of the land in Scotland are a great deal worse than robbers' titles. They are not titles won by the strong hand or by conquest. They are rather the titles of the sneak thief – or worse. These Highland chiefs betrayed their brethren – took advantage of a language and a law that they did not understand. They were won by treachery and treason.

I don't propose to go back into inquiries of that sort, because, to my mind, it makes no difference how a man got the land. It may be said he bought it. Supposing he bought the sun? Could he buy it from anyone who had the right to sell it? But where do these titles come from? Has one generation, supposing they were all united, the right to sell the rights of the coming generation? This earth belongs to all generations. You men have carried in a certain direction compensation to the extreme of absurdity, but it has always been compensation to the ruling classes. You paid the descendants of Charles the Second's[52] bastards compensation for hereditary pensions and taxes, and you paid enormous sums to buy out the hereditary jurisdiction of your Highland chiefs. For every sinecure held by one of the ruling classes he gets compensation, but you never hear of a poor man being compensated. How much were the people compensated when the taxation was taken off the land and put upon labor? Why should you compensate the landlords? The only reason is that you have been doing it for a long time.

Nobody proposes to take anything form the landlords. I would give everyone his full equal share. It is not proposed to take anything from them; it is merely to stop them from taking from other people. (A voice: "What about recently acquired land?") Treat it in the same way. Supposing the land was acquired, is it not the principle of law that the buyer can get no better title than the seller has to give? If a man has no right to the land, how can he give another man the right to it? As a matter of fact, you would do no injury by laying down that principle. No one could be hurt by the resumption of the land as common property, save those who could well afford to have their incomes lessened.

The man of small means who had got himself a house and lot would be the direct gainer by the change which would exempt houses from taxation, and put it upon lots, while he would be an enormous gainer by the increase of wealth and the rise in wages. Then the businessmen who are landowners would profit by the improvement and stimulation of the productive energies of society far more than they would lose as landlords.

The typical landlord is like a landlord in Dublin they call Cosey Murphy, who stayed in bed eight years; the typical landlord is the man who goes to the Mediterranean in a yacht and spends the money which he draws from the toil of the people here. Consider, the real thing that would be taken from the people who demand compensation is not land, but the power which the possession of land now gives them of levying toll upon the labor of others. What does the Duke of Sutherland want with his twelve hundred thousand acres; or the Duke of Westminster with his London estates? No more than the Earl of Airlie wants with the water that he sold. They want to have the privilege of taking the wealth of the people who have produced it. That is a right that no one can have. That is a power that can be sanctioned by no purchase, and that no one can justly ask compensation for. (Applause.)

A Gentleman in the middle of the hall stood up and said: Suppose a man was induced by our Land Laws to invest £100 in land. He might have invested the money in any other commercial enterprise. Would Mr. George compensate the man who had lost his money by the so-called pernicious Land Laws?

Mr. George: I would not. (A voice: "You will not do for Scotland." Second voice: "Keep quiet, you fool! Do you speak for Scotland?") If a man invests a gold sovereign in a bad Bank of England note, I would not reimburse him. If a man invests a hundred pounds in slaves, I would not reimburse him. (A voice: "We compensated the West Indian slaveholders.")[53] A very wicked thing it was. I hope you will not do so again. You shunted the loss which the slaveowners ought to have incurred upon the backs of the working classes of this country. You did worse. You strengthened slavery all over the world; you taught the American slaveholder to believe that, if abolition should come, he would get a price for his human property. Up to the verge of the [American Civil] War slaves commanded as high a price as ever they did. If, on the contrary, that agitation had gone on on the basis of absolute emancipation, the thing would have been gradual. The value of slaves would have declined. Men would not have bought and sold them.

Now, the same is true in this. I want to do this at ten o'clock tomorrow morning; but if we all wanted to do it, it would take a good while. It necessarily must be a progressive step. We must necessarily, on account of the resistance, move step-by-step. And as we do this the landowners will have a chance; your recent purchase will have a chance not to purchase. (A voice: "Why not begin at home?") The decline would be slow and gradual. Why not begin at home? I am beginning at home. I don't come over here to preach anything I have not preached in my own country. The very conditions that I have been speaking to you about I have seen growing upon new soil, and it was because of that that my eyes were opened to it. Why not begin at home? I am here beginning at home. We who speak this language are on both sides of the Atlantic but one people – becoming everyday more one. (Applause.) This agitation must go forward on both sides of the Atlantic – by action and reaction. America must be affected through England and Scotland, and England and Scotland will be affected through America. Whatever we do, we do for this whole, great imperial race – the race to whom the destiny of modern civilization is entrusted. (Loud applause.)

A Gentleman: Would you confiscate all rent?

Mr. George: I would confiscate all rent in the economic sense.

The Questioner: Then, would you give compensation for improvements?

Mr. George: Let the improvements stand. Certainly I would. I don't propose to take the improvements, but to let everything stand as it is now. It is the present system that is confiscatory. It is confiscating labor every day. It is not a robbery that is done and passed away; it is robbery that is going on every week and every month, every day and every hour. It is a fresh robbery that is committed on every child that comes into the world.

Now, to go back to this matter of compensation. Some people do propose to compensate. There are some who propose to compensate all who can show that they have purchased the land at the price they gave for it, *minus* the net rent that they have received. Then there is Miss Helen Taylor, the stepdaughter of John Stuart Mill. She is also in favor of compensating everybody who can show that they have purchased their land with the proceeds of their labor. She proposes to make the landowners pay up with interest, and compound interest, all the back taxes from the time of Charles the Second, and then to take part of that money and compensate the people who could show that they had purchased with their own earnings. (Laughter and applause.)

There are people who believe in compensation – compensation not to the landowner, but to the people who have suffered. I would cut the whole thing now. I should be perfectly willing to draw the line at "let the past be the past." If anyone wants to compensate landholders, they have a perfect right, so far as they are concerned themselves, to give compensation. They could make a collection for them. You have a perfect right to do that, but I deny the right of any individual to grant away the natural rights of another individual. Be just before you attempt to be generous. There is only one true basis of social reconstruction, and that is the basis of Justice.

Votes of thanks to Mr. George and the Chairman concluded the proceedings.

MR. CHAMBERLAIN TRANSLATED INTO PLAIN ENGLISH: AN INTERVIEW WITH MR. HENRY GEORGE

(*Pall Mall Gazette*, January 14, 1885).[54]

Mr. Henry George reached Glasgow on his return from Skye in the middle of last week, and late on the night of his arrival a special correspondent of the *Pall Mall Gazette*, then in the west of Scotland, found him in his room at the Cobden Hotel, in the midst of a mass of proofs, papers, and correspondence. He looked little worse for his uncomfortable journeyings in the far north, and he expressed a degree of satisfaction at the substantial progress made by his movement which our representative, with recent declarations of Scotch newspapers and Scotch politicians in his mind, had scarcely anticipated. Mr. George appeared to have been particularly struck with the change which the extension of the franchise will work in the representation of some of the northern counties.

On his first visit, perhaps, the hands of only two electors – these the minister and the schoolmaster – would be held up in support of the Land Restoration resolution at a large meeting; on this last occasion the great body of the meeting were enable to support not merely as men with hands and minds, but as electors with votes. The following record of question and answer will be read with interest:

Has the lot of the crofters improved since your last visit to the north?

There is certainly a great moral improvement. There is an evident growth of manly feeling, and they are rapidly shaking off the slavishness engendered by long oppression.

Has their attitude changed in any respect?

It is more intelligent and more determined. During all this year a rapid process of education has been going on. The God-given rights of men to the use of their native soil, and the necessity of standing together to assert them, seem to have been the great subjects of thought and discussion. If those who think the Skye crofters have no ideas beyond some pitiful reduction of rents could have listened to some of the speeches I heard, they would have been very largely undeceived.

How do they regard Sir William Harcourt's speech and action?[55]

Very much, I think, as that of the fabled crocodile, who sheds tears and then bites. They say, by-the-by, that he was once roughly driven off the island of Rassa by a gamekeeper, when he had presumed to land from his yacht. Personally, they seem to like the marines, some of whom at least are thorough-going land nationalizers.

Apart from your radical remedy, what could you suggest in the way of imme-diate measures of relief for the crofters – that is to say, possible practical measures?

The withdrawal of the army of invasion, the suspension at least as to [the] crofters' holdings of all laws for the collection of rent; the suspension of all laws for the preservation of game, and of the law of requiring gun licenses. The enactment of a short bill of this kind would greatly relieve the crofters while larger measures were being considered, and would obviate the necessity for any charitable fund, such as the Earl of Breadalbane and the Rev. McDonald, of Inverness, are raising, which could be turned to the relief of the landlords if any of them really suffered by not getting rents. This suspension of the gun license and game laws would enable the crofters to protect their crops, and vary their diet, while accustoming them to the use of arms, a thing in itself much to be desired among a free people.

How would you regard a measure which established in the crofter districts Government Commissioners having the confidence of the people, who would negotiate with the landholders for fair-sized lots incapable of division, at a fair rental, the State to lend the crofter money up to two-thirds the amount of rent in order to stock his croft?

As a piece of temporizing patchwork which would settle nothing and please nobody. I am a free trader, and do not believe in governmental fixing of prices. I recognize the obligation of the Eighth Commandment,[56] and do not believe in robbing Peter to pay Paul. And besides, if there are to be Commissions to fix rents in the Highlands, why not Commissioners to fix rents in London? If Government money is to be lent to the crofters, why should it not be lent to agricultural laborers and unemployed workmen.

But would not such a measure satisfy the crofters?

No. It might have done so a year ago, but that day has passed. The crofters will of course take whatever advantage they can of any measures that are adopted, but the movement has gone too far for them to be satisfied with anything short of the complete restoration of their natural rights in the land. My recent visit proved conclusively what I knew and stated before, that the crofters really want land nationalization, not tenant-right tinkering. They have got a firm grasp of the truth that God made the land, not for the landlords alone, but for all the people. They have the Bible at their tongues' ends, and the power of the ministers who have long been preaching a slavish submission in the name of Christianity is rapidly passing.

A good exposition of the ideas current among the crofters is the speech made by the Rev. Mr. McCallum at the Anderton Hotel conference, which has been reprinted in full in the principal Highland newspaper, the *Oban Times*. It is a pity the London papers did not print it for the information if not for the instruction of their readers. Its grasp of first principles as religious truths, its Biblical expression and illustration, are extremely characteristic. It is the "trust-in-God-and-keep-your-powder-dry" spirit that is wakening in the north.

How do you regard Mr. Chamberlain's latest speech? Do you agree with the *Times* that "it was apparently something of the same kind as land nationalization, that he advocated with judicious vagueness?

It seems to me about as clear a declaration of true principles with regard to the land as a Liberal politician and Cabinet Minister could just now be expected to make, and will, I should say, much strengthen Mr. Chamberlain's position with the new forces which are soon to revolutionize English politics. Translated into plain English, I take it to be that the land belongs of natural right to the whole people, but that as it would be impossible to equally divide land, it should be left in form to private owners while the people take the rent.

When Mr. Chamberlain says that in the origin of things men were born with certain natural rights, he of course implies that they are born today with the same rights. When he says that these rights have passed away, he of course does not mean that they have ceased to exist, but that they have been ignored. This speech is to me evidence of Mr. Chamberlain's political sagacity. He has stepped forward at the right time. Mr. Chamberlain seems everywhere regarded as the man who is to lead the great Democratic party which is to spring into the political arena as soon as Mr. Gladstone retires, and his strength seems to have been the belief that he was far more radical than he had yet seemed, and was only waiting opportunity.[57] To delay moving forward would be to lose that confidence which is a political leader's greatest strength.

What has most impressed me during my recent trip is how much in advance of their leaders are the Radical rank and file. For the earnest, active men Mr.

Chamberlain cannot go too fast or too far, and the more moderate, who are disposed to wait for the word of command, will be swept along by the current whenever that is given. What is called the Liberal organization seems to me up to this time to have been largely a machine for repressing Radical sentiment.

You think, then, the land question is coming into practical politics?

It can't be kept out. The land question is simply the great social question – the question of work and wages; of food and raiment. Everything is contributing to force it into politics: the crofters' revolt and the misery in London; the losses of farmers and the distress among artisans; the work of the Land Leagues and the Fair Trade propaganda. And now the dyke that held back the flood is broken. I never fully realized what the extension of the franchise meant until I found around me great audiences of men imbued with the most Radical sentiments who have never yet had a vote.

Politicians who profess a desire to be kind to the people must soon give place to those who will assert equality and demand rights. Those who do not realize this, count, I think, far too confidently upon the stupidity of the newly enfranchised millions. It has been said, for instance, that crofters and agricultural laborers could not understand the beliefs I hold. I know from experience that this is the very reverse of true. They are as logical as children, and are not perplexed by considerations of what would happen to this or that if Justice be done. "God made the land for one man as much as for another." No rustics are so stupid as not to understand that, and to carry their "betters" they may continue in the old ruts a while longer; but when they are appealed to in the name of natural rights, I believe that even the English laborer, of whose sluggishness I have heard so much, may be roused as he has not been roused since the times of John Ball and Wat Tyler.[58]

And whenever such practical politicians as Mr. Chamberlain bring the organization and means they can command to the work of appealing to these classes, the squires, Whig and Tory, will soon be left just where the same class have been left in Ireland. In short, everything I have seen in Scotland has convinced me that Patrick Dove[59] was right when years ago he predicted that the extension of the franchise would at once bring the land question to the front, in the struggle over which the whole system of aristocratic government would be overthrown. It seems to me that the great English revolution, which means a social revolution throughout the civilized world, is already begun.

But most of the Scottish journals say that your second campaign in Scotland has been a failure?

That is what they said of my first campaign. This has been just such a failure as that, only very much so. I am satisfied, the Scottish Land Restorationists are satisfied, and if the journals who are trying to boycott the discussion of the land question are satisfied, there is satisfaction all around. "Let him laugh who wins." My trip this time has been much more telling than the last. I did not go into the Highlands, except to Skye. They can be safely left to the lairds, the factors, and my efficient coworker, Mr. Winans. But I have been speaking night after night in the Lowland districts, where the political strength is.

And wherever I have been I have left discussion behind me. All political and social powers combined cannot keep the movement down in Scotland. It enters into practical politics at the next election, and will stay there until finally settled in the only way it can be settled – the full acknowledgment of equal rights, the restoration of land to the people. I, of course, do not expect any political party to come forward on restoration lines, though here and there there will be men who will do so. Great questions always enter politics in moderate form – the thin end of the wedge.

What pleases me more than anything else is the attitude of the moderate men – the men who declare that they will not go for extreme measures, and who denounce me as advocating confiscation, but in the same breath declare themselves for the tax of 4s. in the pound, or for such measures as that proposed by Sir George Campbell[60] to break up the monopoly of building ground around the towns. The Financial Reform Association is doing splendid work. Besides the *Almanac* and the tracts it has issued a pamphlet by Mr. Heywood on the land tax, than which nothing better adapted to popular instruction [could be found].

Does Mr. Chamberlain's speech point to the remedy of the Financial Reform Association – the reimposition of the 4s. land tax?

I should say it pointed to 20s. in the pound, though of course this is not the thing for the practical politician now to avow. The propagation of ideas is one thing, the carrying of those ideas into politics is another. The practical proposition should always be the lowest which embodies the principle, as the greatest strength can thus be combined against the least resistance, and the men who will only support a moderate proposition lose their moderation as they get warm in the struggle. One shilling in the pound would be a good enough line on which to fight the political battle, but the 4s. tax has historical associations which peculiarly adapt it for a rallying point.

Do you seek the abolition of rent?

Certainly not. One might as well seek the abolition of gravitation. Rent arises whenever social development reaches a certain point, and increases with the progress of society. The only question is: Who shall get it? Let it go to individuals, and material progress tends to greater and greater inequality. Take it for common uses, and the progress of society is towards a greater equality of conditions. As declared in the *Anti-Corn Law Catechism* (a book which the half-way free traders of the Cobden Club might read with profit), no tax should be levied until rent is exhausted.[61]

Are you prepared to listen to the cry of many sympathizers for a scheme of compensation? How would you regard a scheme of graduated compensation – a scheme which would return to a man who bought an estate yesterday the full value of it, to a man who bought one fifty years ago two-thirds its value, to a man who purchased one two hundred years ago one-fourth, and so on; land in cases in which it can be proved that the present holders stole it hundreds of years ago to be taken without compensation? Granting that landowners should be compensated, your scheme surely is upside down.

The only ground, in my opinion, on which any claim to compensation could be based is that of ignorance and surprise. The man who purchased fifty years ago might urge that he had never heard of the right of the people to the land; but how could a man who purchased yesterday? [If] you passed such a measure, unless you made it retroactive, you would find that all the land in the three kingdoms had been purchased yesterday. If any discrimination is to be made between those who buy and those who inherit land it should be in favor of the latter, since purchase is an act of deliberate volition, but inheritance is not.

But it is needless to talk of compensation. My judgment of the British people, formed before I had set foot on British soil, has been abundantly verified by observation. I was told that the British people were so Conservative that they would never hear of the resumption of their rights in their native soil without compensating the landlords. I did not believe it then, and have seen enough of the drift of opinion to know that I was right. To my own knowledge, man after man who a year ago was in favor of compensation scouts it now.

To be sure, all English precedents are of the compensation of the ruling class. Scotch lairds were compensated when the privilege of hanging their fellow-countrymen was taken away form them. Irish lords were paid the capitalized market value of their rotten boroughs when the Irish Parliament was abolished, and underfed British workmen were taxed to make up to West Indian planters the value of their human chattels, while no sinecure could be abolished nor an hereditary pension be got rid of without compensation.

But that day has passed. Not only is "the schoolmaster abroad," but political power has passed out of the hands of the privileged classes, who have been so ready to compensate each other at the expense of the people. If on the settlement of the land question any compensation is exacted, it will be from, not to, the landholders. Further than this, the resumption of the people's rights in the soil must be a gradual process which will injure no one. Stupid people persist in talking of it as though it could be accomplished at the drop of a handkerchief.

How do you regard Mr. Hyndman's argument, that if Mr. Chamberlain has become a land nationalizer he cannot stop at the point that he has reached? – that "the change he implied and advocated in dealing with the land can only be brought about with any advantage to the people at large by first taking possession of that which Mr. Chamberlain has so ably represented – money?"

I think that those who, like Mr. Hyndman, believe that to secure any benefit to the people capital as well as rent must be made common property, have never really thought out the social problem. The evils they attribute to capital really spring from monopoly, of which the monopoly of the soil is the most important. Mr. Hyndman, on some important points at least, seems to have made a convert of the editor of the London *Times*, but Mr. Chamberlain I trust, has a more logical mind.

You do not then regard the capitalists as "greater robbers" than the landowners?

A capitalist may or may not be a robber. As things go at present many of them are, for as society is at present based it is very much "rob or be robbed." Many things are commonly spoken of as capital which are not capital at all, but capital in itself is an aid to labor, not an injury. It is not capital which forces down wages, but the monopoly of the land which compels men who have been deprived of the natural means of employing themselves to compete with each other under the pressure of starvation.

What is your attitude in regard to the nationalization of capital?

Insofar as by here the nationalization of capital is meant the undertaking by the State of businesses that are in their nature monopolies – such as the telegraph, railways, etc. – and assuming functions that are in their nature cooperative, I am in favor of it as fast as practicable. But to go further than this, would be to strike at the springs of individual well-being and national wealth. Instead of repressing enterprise, and discouraging thrift, our effort should be to encourage everyone to produce and accumulate all he can, by removing all obstructions and sacredly guarding the rights of property.

It grieves me to see such a respectable journal as the *Times* undermining the rights of property and of social order by teaching that there is no other basis of right than the will of the majority. To say nothing of the blank atheism of such teaching, it is dangerous in the highest degree, and I do not think the editor of the *Times* can realize the full import of such words as these which are contained in his article on Mr. Chamberlain's speech:

> Property has no rights except such as are conferred by law. . . . Rights are the creatures of
> laws, and laws rest upon the assent, explicit or implied, of the majority.

HENRY GEORGE IN SCOTLAND
(The Standard, May 25, 1889).[62]

I finished my Scottish trip, or at least the speaking part of it, at a great meeting in the Glasgow City Hall on Thursday night, and before the afternoon closes will be over the border on my way south. I have not been as far north as I expected, the visit to Wick, which had been on the program, having been abandoned by the committee in order to permit of [sic] some other meetings in the Lowlands. So I missed the walk along the beach by John o'Groat's house that I expected to take with [the] Dean of Guild Brims, and the greeting of their other friends I count in Caithness and Sutherlandshire. But I think the committee were wise, and that the time that would have been consumed by the trip to the top of Scotland has been better utilized nearer the border. Caithness and Sutherlandshire are already represented by Dr. Clark and Angus Sutherland, and can be depended on in the next election, and it is in the districts that are not sure that time thus saved has been spent. Of the three weeks that I was to devote to Scotland the committee originally promised me one as a holiday, for they were patriotically anxious that I should see "bonnie Scotland" when the sun was on it and the leaves were coming out, and all my previous trips have been in winter and during the shortest of the short days. The week's holiday in Scotland gleamed before me like an oasis, but when it came to the reality it was only by strenuous efforts that I got one day of it, and that I had to put in a futile effort to catch up with pressing arrears of personal correspondence. But I think good work has been done.

All my speaking – with the exception of one meeting at Campbelltown, on the Cantire peninsula, in the Duke of Argyll's dominions, has been in the Lowlands; and could the requests that have come to the committee have been possibly acceded to, I could have put in effectively many weeks more, in spite of the fact that the days are now so long. The Highlands are all right. Their people will be all ready when the south is ready. This I learn from many sources. The reductions of rent and the sweeping away of arrears by the Crofter Commission are only whetting the appetite of the crofters for more. But they are doing a still more important work in raising the spirit of the people. It is a great thing for the men who have hitherto stood in dread of the power of landlords and factors, and been compelled to look up to them almost as a superior race, to sit at the same table with landlord or factor – to tell their story and hear landlord or factor tell his – and then to have the commission decide against the "higher orders" – as it does when it decrees a reduction of rent. It is a new experience, and one that bodes no good to Highland landlordism.

One of the little "straws" from the Highlands that has much pleased me is a bit of news that has come to me first from Donald McCrae, the schoolmaster of Balallen, who organized the deer raids, and afterward from D. C. MacDonald of Aberdeen, the counsel for the Lewis crofters.[63] It tells of the promotion of the Rev. Donald McCallum of Waternish, Skye, to what is considered in the Western Isles a very fat living at Tiree, [on] the Island of Lewis. The Rev. Donald McCallum, who has been foremost of all the Highland ministers in preaching the gospel of land for the people, and who when he and I spoke together, when I was in Scotland before, seemed to me about the most telling speaker I had ever listened to, had only a *quoad sacra* charge in Waternish.[64] That is to say the church of which he was minister, though belonging to the Established Kirk of Scotland, was not one of the legally established parishes with a revenue drawn from the "teinds" – the stipulated rent charges reserved for the support of the Kirk when the immense property of the Catholic Church in Scotland was, in spite of the protests of John Knox, turned over to individual landlords at the time of the Reformation.[65] The *quoad sacra* churches are supported, and their ministers' stipends are paid, from a fund accumulated by contributions, legacies, etc. – a sort of Established Church home mission arrangement – and the positions and salaries of the ministers are generally not so good as those of the regular incumbents, who correspond to the English parish rectors.

The last incumbent of the Lewis Parish was the Rev. Ewen Campbell. He had been a sailor in his youth, and had afterwards studied theology, been ordained, and obtained this parish in the Hebrides, where there was £200 a year to get and nothing whatever to do except to preach on Sundays and fast days to his servant maid and sexton – for the Rev. Ewen Campbell, being a bachelor, had neither wife nor children to swell his congregation. In Lewis, as in most parts of the Western Highlands, the people had walked out of the Established Kirk in a body at the time of the "Disruption," and had got to regard it as not much better than popery.[66]

Though in his solitary life Mr. Campbell became a good deal of a recluse and eccentric, he was a man of considerable energy of character, and the landlord must oft have been tempted to couple his name with Dr. Watt's line: "Satan finds some mischief still for idle hands to do."[67]

For the Rev. Ewen Campbell having only two of a congregation and plenty of leisure for documentary research and ecclesiastical law investigations, set himself to ascertaining and asserting the rights of the Church. The parish had been held for several incumbencies by ministers who were tacksmen (large tenants) of the great landlord, and the glebe lands[68] had gradually been merged and lost in the larger areas they rented for sheep farming. Rev. Mr. Campbell set himself in the first place to discovering and proving up the boundaries of the ancient glebe, and by threat of proceedings in the ecclesiastical courts compelled the landlord to concede it. Then he addressed himself to the condition of the kirk. The landlord protested that there was no congregation, but the Rev. Ewen Campbell replied that that made no difference, as there might be a congregation sometime; and anyhow, it was the landlord's business to repair the kirk. And the landlord finally had to do it.

Then the Rev. Ewen Campbell addressed himself to the manse.[69] This was in a very dilapidated condition, and as there was no legal escape, the landlord had finally to build a virtually new and commodious manse. At this point the landlord naturally thought he had satisfied his obligations to the church. But, no; in drawing the plans for a manse, to be occupied by a bachelor minister, the architect had left out the nursery, which is, it seems, embraced in the legal regulations for the proper housing of a minister of the Established Kirk of Scotland. As soon as the Rev. Ewen Campbell had got well rested he called attention to the fact and demanded the nursery. This seemed to the landlord like adding insult to injury, and he warmly protested against the manifest absurdity of building a nursery as an addition to the house of an old bachelor. But the Rev. Ewen Campbell was deaf to all such remonstrances, or, rather, they only made him more determined to insist on his legal rights. Even if he was a bachelor, he insisted, he had the right to get married if he chose, and when he did get married he wanted a nursery for his children and a room for the children's governess. And besides, even if he himself never got married, his successor might get married and might badly need a nursery and a room for the governess. Finally, he had his way. And so, having secured the eight square miles of glebe land, having had the church repaired, a proper manse built, and the nursery added to the manse, the Rev. Ewen Campbell rested from his labors, and about four months ago closed his eyes for the last sleep.

Now according to Scottish ecclesiastical law – I believe it is a new law – a majority of heads of families, regular members of the Established Kirk, and communicants of the parish, can present a minister when a vacancy occurs. All of which had been pondered over by the schoolmaster of Balallen.

So, perhaps, it came that with the news of the death of the Rev. Ewen Campbell the Presbytery to which he belonged received a much more astonishing notification. During the years of his charge the old bachelor minister had got eight square miles of glebe lands, a repaired kirk. a new manse, and a nursery to the manse, but he had not added a single soul to the congregation. His death, however, had proved more potent than his life, for the document which the astonished Presbytery received with the notice of the death of the Rev. Ewen Campbell, was a declaration signed by eighty heads of families in the parish, setting forth their earnest desire to return to "the kirk of their fathers" and to partake of a sacrament according to its form, and requesting that a minister be sent by the Presbytery to examine them.

Of course, the news of this reverse disruption that threatened to take the whole parish back from the Free Kirk to the Established Kirk was not long in coming to the ears of a Free Kirk minister. He was naturally aghast at a prospect of having the death of the Rev. Ewen Campbell reduce him to such a congregation as that to which the deceased gentleman had ministered in his lifetime, and at once bestirred himself to prevent such a disaster. But finding protestations to individuals useless, he at last got the whole eighty together, and after unavailing remonstrances finally told them that if they took such a course neither he nor any other Free Church minister would baptize their children. The people had merely listened till then, but at this threat one of the crofters got up and said that the schoolmaster of Balallen could preach a good sermon – in fact, they liked his sermons better than any they had ever listened to, and if they could get their children baptized by no one else they were quite content to get the schoolmaster to baptize them. The Free Church minister resigned himself to his fate and left the eighty to theirs.

In due time, the delegate of the Presbytery came to the parish, and the examination of the applicants for admission to the Established Church began. The examination was vigorous, and more as if it were desired to keep these yearners after the faith of their fathers out of the established pale than to let them in; but they were all fully up in the prescribed questions, and the delegate had no alternative but to admit the whole eighty. And then, as if in imitation, the gamekeepers, gillies[70] and other attaches of landlordism in the neighborhood also began to be seized with the desire to return to the faith of their fathers, and it began to look as if the election to the place of the Rev. Ewen Campbell would be a warm one.

But a day or two before the appointed time for the church members of the parish to express their preference it began to be whispered in game-preserving circles that the whole thing was a ruse to cover a deer raid on a larger scale than had yet been attempted, and the movements of crofters seemed to give color to the suspicion. So strong did this at last become that by the hour appointed for the voting every gamekeeper and gillie was at his post to protect the sacred animals, and the crofters, executing a flank movement, marched to the kirk and triumphantly presented the Rev. Donald McCallum, *quoad sacra* minister of the Waternish and Tyree, as the choice of the parish for the incumbency made vacant by the death of the Rev. Ewen Campbell.

Mr. McCallum is not yet installed, but that is only a matter of a month or so. He is sure of the place in that time if he lives, and will soon be in possession of the eight square miles of glebe, the comfortable stipend, and the new

manse with nursery attached. Mr. McCallum is as yet a bachelor, like his predecessor, but he will probably soon correct that error.

When he is inducted the crofters of Lewis will have eight square miles on
which to build huts in case they are evicted, and the possession of this tract
by a clergyman of the Rev. Donald McCallum's sympathies opens another
possibility. The glebe lands abuts on an arm of the sea up which the salmon
of Lady McDonald (who is now the owner) come in great numbers in proper
seasons. There is one point from which a net properly stretched across the inlet
would swoop in pretty much all the salmon that come in and leave the gentleman
to whom Lady McDonald has sold the right to fish salmonless. Whether the
incumbency of the glebe land carries with it the legal right to fish is a law
point not yet determined, but the Rev. Donald McCallum is a man who is likely
to let the onus of proving that it is not devolve on the landlords.

There are so many clergymen who find that the road to promotion is the
betrayal of Justice and the preaching of a gospel suited to the tastes of the very
class that Christ denounced, that it is refreshing to hear of a clergyman who
has been called to a better position and a higher salary, because of his advocacy of the equal rights of all men, and we can all wish for the Rev. Donald
McCallum that his predecessor's foresight in making the landlord erect the
nursery may in due time be justified in his case, and that the fishing on his
new estate may be good.

Speaking of salmon, a carpenter was recently heavily fined in Newcastle for
infringing the rights of the Duke of Newcastle by catching some of them off
the Northumbrian coast. The theory, if there was any theory, must have been
that the salmon were the private property of the duke, not that the carpenter
was trespassing on the duke's ocean. But in the cases of Donald McCrae and
the other crofters who were prosecuted for killing deer, it was held that deer
were *fera naturæ*[71] and could not be subject to property rights until taken.
Therefore, the deer raids involve no legal offense in killing the deer, but only
that of trespassing on the deer parks.

Donald McCrae gave me information of hunts that are contemplated when
the hunting season begins that will, if properly carried out, attract a great deal
of attention. The crofters like to wait till the proper season for killing deer,
even when their purpose is to get rid of them.

I began this letter with the intention of giving something like a review of
my trip in Scotland. But a chapter of accidents has made it impossible now to
do so in time to catch a mail. I can only say at this time that it has been very
successful and exceedingly gratifying to me. My audiences have been large,
my reception of the warmest kind. Our friends everywhere are awake and full
of hope, and I have everywhere found evidences of the greatest progress.

My hardest work is now, I hope, pretty nearly over, and by next week I will be able to send a full letter.

ADDRESS OF WELCOME TO MR. HENRY GEORGE FROM THE DUNDEE AND DISTRICT UNITED TRADES' COUNCIL
(The Standard, May 25, 1889).[72]

Sir: The Trades' Council of Dundee, representing the working classes of the city, on the 16th of April, unanimously resolved to ask you to receive an address of welcome and of thanks for your great services in the cause of labor; and in accordance with that resolution, we desire to assure you that you have the council's warmest welcome among us and their most sincere thanks for the disinterested devotion with which you have pled the cause of the disinherited of the world.

Ever since you commenced to think you have had the industrial problem before your eyes, and you say somewhere that it gave you no rest till you solved it, your solution was to replace a political economy that was truly called the "dismal science" by a system which is radiant with hope for the masses and impossible to call unjust to the classes. The great difference between yourself and preceding economists lies in the extreme simplicity and beautiful correlation of your laws of distribution as contrasted with the confusing unconnectedness of theirs. Three parties unite in production – the laborer, the capitalist, and the landowner – and the shares of each are wages, interest, and rent.

The existing law of rent you accepted; it was with regard to the law of wages and the law of interest that you differed from the current political economy. If "the rent of land is determined by the excess of its produce over that which the same application can secure from the least productive land in use," surely it is clear that the recompense left to labor and capital is all production minus this excess. That conclusion of yours is unavoidable, and it has an infinity of meaning. As far as labor and capital are concerned, it is as though all production yielded no more than what is secured at the lowest levels; it is as though there was no better land than the worst. That is a sufficient reason why wages are low, and why the reward of labor will remain low until it learns to ask that this excess shall be equally divided.

This excess is produced not by any limited body of individuals, but by the whole community, and should belong to the whole community. "On the land we are born, from it we live, to it we return again – children of the soil as truly as is the blade of grass or the flower of the field;"[73] and it is the most reasonable proposition in the world that everyone has an equal title to land. It is instructive to analyze the motives of most landowners for monopolizing land. It is not to labor on it or to dwell upon it – both blameless motives when they covet no more than their equal share; they monopolize land almost solely that they may be able to intercept this excess – making others suffer that they may live in idleness and luxury.

The significance to labor of that part of your argument which deals with the effect of labor-saving contrivances cannot be overestimated. Inventions take the place of men. It is all very well to say that they cheapen the cost of production, and that so the working classes are benefited. But let those who administer such comfort remember the law of rent. The rent of land is the excess of its produce over what is secured on the worst soil in use – otherwise called the margin of cultivation. It is undoubted that cost of production is lowered by inventions; but it is only insofar as it is lessened on the margin that the working classes share in the saving: it is rent which reaps the great increase.

With regard to that other branch of your subject – the relation of labor to capital – let it suffice to say that we fully appreciate the fact that the natural and just distribution of wealth by conferring the power effectively to demand would enormously increase the request for labor, and that this increase would enable labor to meet capital on neither more nor less than equal footing. To put it shortly, it would be as easy for the employe[e] to find the employer as for the employer to find the employe[e]; under these circumstances capital would be shorn of its power to oppress.

In one of the most eloquent passages of your eloquent book you speak of man when his nobler nature is developed.

> He works for those he never saw and never can see; for a fame or it may be but for a scant justice, that can only come long after the clods have rattled upon his coffin lid. He toils in the advance where it is cold, and there is little cheer from men, and the stones are sharp and the brambles thick. Amid the scoffs of the present and the sneers that stab like knives, he builds for the future; he cuts the trail that progressive humanity may hereafter broaden into a highroad.[74]

True as this always is in the case of men who advance in thought and try to teach the world new truths, we think you are especially fortunate as a thinker in that the cause you advocate has made a progress that must astonish even yourself. In your native country, America, the New Crusade must be an uncomfortable fact for the mechanical politicians of other parties, and it is inevitable that the best of the Republicans and Democrats will gravitate in your direction. In this country the most humdrum village does not escape a discussion of your theories, and you are no doubt aware that the Trades' Congress – which meets this year in Dundee – has already resolved that the land must be nationalized. There is not a civilized country in the world in which your name is not known, and in which your adherents do not spread the light kindled at your flame.

In conclusion, let us express the conviction that if one-half of the working classes of these islands had a decent grasp of the laws of distribution as taught in *Progress and Poverty* it would go hard with our present state of industrial slavery: and let us again assure you of our deep gratitude for the noble work to which you have dedicated your life.

Signed on behalf and by authority of the Dundee and District United Trades' Council, at Dundee, this sixth day of May, 1889. . . .[75]

HENRY GEORGE PREACHES IN DUNDEE:
Gilfillan Memorial Church Filled to Overflowing;
Devotional Services Conducted by the Pastor,
Rev. David Macrae
(*The Standard*, June 1, 1889).[76]

Mr. Henry George appeared in the Gilfillan Memorial Church last night and delivered an address on the land question. The church was crowded to excess, chairs, and forms[77] being placed in the passages, and all available space occupied. The Rev. David Macrae conducted the devotional services, and in commenting on Psalm CXIX, said it was interesting to note what the Bible said on the great question of the land. Isaiah said: "Woe unto them that join house to house and field to field till there be no room for others, that they themselves may be placed in the midst of the land." Amos spoke about judgment being pronounced against those who stored up robbery in their palaces, "who swallow up the needy and make the poor of the land to fail," who "grind the faces of the poor and have the spoil of the needy in their houses.[78] One might, Mr. Macrae said, suppose the prophet looking at some of our own landholders clearing away the inhabitants to turn the land into deer forests, or appropriating to themselves without compensation their tenants' improvements.

He wished to welcome Mr. George to Dundee that night not only as a reformer, but as a Christian reformer. Mr. George's name was known all the world over, and identified with the great question of land reform. Land reform was a question which had a special claim upon the attention of Christian people. Land monopoly had been the source of vast and innumerable evils, and if Christianity had within its province seeking to remove evil effects it had within its province seeking to remove evil causes. He could predict that what Mr. George said that evening would be worthy of profound attention, and would be calculated to stimulate others to study the great question of land reform. Mr. George at the outset quoted a verse of the psalm which the congregation had just sung:

> Blessed are they that undefiled
> And straight are in the way,
> Who in the Lord's most holy law
> Do walk, and do not stray.[79]

Blessed are they! What did it mean? The blessing of a God who left men in heaven and allowed them to starve on earth? The blessing of One in whose house were many mansions, but whose children on earth – whole families of them – live in one-roomed houses? No! All through that book ran the promise of peace and wealth and length of days to those who obeyed His laws. When Mr. George was in London four years ago a clergyman of the Established Church of England came to him and said: "I am in great mental trouble. I am a university graduate. I took orders and got into the navy as chaplain and most of my life has been spent abroad. I have been up to this time an absolute believer in the Bible. I have read the text which says: 'I have been young, and am now old; yet have I not seen the righteous forsaken, nor his seed begging bread.'[80] I believed that until I came to live in the East End of London. I do it no longer. I cannot believe it. I find that my faith is going as I thought it never would go."

Mr. George replied: What is the promise of the Bible? All through the Bible there are promises to the righteous. You never find anywhere that God promises to destroy the righteous – yet if there had been nine righteous men in Sodom they would have been burned up. The demands of the law applied not only to individuals, but to communities. Nations must obey the law as well as individuals. There were great masses of the people living in the midst of wealth condemned to poverty, and great masses of people who must work all their lives for a bare living, and then at last many of them will fill paupers' graves. The people who fill those twopenny and fourpenny lodging houses, or sleep in sheds or on walls, those little children who die by the thousands in the poorer quarters of our great towns, are they unrighteous?

All through our communities today there were many thousands who had never enough and to spare of the necessities of life, and yet there were many who said that the poverty and vice and crime that were born of want were in accordance with God's will. What was the cause of that great and constant stream of benefactions given by the rich to the poor but the uneasy feeling that though there might be some who deserved poverty there were great masses who did not.

The One Cure For Poverty

That stream flowed on, and poverty continued; it might be doubled, and yet poverty would remain. Charity never could cure chronic poverty; there was but one thing that could cure it, and that was Justice. Men should be just before they are generous. Take one of the shortest commandments in the sacred book – Thou shalt not steal. Was not the accepted version of that: Thou shalt not get into the penitentiary. (Laughter.) If they stole a shilling there was a good chance of them going to prison, but if they stole £1,000,000 there was a chance of them becoming what they called in America one of our first citizens (laughter) – that was if they got off with it. (Renewed laughter.) What did the words "Thou shalt not steal" really mean? Not merely that they were not to commit petty larceny nor burglary, but just what it said – they were not to take from any man that which was his. They had just to look round human society in their so-called Christian community, when they saw that there must [be] a good deal of stealing going on that was not accounted for by the people who filled the prisons. (Laughter.) When they spoke about the country having increased in wealth they meant that it had increased in those things which were properly called wealth in economic language, and these had been produced by labor. It was not the things they found in the world – not the things that Adam when he first came found. Labor – human labor – was the only producer of wealth. If that were the case, then it was perfectly true, as a British writer had said, that all mankind might be economically divided into three classes – the working men, the beggar men, and the thieves. (Laughter.) Looking around, they saw that instead of those who labored, who produced the wealth, having most wealth they were the poorer class, while the class who enjoyed the largest amount of the produce of labor were a class that did not labor at all.

The Working Class Robbed

If that were the case the working class must be robbed. There were other means of stealing than by forcibly taking from a man that which belonged to him. To illustrate, he would tell them a story which he thought was proper enough to tell in a church. There was once a good Christian who got among some wicked Arabs, and he started to Christianize them. He did pretty well, as he thought. They were traveling, and passed a rich caravan away out in the desert, where there was no water for miles. The Arabs and the old Adam began to come up again in their hearts in spite of the Christianization. They proposed to go back at nightfall, set upon the caravan, and take all that there was. But the Christian set himself against that, saying it would be a violation of the commandment "Thou shalt not steal." But, said he: "I'll show you an easier and a better way. That caravan is going evidently to get water at the spring we passed – let us go back there, take possession of the spring, and sell these people the water." (Laughter.) They went back, took possession of the spring, and then when the caravan came up they were in possession – the spring was theirs. (Laughter.)

The caravan people could not go farther, and they finally offered to buy the water. (Laughter.) The consequence was that in a little while all the contents of the caravan, as well as the animals, had changed hands. Now, what was the difference? Was it not stealing? (Applause.)

That was what was going on in this country, and going on in America, all the time, and by the most respectable citizens. The root of the present day evil lay in the fact that where the population had gone men in advance had seized the land and the water, and those who followed had to pay a price for God's bounties. The people had to go up and beg for the privilege of using God's earth, for the privilege of availing themselves of His bounties. When one of the Indian agents a while ago was found selling the blankets to the Indians on the reserves which the Great Father at Washington,[81] as they called him, had given them, he was sent to prison. But the people permitted many men to sell to their fellows what the All-Father in heaven had sent to his children. (Applause.) The Indian agent they sent to the penitentiary: men who did the other thing were sent in many cases to the Senate of the United States, just as in Britain they were frequently sent to Parliament. (Laughter.)

Land Created For All

God created the land, not for some men, but for all men. It was never "thy land," or "the land which thou bought," but "the land which the Lord thy God giveth thee." The right to the land was just as clear as was the right to the light that streamed around. The evil was at the bottom of the civilization of today. It was the fundamental curse that was giving to the whole civilization of ours so unstable and one-sided a development, and converting what ought to be blessings into curses. All God's creatures were entitled to those natural elements that were indispensable to life. Daily they thanked God for His bounties. But how did God provide these? He did not place them upon the table. His provision was in the earth, and He had given man the power to labor and the power of producing these bounties. Some people said that all that might be quite true, but it was utterly impracticable to amend. Was it utterly impracticable to do God's will? Let them ask themselves the simple question: Did God intend this world for some people or for all his people? Did He intend that one little naked child who came into the world should have the right to 100,000 acres of the surface of the planet, to the minerals that were embedded in its bowels, to the birds that flew over it, to the salmon that came to its shores from fathomless deeps of the far-reaching ocean, while another little naked infant came who had not the right to one square inch of the world's surface – was it not monstrous?

Someday men answer for what they have done in this world. Supposing God asked about these thousands of little children He has sent upon the earth, and who are perishing in the city slums, supposing He asked about these children who are growing up under conditions in which only a miracle can save them; supposing He asked about the women who ought to be happy wives and mothers who were by thousands prowling the streets of our great cites at night; supposing He asked about that bitter misery and want that may been seen even in the centers of wealth? Did they think they would get off by saying that these were none of their affairs, and that they did not make the world? No, God made the world, but He made it wide enough for all the human beings He has brought upon it.

If there were today sin and distress, and uncleanness and drunkenness, and the vice and crime that were born of want, it was not God's law. God is all bountiful. He has given enough and to spare. Injustice of man was at the root of it all. It was impossible to imagine heaven treated as we now treat this earth without seeing that no matter how salubrious its air, how bright its light, how magnificent its vegetative growth, there would be poverty and suffering if heaven were parceled out as we have parceled out this earth. How could God Himself relieve the vast poverty? If He were to rain down wealth from heaven, though He was to cause it to gush up from the bowels of the earth, to whom would our laws say it belonged?

Through the present system, no matter how bountiful God would be, someone would grab these bounties. Some said that was not their affair. The text did not mean men should not steal themselves, but it meant also that they were to be accessory to stealing, either by helping or by refusing to prevent it as much as in their power. An ignorant public sentiment of today refused to act, was content to stand and see injustice done. It was their duty to look into these questions, and to help bring in that kingdom of Justice and of love for which Christ taught us to pray. (Loud applause.)

The Rev. David Macrae, before concluding the service, conveyed to Mr. George the thanks of the congregation for his able address.

ADDRESS BY HENRY GEORGE IN THE CITY HALL, GLASGOW ON AUGUST 20, 1890[82]

Mr. George on rising was received with great enthusiasm. He said: Friends! It is a pleasant thing for me to be thus welcomed back to Glasgow; and it is a pleasant thing for me to see on this platform so many men that I know as friends of mine and know as friends of our cause.

It is a pleasant thing for me to find that ex-Provost Cochran of Paisley (great cheers) with his honorable crown of white locks is still hale and still in the front. And it is a pleasant thing for me to stand on this platform again with the man under whose auspices and under whose presidency, I first stood here nine years ago. And there are other feelings. One man – one good man, one man whose heart was warm and beat truly – William Forsyth, the first President of the first society formed in Scotland for the advancement of our principles has passed away since I was here last: passed away, though, I know, in full clearness of faculty, in full ardor of conviction, with the trust and the faith that, though he should never see it, yet, ere long the right for which he had struggled would be triumphant in Scotland. (Applause.) Some this year, and some next year, and some the year after. Thus it will be to us all. But for me, I feel – and I think every man who has really enlisted in this cause, will feel – that after all, the best that has been given to us in life, the thing that in truth we will most cherish is the opportunity of doing something, though it be but little to aid in the strife of Good against Evil, in making the lives of those who come after us higher and brighter and nobler and better; in doing some little to bring on earth that Kingdom for which the Master had told us to hope, to struggle, and to pray. (Applause.)

I have traveled a good deal since I was in Glasgow. I come here from Australia, having made the circuit of the world. Let me briefly tell you how I found the good cause in the United States. Our movement is going forward with accelerated rapidity, exceeding the hopes of even the most sanguine among us. It is not the growth of an organized movement; it is not the building up of a party; it is an elevation and an education of thought. Our principles are diffusing, they are in the air. They are beginning to permeate all parties, and especially that part which at last is beginning to take the side of industrial freedom, and beginning to struggle against the curse of Protection. From New York to San Francisco I stopped nowhere without finding earnest and ardent men all alive in the cause, all telling the same story of its progress.

Leaving San Francisco – in Australia I found the same thing. As I said the other night to a little meeting in London, if we of Scotland and England and Ireland and the United States, if we Land Nationalists, or Land Restorationists, or Single-Tax men as we call ourselves in America and in Australia were all swept away and every trace of us forgotten, there are men in Australia who would begin again to fight and carry it to victory among the English-speaking peoples. (Loud applause.)

Australia has made the first start. In South Australia the principle of taxing land values, irrespective of improvements, is already in force to a small degree.[83] In South Australia there is a tax of a halfpenny in the pound on unimproved land values – very little it is true – but little as it is, it is enough to show that the thing can be done, enough to end the objection that there is no possibility of distinguishing between the value of improvements and the value of bare land; enough to make a start. In the colony of South Australia, a proposition made by the late Government to increase that tax by progressive steps until it amounted to 3d. in the pound on unimproved land values, has first been defeated in the lower house by a majority of four.

That was a measure that did not have cordial support of our people in Australia. They did not believe nor do I believe, in the principle of making fish of one and flesh of another; in the principle of taxing a man higher because he has much and letting him go because he had little. They contend, as I believe all of us here will contend, that the man who has valuable land holds from the state, a state privilege and if he has much of it, then should he pay much; if he has little, then should he pay little. But in any case, if he has an advantage greater than that of his fellows then should he return an equivalent to the state. (Applause.) But the mere introduction of that bill by the Government, the pressing of it to a vote, the narrow margin by which it was beaten shows at once the advance that the principle of taxing unimproved land values is making. The larger bill (a better measure remained behind) has been twice passed in the lower house and as I was informed on all sides will be passed by both houses this year, for the Colonial "House of Lords" will not dare much longer to block it. This bill gives the municipalities the power to put all their rates on unimproved land values. That will bring the great principle before every municipality.

In New South Wales, a similar bill has been proposed by the Government and will in all human probability be passed in this session of the Legislature.

The Australian colonies, I was going to say, offer special and striking evidences of the inevitable result of putting land into the hands of individuals, of treating land as we rightly treat the things which are produced by human labor. But I do not know that I could really say that. From Glasgow round the world to Glasgow again where can it be said that there are the most striking evidences, the same wrong everywhere shows the same results. ("Hear, hear.") But there is a new country with an exceedingly sparse population – a country where the production of wealth is probably higher, man for man, than anywhere else in the world! Yet, with all those advantages you find symptoms of the same disease as is to be found here – the cry of unemployed, the strike and the struggle; the pressure of a one-sided competition.

Starting with Victoria – great masses of the people have been led onto a wrong track and have attempted by what is called Protection to do something to improve the condition of the workers. Even in New South Wales, that held out [the] longest, Protection came at [the] last election within one seat of carrying the country and making the Government. Today I am glad to say on the authority of the protectionist papers that there is no possible opportunity, no possible chance of Protection at the next election carrying the day in New South Wales. (Applause.)

And it is because our people have come to the front; it is because the standard of real Free Trade has at last been raised. (Applause.) Very largely the Protection movement there has been a Democratic movement, the party of so-called Free Trade being the party of the large squatters of the landed interests; the Free Trade that existed being a mere make-believe Free Trade like the Free Trade that you have here. (Laughter.) But now in the name of Democracy the banner of real Free Trade has been raised. Now an active, energetic devoted band of men – men of the highest ability and cleanest character – are devoting their energies to propagate the gospel of freedom, to demand the equal rights of all citizens, and propose a Free Trade that means the sweeping away of every restriction, that means the doing away with every monopoly. For, in Australia they are ahead of us both in Great Britain and the United States in two or three things. There the railways belong to the colonies. (Applause.) There is no such thing as a railway syndicate or a Railway King, and there the great monopoly is clearly the fundamental monopoly – the monopoly of land, the monopoly that is the parent of all the little monopolies. ("Hear, hear," and applause.)

I come back from Australia not merely with a sense of the greatness of the nation that is growing up there, not merely with a higher idea than I ever had before of the great possibilities of that country and that people, but with the feeling that the fight in which we are engaged is so far forward there that it is only a matter of time – a matter of successive steps when it will be triumphant (Applause.) And coming back here to the old country, I find the same evidences of progress. I do not think it has been here, nor has it been anywhere, a progress in organization, but a progress in thought. Our ideas are in the air. Men are beginning to get them and to hold them without knowing they have got them. ("Hear, hear," and laughter.) Men are getting to believe in these ideas that do not believe in us, and that is the way in which all great movements must go forward. The real thing, the real work we have to do is the work of education; the real thing that we have to look to is public opinion. Thought always must precede action, and right thinking always will produce right action. And, as I have said in this hall before, if the masses of the people are anywhere oppressed where the English tongue is spoken, it is not because of any class of oppressors who hold them down, it is not corrupt politicians, it is not because of grasping capitalists, it is not because of tyrannous landlords – it is because of the people's own ignorance, because of the people's own selfishness.

I believe today that in Great Britain, the fact that something is wrong is so clearly apparent. Aye! I believe today that the idea in Great Britain, the feeling that all men have natural rights in the land, has gone so far that if energy could only be concentrated, the work of carrying that principle into effect could not by any power be long delayed. We are fought, not now so much directly, as by diversions, by calling "Lo Here" and "Lo There" and the leading of men, anxious to do something, into bypaths, into lanes where there is no ending.

When I first stood on this platform with Mr. Ferguson, the idea of the British taxpayers buying the land of Ireland and giving it to the agricultural tenants – why that was the most radical demand of men who really thought. They were radicals, and it was not within the range of practical politics. There is one measure of the advance. So great has been the advance of thought in these years that what was then thought so radical is now the last ditch of the Irish landlords. And I am told that a bill to appropriate thirty-three millions of your money, or of your credit is to be forced through your next Parliament, or your present Parliament at its next meeting. And a gentleman, one of your Glasgow Councilors told me this evening to my astonishment that your council of the municipality of Glasgow is going down to ask permission to raise the water three feet in Loch Katrine. Nay! The fools are not all dead yet.

Look at what we have been doing in the United States. Let me say it to you that absurd [and] profligate as has been the action of our Congress – nothing could be better conducive to the spread of right principles. Protection in the United States is in its death gasp. In bringing forward this McKinley Bill,[84] as it is called, in proposing to put more fetters upon fettered commerce, in carrying to the most profligate extreme the principle that it is the business of Government – to tax some people in order to enrich other people, the party in power is preparing the way for being hurled from power at the next election. It is helping our people to carry on a most effective propaganda.

But this is a digression. Speaking of this Land Purchase Bill[85] – that it should be introduced, that such forces should be united to pass it shows the advance in a certain way that has been made in these years. But in itself it is not a measure that leads to good. It is a measure that leads the other way. It is not a measure that goes with us. It is a measure designed to block us. It is not merely, as Davitt styled it, a "Landlord's Relief Bill," but it is a bill designed to create in Ireland a larger class of landowners, a bill designed to bribe a certain number of the people to stand on the side of escaping landlordism. Whether it passes or not, its mere introduction, the discussion that must take place, the feeling of injustice that will be engendered must help forward right principles, must advance our cause.

Aye! We are so far now that every wind that blows, every current that sets, must help us on. That is the principal property of truth. Truth need only fear to be ignored. It can never fear scrutiny. It must always hail discussion. If it be truth, you have only to get men to talk, think, and feel about it, and everyday its advocates must increase.

As I said, there is a rapidly growing feeling of social injustice sweeping over this country and the civilized world. Things cannot continue as they now are. There must be an advance. The only question is whether it shall be on the straight line, or whether it shall be diverted into the bypaths. That can only hinder for a time, but it will hinder for a time.

But whenever I get before a Glasgow audience now, I feel as though I were talking to an audience of friends, an audience who understands precisely what I mean by "our cause" and "our plan," to an audience to which I have to explain nothing. But probably I am wrong in that, probably there are in this audience women and men who do not clearly understand for what we contend. Let me therefore explain.

We see everywhere evidences of the social disease. We see that all over the civilized world labor does not get its fair reward. We see that there is every-where, where social development has gone even a little way, a difficulty of finding employment a fringe, more or less deep, of unemployed men. We say that these phenomena indicate some general cause, a cause that is not peculiar to any country or to any institutions. We trace that cause to this great wrong – that everywhere the element which is essential to human life, the element which is the raw material of all production, the land, on which men must stand, on which men must work, the land that, to a man, is the whole material universe so far as he can get access to it, the land which is vitally necessary to every human being, is made the property of some. There, we hold is the fundamental cause. It lies not in competition; it does not result from any inevitable conflict between labor and capital; it is not a legitimate outgrowth of the wage system. It is simply the inevitable result of making that element necessary to all men the private property of some men, the inevitable result of disinheriting men of their natural rights, of compelling them to pay tribute for the good gifts of their Creator. There, we say, is the fundamental wrong.

There can be no remedy, no matter what it is, there can be no remedy until it goes to that wrong and rights it. We hold it as a self-evident proposition that there never lived a human being who had any right to give away this world. We hold that every child who comes here is, in the moment of its birth, seized with an inalienable and indefeasible right to the element which is necessary to its life. We hold that there is an absolute right of private property in all things which labor produces from land – *an absolute right*. We do not hold that men have equal rights to wealth. We do not hold that there is among men an absolute equality as to power or capabilities. But we do hold that, since the land was given by God, since it was given to all His creatures, since, under His law, and by His providence, it is made indispensably necessary to each human being, since to deny the rights of some, and to extend the rights of others in land, to make some the masters of others, we hold that men's rights in land are, and must be equal. *And what we aim at is none of these two-penny [or] halfpenny reforms that might alleviate things a little, or for a small class.*[86] We aim at nothing less than the securing to every man, woman, and child in every country, their equal rights to live and to work on the land of their country. And our means are as simple as our aims. We do not propose to buy land from some people and sell it to other people, as is proposed in the Purchase Scheme.[87] We do not propose to buy a piece of land here and a piece of land there, and to make it the property of the people, and have it rented out. We do not propose either to buy or take the land and divide it among individuals. We simply propose to assure equal rights by having the people in their collective capacity, in their Municipalities or Councils or States act as the landlord and collect the rent. We do not propose to do that under that form. We do not propose to call the people the landlords; we do not propose to demand any payment as rent. There is an easier way than that. We propose simply to let things stand as they are, not to take a square inch of land from anybody, to let things stand as they are and to collect rent under the form of taxation. "The rose by any other name." (Loud laughter.)

What we propose to do is to abolish all taxes now levied on the products of human industry, and to put taxes on the unimproved value of land – the value exclusive of the values of all improvements in or on it. And we propose to raise that tax until we come as near as possible to taking the whole of economic rent. I think that is so much the easiest way, that we may truly say it is the only way. All our English-speaking peoples are accustomed when Governments take anything like land directly, to think that they should pay compensation, for it is always proposed to compensate landlords for taking away their land. The idea of compensation always arises whenever Land Nationalization (in its narrow sense) is proposed.[88] But no one ever heard of compensation for taxes.

There is a method that may be pursued without the slightest difficulty of that kind; a method that harmonizes with all our thoughts; a method which offends none of our prejudices; a method that is but agoing back on the way we have come. For really the way land has been made the property of individuals is mainly by a series of "No Rent Manifestoes," by which the holders (not the owners) of the land got abolished the taxes which used to be levied on land values. There is a method that harmonizes with Free Trade. Nay, it is clearly and plainly the carrying of the principle of Free Trade to its logical conclusion – the carrying of it to the only conclusion that will enable Free Trade to do what Richard Cobden hoped it would do, to utterly abolish pauperism, and in every way elevate the condition of the worker. Putting our taxes upon the proper place for taxes, we are then able to sweep away all taxes that lessen the reward of labor – all taxes that discourage energy and thrift – all taxes that increase prices, and build up monopolies. That is the way we in the United States and in Australia call the way of the "Single Tax." We do not want to take land, we simply want to abolish taxes on the products of labor and levy them on the value of land irrespective of improvements.

Now compare that with some of the methods on which stress is being laid and energy expended. Take, for instance, the organization of labor. The organization of labor and establishment of trades unions is, in many cases, a good thing. It does something. It enables men to raise their wages, at least for a time, and over a certain area; but it can be no final settlement of the Labour Question in that way. You know the Docker's Strike, a struggle which raised an enthusiasm and brought out a substantial help that makes it famous in the annals of such movements.[89] The Docker's Strike has done a great deal for a downtrodden class, made them more independent, gave them greater self-reliance, greater capacity; but it has only done so for a certain class of men. Here is the final report of the Docker's Strike, *The Story of the Docker's Strike*, written by Mr. L. Smith with an introduction by Sidney Buxton.[90] These men say:

> The first and greatest of the results of the new agreement is to place dock labor for the first time in the position of an organized and regular industry. If the settlement lasts, work at the docks will be more regular, better paid, and carried on under better conditions than were before. All this will be exclusively given to those who get the benefit from it; but in other results will undoubtedly be to contract the field of unemployment and lessen the number of those for whom work can be found. The lower class [and] the casual [workers] will in the end find [their] position more precarious than ever before. In proportion to the increased regularity of work, that the [ ] of laborers will secure, the effect of the organization of dock labor, as of all classes of labor, will be to squeeze out the residuum, the loafer, the cadger, the casual follower in the industrial race. The members of Class B of Mr. Booth's [designation] of social classes will be no gainers by the change, but will rather find another door closed against them, and this in many cases, the last door of employment. Hitherto, all these grades of labor have jostled each other at the dock gates and the standard of life of the lowest has set the standard of all. Now they will be more sharply divided. The self-respecting laborer will no longer be demoralized into the lower. Thus we may look in the immediate future for a Class B, diminished in number, but in a more hopeless condition than ever. The problem of dealing with the dregs of London will thus loom up before us more urgent than ever; but it will be simplified by a change which will make it impossible, or at least unpardonable, to mix up the problem which is essentially one of the treatment of social disease, with the radically different question of the claims of labor.

Here is an utter misapprehension. Social disease and the question of the claims of labor are not two diverse things, but one thing. Ill-paid labor is a symptom of social disease – of the same social disease which produces the casual [worker] and the tramp and the pauper. But here is an admission of the utter failure of organization as a settlement of the labor question.

Nothing can permanently elevate the condition of labor, nothing can solve the labor question, unless it goes to the root of the subject. For organize as you may, there must always be a class outside of organization. Work as you may, with invention succeeding invention, new industrial processes coming out every day, men must be drawn out of their former class and pursuits and must fall to the bottom. Therefore, it is that there can be no hope of any full solution of the labor question in that way.

Nor can there be any hope in charity. Look at the enormous sums that are raised in this country, besides the preparation of Government. Take the enormous sums that are raised to help the poor in various ways. What does that do? They may help individuals for a time, but that is all. The poor continue and the poor increase. Charity clearly increases poverty by diminishing self-respect. So these charity workers tell us. Neither can any effort that does not go [to] the bottom, that does not strike at the root avail.

You know here. Aye! I think they know in every part of the world, the organization that goes under the name of the Salvation Army.[91] I came back from this trip round the world with a higher and deeper respect for the Salvation Army than ever I had. I have found no town in all the Colonies where it does not have a station; where I did not hear the voice of the Salvation lads and lasses. And everywhere, I was told, they had done real and substantial good. And even in Colombo, [Ceylon] when driving from a Buddhist temple, what did I run across but a station of the Salvation Army? The Salvation Army was started to save men's souls. There is already dawning upon the Army this great truth – that unless men's bodies can be saved, it is a hopeless task to try to save men's souls. And in this way it is even beginning to resolve itself into a kind of "Anti-Poverty Society" and to make war against want as well as against sin.[92]

And growing out of this Docker's Strike, and out of its organization and machinery for feeding the strikers, the Army has gone on to attempt to take up the broken-down wanderers of London, and to make them men and women again. I visited the other day one of these stations, and it is a visit that I shall long remember. There, picked up out of a London street – there as we entered – were a lot of men in whose faces you could see traces of the hard times they had undergone: And how crushing it had been to soul and body! There they were upon their knees praying and singing before they took their noonday meal. And I asked what the Salvation Army found for these men to do. I learned that they were teaching these men to make mats, to make brushes, to split up kindling wood. The next day I read in a London paper of a great protest by the splitters of kindling wood against the Army for diminishing the work of outside men who were trying to earn a living by making kindling wood. It will be the same way with the brushes and the mats. They were not made by the laziness or viciousness of these men. The trouble with them is the one of finding employment.

Would there be any strikes were it not for this difficulty of finding employment? That is at the bottom of the social problem. Why should the Salvation Army need to find mere employment? Is there any searching of work where the great majority of the people do not have enough of the things which work produces? Human labor is the active force which produces all wealth. How in a country where the mass of men are ill-clad and certainly need houses if they are not unfed. How in a country and world where the great majority of men want more of the things which labor can produce, and might live higher lives if they had more? How in such a country, or in such a world, can there be any search of unemployment? This underlies all these ideas of Protection which still linger under your limited Free Trade and come out in Australia or the United States – this is the root of many of these half-going remedies that are in their nature restrictive – the idea that work is a fixed quantity, and therefore to let one class of men get work [while] other men must be debarred from it. But how can that be? Every man has two hands, and has a demand for the labor of those hands.

The want of work! There is no want of work! There is a want of the opportunity of working. But that can only exist where the natural opportunity of work is shut up; can only exist where labor is debarred from land. Do you want to know why wages are so low? Do you want to know why thousands of men in your great cities sleep where they may at night, and rise in the morning without knowing where they are to get a meal? Why have you in Glasgow – that ought to flourish by the preaching of the word – to erect lodging houses? Do you want to know why women are making matchboxes in London for four-pence halfpenny a gross? Man, though he has in him something of the likeness of his Creator is physically a mere land animal. All his work is merely the changing in form or in place of what he finds in land. All human production, in the last analysis is but the union of labor with land.

Consider that fact, and look around. See twenty-six and a half millions of acres unused on this small island. See around your crowded cities the great tracts held unused or hardly used while in the cities themselves human beings are crowded as they are. That is the curse of civilized communities. Open the land to labor and there can be no such thing as scarcity of unemployment – open the land to labor, give men equal rights to the primary element – then you need not trouble yourselves about competition except to sweep away every restriction to its free play. Open the land to labor and you need not trouble yourselves about finding work for anybody – they will find work for themselves. Why, consider the Salvation Army: what does it tell you constantly? One of its endeavors is to send people back to the country. Why did they drift from country to the town? Simply because other men owned the land on which they were born and on which they must live, if they would live at all.

There is the cause of low wages; the primary occupations of men are those occupations which directly extract wealth from land. Where a man's labor can make one-pounds worth of wealth a day from the land his wages will be £1 a day. But when somebody else owns the land, and a man has to pay 10s. a day for permission to use the land, then his wages can only be 10s. a day; and if he has to pay 15s. a day for permission to use the land, then his wages can only be 5s. a day. And so on, down to the point where . . . you find men working on the land for 10s. a week, and thinking themselves privileged at that. There is no mystery about this social disease. The simple facts that the smallest child knows, if put together will explain it. And the remedy is as clear. What we have to do is to lift all taxes from production; to cease fining a man who builds a house or otherwise produces wealth; and to concentrate taxation upon the unimproved values of land, taking what political economists call "rent" or what John Stuart Mill calls the "unearned increment;" thus making the land useless to the mere landowner (did you ever think how utterly useless a being the mere landowner is?) – making land useless to the mere landowner – the man who wants to reap without sowing; but making it more useful than ever to the man who wants to use it, and making it easy for the man who wants to use it to get it.

Here I am more and more convinced is the true line. And it is the easy line. Every step taken in that direction will make the next step clearer. Every step that you take in that line will create, will open opportunities for labor; raise society from its foundations; and lessen resistance to the next step.

NOTES

1. The "Land and People" is an abridged and retooled work containing parts of some chapters from George's 1881 work "The Irish Land Question" (subsequently renamed "The Land Question"), to adapt to a Scottish readership; ILD. See: Henry George, "The Land Question," in *The Land Question*, pp. 35–55, 65–69, 70–71, and 76–86.

2. A cottar or cotter is a peasant who exchanges his labor or other services so that he may live in a cottage and usually with a small parcel of land for personal use.

3. Herbert Spencer (1820–1903), an English philosopher who attempted to correlate natural sciences with philosophy, was best known for a strongly capitalist economics, which became associated with Social Darwinism. (See footnote no. 29.) George wrote *A Perplexed Philosopher* (1892) as a rebuttal to Spencer's recantation of a progressive view in *Social Statics* (1850) of the land, including a single tax.

4. Julius Caesar and his Roman legions landed in Great Britain in 55 B.C.

5. William Kidd (ca. 1645–1701) was commissioned as a privateer but was arrested and then subsequently hanged for piracy. He is associated with legendary exploits and hidden treasure.

6. Foot-binding of women was a practice among the aristocracy of China considered to enhance beauty. The Flathead Indians originally lived in the Bitterroot River valley in Montana. After the introduction of the horse they adopted a life on the Plains but suffered at the hands of the Blackfoot Indians.

7. Article 1, Section 9 of the Constitution stipulates that: "The migration or importation of such persons as any of the states now existing shall think proper to admit, shall not be prohibited by the Congress prior to the year one thousand eight hundred and eight, but a tax or duty may be imposed on such importation, not exceeding ten dollars for each person." (H.G.).

8. References to the usage of property rights in regard to slavery by abolitionists prior to the Civil War.

9. Henry George was born in Philadelphia, PA in 1839.

10. The original chapter heading in "The Irish Land Question" is: "The Only Way, The Easy Way."

11. The "three F's" are Prime Minister William E. Gladstone's "fixity of tenure, fair rents, and free land sale" as part of the Land Act of 1881 that he pushed through Parliament.

12. Formatting has been changed by the editor.

13. "Social Statics," Chap. ix, sec. 8 (H.G.).

14. *Social Problems* was published in 1883 and *Progress and Poverty* in 1879.

15. The free-trade issue was important in nineteenth-century Britain. Major milestones were: endorsement by Adam Smith, repeal of the Corn Laws in 1846, and their reestablishment by William Gladstone in 1869.

16. The Stuart Charles II (1630–1685) reigned from 1660 until his death.

17. The Industrial Revolution once dated from about 1750 in British history, has more recently revealed evidences of itself in earlier modern times. Historians agree, however, on its eventual transformation of society from a predominantly agricultural to an industrial base. This change was fostered by new inventions in manufacturing, communications, and transportation. The Industrial Revolution in the United States took off after the Civil War with much the same patterns of change, although these elements appeared earlier.

18. Italicization and formatting of this phrase are by the editor.

19. A crofter is a tenant of a small piece of land who for the most part lived in poverty at the whim of a landlord. The Old English word croft originally meant an enclosed field.

20. Henry Peter, Baron Brougham and Vaux (1778–1868), was a noted British statesman, jurist, orator, scientist, and reformer.

21. I of course, use the word "rent" in its economic, not its common sense, and mean what is commonly called ground rent (H.G.).

22. John J. Astor (1763–1848) was born in Germany but made a fortune in the United States as a merchant and fur trader. At his death he was the richest man in the country. Some of his descendants, known for their wealth, were prominent in public affairs.

23. Louis XIV (1638–1715), known as the "Sun King," reigned from 1643 until his death and was succeeded by his great-grandson Louis XV (1710–1774). The tsar of

Russia at the time of this writing was probably Alexander III (1845–1894). He succeeded his father Alexander II (1818–1881), who was assassinated in March, 1881.

24. Henry George, "Scotland and Scotsmen" (Glasgow: "Land Values" Pubications Dept., 1884): ILD. This speech was given on Feb. 18, 1884.

25. The first time George visited Scotland was in 1882.

26. For John Bright see footnote no. 96, page 103.

27. Guano is the excrement of seabirds used for fertilizer.

28. The Tierra del Fuego (Land of Fire) is an archipelago at the southernmost part of South America with an extremely harsh climate. The Indians were quite poor. "Black fellows" refer to aborigines. Digger Indians included Indian peoples living on the far west of North America who dug up roots as part of their diet.

29. Many people in Europe and America, especially with a strong religious bent, believed industrialization and urbanization, which drastically changed rural life, to be detrimental. The newer forms of weaponry shocked some of them. Social Darwinism used principles from the natural sciences such as the doctrine of evolution and the struggle for the fittest to become a unifying factor of knowledge and by extension human affairs. Such an application was regarded by these critics with horror, for in its extreme forms, poverty, war, and suffering, could be justified.

30. Feu (or fee) has a number of definitions: an inherited estate of land without limitation to any particular class of heirs (fee simple) or limited to a class of heirs (fee tail); an estate or territory held by a feudal lord on condition of service, or; full ownership in land.

31. References to the rapacity of landlordism and the forced exodus of peasants.

32. Laird is a term used in Scotland for a landed proprietor.

33. I have not been able to identify Mr. Winans.

34. Trench here means to encroach upon.

35. A hook has a crescent-shaped blade with an inner cutting edge. A flail is a hand instrument used for threshing.

36. Probably a curved-shaped spade.

37. Kail is a spelling variation of kale.

38. John o'Groat's is the northernmost point of the United Kingdom, located in Scotland.

39. Buckie groat is a general term for a food grain, which is then stripped of the hull and crushed, and is larger than grits.

40. Tacksman has two definitions: A holder of a lease or a tack of land, tithes, watermill, fishery, or anything leased or farmed; or most probably here a middleman who has leased land from a proprietor and sublets it in small farms. Tack in this sense can mean a leasehold tenement, the tenure of tenancy of a piece of land, or the period of tenacy.

41. Royal and parliamentary commissions were used in Great Britain to investigate various abuses and problems, and to suggest methods of amelioration including the passage of laws.

42. A stirk is a cow or bull between one and two years of age.

43. A stone is equivalent to fourteen pounds.

44. A factor is either a person who transacts business for another (a commission merchant), or a lender of money to dealers or producers.

45. Nicodemus was a leading Pharisee who visited and later helped bury Jesus. See the Christian Scriptures, John 3.1–21, 7.50, 7.51, and 19.39–19.42. There is the Gospel of Nicodemus in the Pseudepigrapha.

46. Polander is an old term for an inhabitant of Poland.

47. National or royal imposts are known as taxes, but local levies as rates.

48. The highland regiments were first raised in 1725 as a military police force.

49. The thistle is the national symbol of Scotland.

50. See Henry George, "Moses" in Wenzer, *An Anthology of Henry George's Thought*, 14–25.

51. The eighth Duke of Argyll, George Douglas Campbell (1823–1900), vehemently opposed Henry George's political economy. His attack of 1884 entitled "The Prophet of San Francisco," which originally appeared in the *Nineteenth Century*, can be found in Henry George, *The Land Question and Related Writings*. George's rebuttal, the "Reduction to Iniquity," is also presented.

52. Charles II (1630–1685) reigned between 1660 and 1685.

53. Slavery in the British Isles was abolished in 1807 and slavery in the British West Indies in 1833.

54. Henry George, "Mr. Chamberlain Translated into Plain English: An Interview with Mr. Henry George," *Pall Mall Gazette* (Jan. 14, 1885): PDM.

55. Sir William Harcourt (1827–1904) was a noted English statesmen and leader of the Liberal Party.

56. The eighth commandment is: Thou shall not steal.

57. For a definition of radical see footnote no. 33, pages 302–303.

58. John Ball and Wat Tyler (both d.1381) were English rebel leaders of an unsuccessful popular movement and revolt protesting fixed wages and increased taxes during the reign of Richard II (1377–1399).

59. Patrick Edward Dove (1815–1873) was an English theoretician of the single tax and the author of *The Theory of Human Progression*. See chapter 6 by Julia Bastian in Wenzer, *An Anthology of Single Land Tax Thought*, pp. 159–202.

60. Sir George Campbell (1824–1892) was a noted administrator, author, and member of Parliament from 1875 until his death.

61. Richard Cobden (1804–1865) was an English statesman influential in diplomatic affairs and in the repeal of the Corn Laws, which relieved food prices by permitting a greater import of grain.

62. Henry George, "Henry George in Scotland," *The Standard* (May 25, 1889); GRD. The subcaption reads: "Glasgow, May 11, 1889." Capitalization has been added in the appropriate places.

63. D. C. MacDonald, a land reformer, edited and wrote the biographical notes and preface to William Ogilvie's *Birthright in Land: An Essay on the Right of Property in Land* (London: Kegan Paul, Trench, Trübner & Co.,1891; reprint; New York: Augustus M. Kelley, 1970). See also chapter 4 in Wenzer, *An Anthology of Single Land Tax Thought*, pp. 73–132.

64. *Quoad sacra*: As regards to the holy things.

65. The Reformation traditionally began with Martin Luther's nailing of the ninety-five theses on a church door in Wittenburg on Oct. 31, 1517.

66. About a third of the prelates and many Scots left the Scottish Presbyterian Established Church in 1843 to form the Free Church over patronage and other issues.

67. Isaac Watts (1674–1748) was an English hymn writer and a dissenting clergyman. This line comes from his "Against Pride in Clothes" which appeared in the 1715 work, the *Divine Songs for Children*.

68. A glebe is land yielding revenue or belonging to a parish church or any ecclesiastical office.

69. A manse is the residence of a clergyman.

70. A gillie can be either an attendant on a Highland chief or a hunting and fishing guide.

71. The Latin term *fera naturæ* means of a wild nature, or not domesticated.

72. "Address of Welcome to Mr. Henry George from the Dundee and District United Trades' Council," *The Standard* (May 25, 1889): GRD.

73. George, *Progress and Poverty*, p. 296.

74. Ibid., p. 136.

75. Six names have been deleted by the editor.

76. Henry George, "Henry George Preaches in Dundee," *The Standard* (June 1, 1889); reprinted from the *Dundee Advertiser* (May 6, 1889): GRD.

77. Form here is a seat or bench.

78. In the Hebrew Scriptures, Amos 8: 4, Isaiah 3:15, and Isaiah 3:14.

79. These lines could be a rewording from the Hebrew Scriptures, Psalm 119:1.

80. In the Hebrew Scriptures, Psalm 37: 25.

81. "Great Father at Washington" refers to the president.

82. Henry George, "Address by Henry George in City Hall, Glasgow on August 20, 1890;" #9, HGP.

83. Land-value taxation was introduced in Australia in the early 1870s. For more information consult Wenzer, "The Degeneration of the Georgist Movement: From a Philosophy of Freedom to a Nickel and Dime Scramble," *The Forgotten Legacy of Henry George*, pp. 71–72.

84. The McKinley Tariff Act of 1890, sponsored by the future president William McKinley (1843–1901), was highly protective and pushed rates up to their highest levels at this time.

85. The Land Purchase Bill probably refers to the extension of land purchases in 1891.

86. Italicization added by the editor.

87. Purchase Scheme probably refers to the purchase clauses of the Land Act of 1881 and its subsequent extension through other legislation in the 1880s.

88. See pages 263–281 for George's ardent anti-compensation stance in a speech.

89. For the Docker's strike see pages 238–249.

90. Possibly Sidney C. Buxton, first earl of Buxton (1853–1934), who was at this time a member of Parliament from the London borough of Poplar.

91. The Salvation Army was founded in London in 1865 by William Booth (1829–1912).

92. The Anti-Poverty Society, founded by George's follower and critic, Reverend Edward McGlynn, was established in 1887 and lasted for about a year. See footnote no. 92, pages 102–103.

SECTION III: ENGLAND

PROGRESS AND POVERTY: AN EDITORIAL

(John Milne, London *Times*, September 14, 1882).[1]

Mr. Henry George, having left the United States to study the Irish land question on the spot, was twice arrested by the Irish police as a suspicious character. He has no reason to regret the temporary inconvenience which he suffered. He has been laboring for some time, but with little success, to instill certain ideas into the minds of the English-reading public. Two years ago he published a work entitled *Progress and Poverty*, wherein many accepted economic doctrines were pronounced erroneous and several novel doctrines were propounded with considerable show of plausibility and in a fascinating style. Hitherto, Mr. George has not been regarded in his own country as a prophet. Now, however, he has attracted attention both among our countrymen and his own, and both are curious to learn more about himself and his writings.

Before examining his larger work, we shall devote some space to his pamphlet on Ireland.[2] Unlike other pamphlets which have been written about the Irish land question, this one is antagonistic to Irish landlords and tenants alike. Mr. George holds that both are in the wrong. He puts forth a view of the case which has not received due recognition. Among the wild projects for rendering Ireland prosperous and contented we may include one for annexing it to the United States. Agitators would have the ignorant believe that, if this were effected, tenants' wrongs would all be righted and Irish landlordism would cease to be a curse. Those persons who speak or write in this fashion have probably omitted to inquire what would really be the landlords' loss and the tenants' gain if Ireland became a state in the North American Union and were emancipated from what is styled British domination and oppression. Mr. George asks this question, and he answers it in the following terms:

> The truth is that the gain would be to the landlords, the loss to the tenants. The simple truth is that, under our laws, the Irish landlords could rack-rent, and without any restriction from Ulster tenant-right or legal requirement of compensation for improvements. Under our laws they could, just as freely as they can now, impose whatever terms they please upon their tenants – whether as to cultivation, as to improvements, as to game, as to marriages, as to voting, or as to anything else. For these powers do not spring from special laws. They are merely incident to the right of property; they result simply from the acknowledgment of the right of the owner of the land to do as he pleases with his own – to let it or not to let it. So far as law can give them to him, every American landlord has these powers as fully as any Irish landlord. Cannot the American owner of land make, in letting it, any stipulation he pleases as to how it shall be used, or improved, or cultivated? Can he not reserve any of his own rights upon it, such as the right of entry, or of cutting wood, or shooting game, or catching fish? And, in the absence of special agreement, does not American law give him, what, as I understand it, the law of Ireland does not now give him, the ownership at the expiration of the lease of all the improvements made by the tenant?

What single power has the Irish landowner that the American landowner has not as fully? Is not the American landlord just as free as is the Irish landlord to refuse to rent his lands or his houses to anyone who does not attend a certain church or vote a certain ticket? Is he not quite as free to do this as he is free to refuse his contributions to all but one particular benevolent society or political committee? Or if, not liking a certain newspaper, he chooses to give notice to quit any tenant whom he finds taking that newspaper, what law can be invoked to prevent him? There is none. The property is his, and he can let it or not let it, as he wills. And, having this power to let or not let, he has power to demand any terms he pleases.[3]

Mr. George summarily disposes of the objection that, Ireland being a conquered country and the land there having been confiscated, the titles to the soil are bad, by saying that this has nothing more to do with the question of the hour than the confiscations of Marius and Sylla.[4] He remarks that England is also a conquered country and that her soil has been repeatedly confiscated, and he contends that, despite "all talk about the Saxon oppressor and the downtrodden Celt, it is not probable that, after the admixture of generations, the division of landholder and nonlandholder any more coincides with distinction of race in the one country than in the other." He admits that, in his own peacefully settled country, the titles to land often rest on force and fraud, saying: "Even in our most recently settled states, how much is there to which title has been got by fraud and perjury and bribery – by the arts of the lobbyist or the cunning tricks of hired lawyers, by double-barrelled shotguns and repeating rifles!"

One of the Irish grievances is that many landlords are absentees. In truly Hibernian phrase, a Home Ruler recently complained that Ireland "swarmed with absentee landlords." Mr. George points out that several English and Scottish landholders never live on their estates, and that absenteeism is as common in the United States as in any other country:

In New York, in San Francisco, in Washington, Boston, Chicago, and St. Louis, live men who own large tracts of land which they seldom or never see. A resident of Rochester is said to own no less than 400 farms in different states, one of which (I believe in Kentucky) comprises 35,000 acres. Under the plantation system of farming and that of stockraising on a grand scale, which are developing so rapidly in our new states, very much of the profits go to professional men and capitalists who live in distant cities. Corporations whose stock is held in the East or in Europe own much greater bodies of land, at much greater distances, than do the London Corporations possessing landed estates in Ireland. To say nothing of the great land-grant railroad companies, the Standard Oil Company probably owns more acres of Western land than all the London companies put together own of Irish land. And, although landlordism in its grosser forms is only beginning in the United States, there is probably no American, wherever he may live, who cannot in his immediate vicinity see some instance of absentee landlordism.[5]

While Mr. George maintains that the land system in Ireland is neither exceptional nor in any respect less advantageous than that of the United States, he also maintains that there is nothing exceptional or specially deplorable in the fact that wholesale emigration is one of its results. When continued emigration is proposed as a relief for Irish suffering, the remedy is often referred to as a harsh and intolerable one. Mr. George reminds those persons who take this view that no special commiseration is awarded to the Scots, the English, the Germans, the Italians, and the Scandinavians when, like the poor Irish, they cross the Atlantic in order to better their condition. Moreover, he notes that in the United States there is an emigration from the Eastern to the Western states caused by the same motives which cause the emigration from Ireland. Respecting the last famine in Ireland, he makes some pertinent remarks. That it was severe he freely admits, but that it was unparalleled he denies. On the contrary, he holds that quite as serious distress has prevailed, during the last few years, in many other countries, and even in the United States.[6]

Perhaps none of his statements or conclusions is more startling than that which concerns one of Mr. Parnell's remedies for Irish discontent. This is the establishment of manufactures in Ireland with a view to diminish the competition for land. The opposite result would take place. As Mr. George phrases it, "put a Glasgow, a Manchester, or London in one of the Irish agricultural counties, and, where the landlords now take pounds in rent, they would be enabled to demand hundreds and thousands of pounds." For other proposals made with a view to pacify Ireland, he has some sharp criticism. To establish tenant-right, he says, is merely to carve out of the landlord's estate an estate for the tenant. To fix a "fair rent" by the decision of a legal tribunal is to lessen the landlord's income and enable the tenant to make his successor pay more, so that neither future tenants nor the mass of the people can reap any advantage. To create a peasant proprietary is to run counter, in his opinion, to the tendency of the times, which is in favor of concentrating much land in few hands in order that it may be cultivated in the most remunerative way. He affirms that "peasant proprietorship in Europe is a survival."

Mr. George is most emphatic in censuring the Land Leaguers who, in the United States and Ireland, rail at England and abuse Englishmen for the manner in which Ireland is governed. He justly says: "To denounce Englishmen is simply to arouse the prejudices and excite the animosities of Englishmen – to separate forces that ought to be united. To make this the fight of the Irish people against the English people is to doom it to failure." He holds that to talk about separation from England is folly on the part of the Irish people, being certain "to arouse passions that will be utilized by the worst enemies of Ireland."

How far Mr. George's "appeal to the Land Leagues" will prove effective we cannot tell. We gladly recognize the large amount of sound sense which his appeal contains and we should be still more glad if his appeal bore good fruit. It is true that the remedy proposed by Mr. George for the evils which he depicts and deplores is open to serious question, but it is one which can be discussed without special reference to Ireland. Indeed, it is quite as applicable, according to Mr. George, to all other countries, while nowhere is it more needed than in the United States. We postpone stating the nature of the remedy till we have examined the work *Progress and Poverty* in which the necessity for it is explained in detail.

Mr. George's book has the merit of being a fearless attempt to solve the problem of modern civilization. He explains in his concluding chapter that he entered upon the inquiry without any theory to support or conclusion to verify. When he first beheld the squalid misery of a great city, the spectacle appalled and tormented him, and he could not rest for thinking "of what caused it and how it could be cured." He learned, as he states in the introduction, that, where the conditions under which material progress makes the most rapid strides – where population is densest, wealth greatest, the machinery of production and exchange most highly developed – there is to be found "the deepest poverty, the sharpest struggle for existence, and the most enforced idleness." Again, he says that "the 'tramp' comes with the locomotive, and almshouses and prisons and are as surely the marks of 'material progress' as are costly dwellings, rich warehouses, and magnificent churches. Upon streets lighted with gas and patrolled by uniformed policemen, beggars wait for the passerby, and in the shadow of college, and library, and museum, are gathering the more hideous Huns and fiercer Vandals of whom Macaulay prophesied."[7]

It is of the great and prosperous cities of the United States as much as of the cities of the Old World he writes when stating that "the association of poverty with progress is the great enigma of our times." He does not consider the universal adoption of free trade to be the one thing needful. He is too clear-sighted to be a protectionist. Indeed, he writes with emphasis about "the evident inconsistencies and absurdities" which cluster round "the fallacies of protection." Yet he also perceives that "free trade had enormously increased the wealth of Great Britain, without lessening pauperism," and he believes that his theory "cuts the last inch of ground from under the fallacies of protection, while showing why free trade fails to permanently benefit the working classes."

In order to conduct his argument to the desired conclusion, Mr. George takes nothing for granted. He discusses, and sometimes he controverts, many accepted doctrines of political economy. He does not do this, like his countryman Mr. Carey,[8] because he thinks that the United States ought to have a political economy as well as a Constitution of their own. On the contrary, he repudiates the article of faith held by many of his countrymen in common with some Liberals in Europe, "that the poverty of the downtrodden masses of the Old World is due to aristocratic and monarchical institutions," and he acknowledges that "social distress of the same kind, if not of the same intensity as that prevailing in Europe," exists under Republican institutions in the United States.

He is of [the] opinion, however, that the standard writers on political economy have erred in many of their conclusions, because they lived and wrote in a state of society where a capitalist generally rents land and hires labor, and thus they were led to regard "capital as the prime factor in production, land as its instrument and labor as its agent or tool." Living and observing what takes place in a new country, he maintains that land is the first thing needful, that capital is the result of labor, that "labor is the active and initial force, and labor is, therefore, the employer of capital." His contention is that "the produce of labor constitutes the natural recompense or wages of labor." He says that earning is making; that whenever labor is exchanged for commodities production really precedes enjoyment; that when a laborer receives his wages in money "he really receives, in return for the addition his wealth has made to the general stock, which he may utilize in any particular form of wealth that will best satisfy his desires; and that neither the money, which is but the draft, nor the particular form of wealth which he uses it to call for, represents advances of capital for his maintenance, but, on the contrary, represents wealth, or a portion of the wealth, his labor has already added to the general stock." Having set forth the above principles, he then, as his manner is, illustrates his argument in the following picturesque fashion:

> Keeping these principles in view, we see that the draftsman who, shut up in a dingy office on the banks of the Thames, is drawing the plans for a great marine engine is in reality devoting his labor to the production of bread and meat as truly as though he were garnering the grain in California or swinging a lariat on a La Plata pampa; that he is as truly making his own clothing as though he were shearing sheep in Australia or weaving cloth in Paisley; and just as effectually producing the claret he drinks at dinner as though he gathered the grapes on the banks of the Garonne. The miner who, 2,000 feet underground in the heart of the Comstock, is digging out silver ore, is in effect, by virtue of a thousand exchanges, harvesting crops in valleys 5,000 feet nearer the earth's center; chasing the whale through Arctic icefields, plucking tobacco leaves in Virginia, picking coffee berries in Honduras, cutting sugarcane on the Hawaiian Islands, gathering cotton in Georgia, or weaving it in Manchester or Lowell, making quaint wooden toys for his children in the Hartz Mountains, or plucking amid the green and gold of Los Angeles orchards the oranges which, when his shift is relieved, he will take home to his sick wife. The wages which he receives on Saturday night at the mouth of the shaft, what are they but a certificate to all the world that he has done these things – the primary exchange in the long series which transmutes his labor into the things he has really been laboring for?"[9]

That capital does aid industry by supplying it with improved tools, and that some wages are drawn from capital, Mr. George expressly admits. But he points out that the usefulness of capital in employing labor and increasing wealth depends upon the surrounding conditions. Ignorance, or the tyranny of antiquated customs, may render the aid of capital unavailing. He is certain that giving a circular saw to a [Terra del] Fuegian, a locomotive to a Bedouin Arab, or a sewing machine to Flathead squaw would not add to the efficiency of their labor. Nor does the lack of capital explain why, in young communities, simple modes of production and exchange prevail. For instance, the cloth required for a little village can be made with less labor by the spinning wheel and handloom than in a great factory:

> A perfecting press will, for each man required, print many thousand impressions while a man and a boy would be printing a hundred with a Stanhope or Franklin press; yet to work off the small edition of a country newspaper the old-fashioned press is by far the most efficient machine. To occasionally carry two or three passengers a canoe is a better instrument than a steamboat; a few sacks of flour can be transported with less expenditure of labor by a pack-horse than by a railroad train; to put a great stock of goods into a crossroads store in the back-woods would be but to waste capital. And generally it will be found that the rude devices of production and exchange which obtain among the sparse populations of new countries result not so much from the want of capital as from inability to practically employ it.[10]

Having arrived at the conclusion that wages are drawn, not from capital, but from the produce of labor, and that in such a case wages cannot be diminished by the increase of laborers, Mr. George proceeds to answer the all-important question: "Do the productive powers of Nature tend to diminish with the increasing drafts made upon them by increasing population?"

The answer to the foregoing question involves an examination of the Malthusian theory. The essence of that theory is that there is a constant tendency for human life, as for all animal life, to increase beyond the means of subsistence available for it, so that the increasing pressure of population tends to render subsistence more and more difficult. Malthus[11] himself argued that population increases in a geometrical ratio; hence, by a series of calculations, he came to the conclusion that in two centuries the population would be to the means of subsistence as 256 to 9; in three centuries, 4,096 to 13; and in 2,000 years the difference would be almost incalculable. This consequence has seemed so appalling as to have apparently hindered inquirers from testing the validity of the assumption on which it is based. Mr. George denies the truth of the assumption, and he declares it equivalent to asserting that, because a puppy doubled the length of his tail while he added some many pounds to his weight, therefore there was a geometrical progression of tail and an arithmetical progression in weight adding:

the inference (drawn by Malthus) from the assumption is just what Swift in satire might have credited to the savants of a previously dogless island, who, by bringing these two ratios together, might deduce the very "striking consequence" that by the time the dog grew to a weight of fifty pounds his tail would be over a mile long, and extremely difficult to wag, and hence recommend the prudential check of a bandage as the only alternative to the positive check of constant amputations.[12]

In another chapter, he recurs to the subject, and pronounces the assumption of Malthus quite as grotesque as if Adam had estimated the future growth of his first baby in accordance with the observations made during the early months of its life. Adam might have argued from the fact that as at birth it weighed ten pounds and twenty pounds eight months afterwards, it would be as heavy as an ox at the age of ten, as heavy as an elephant at the age of twelve, and that it would weigh no less than 175,716,339,648 tons when it attained the age of thirty. He sees no more reason for our troubling ourselves about the pressure of population upon the means of subsistence than there was for Adam worrying himself about the growth of his first baby.

Mr. George asks why it is that in no considerable country poverty and want are exclusively attributable to excessive population, why the globe, which is millions of years old, is still so thinly populated? We count the increase in population without remembering how great has been the decrease in historic times. Large tracts are now desolate which were once crowded with inhabitants; wild beasts lick their cubs among the ruins of vast and populous cities. In Peru alone there was once a population greater, probably, than that of the whole of South America at the present day. Southern Europe, Africa, and Asia were at one time more thickly populated than at present. The most marked change has occurred in Asia Minor, Syria, Babylonia, and Persia, where squalid villages and barren wastes are now found where used to be great cities and a teeming population.

Mr. George thinks it somewhat strange that among all the theories put forward, that of a fixed quantity of human life on the globe has not been broached. It would better reconcile and explain some puzzling facts than the allegation that the constant tendency of population is to overrun the means of subsistence. Hitherto, the principle of population has not been strong enough to settle the world, because it is yet sparsely peopled when compared with its capacity for sustaining human life. A broad fact must strike anyone, Mr. George thinks, who extends his view beyond modern society. This is that as Malthusianism predicates as a universal law the natural tendency of population to outrun the means of subsistence, this law must become obvious wherever population has attained a certain density. He asks:

How is it, then, that neither in classical creeds and codes, nor in those of the Jews, the Egyptians, the Hindus, the Chinese, nor any of the peoples who have lived in close association and built up creeds and codes, do we find any injunctions to the practice of the prudential restraints of Malthus; but that on the contrary, the wisdom of the centuries, the religions of the world, have always inculcated ideas of civic and religious duty the very reverse of those which the current political economy enjoins, and which Annie Besant is now trying to popularize in England![13]

The essence of the Malthusian doctrine is that population presses against the limits of subsistence. How is it, then, Mr. George says, that families in which want is not merely unknown, but every luxury enjoyed so often die out? A practical example of the rule in which a family increased under the most favorable conditions, is to be found in unchangeable China. There the descendants of Confucius form the only hereditary nobility and enjoy peculiar privileges and consideration.[14] If the hypothesis of Malthus were justified by results that family, after the lapse of 2,150 years, ought to number 859,559,193,106,709, 670,198,710,528 souls. Instead, however, of any such astounding number, the descendants were but 11,000 males, or say 22,000 souls, in the reign of Kanghi – that is, 2,150 years after death of Confucius.[15] Now the position held by this family prevented the operation of the positive check referred to by Malthus, while his preventive check was in direct opposition to Confucius' teaching and precepts. The increase of this single family is great, yet this increase does not imply a vast addition to population. The web of generations is, as Mr. George puts it

like latticework or the diagonal threads in cloth. Commencing at any point at the top, the eye follows lines which at the bottom widely diverge; but beginning at any point at the bottom, the lines diverge in the same way to the top. How many children a man may have is problematical. But that he had two parents is certain, and that these again had two parents each is also certain. Follow this geometrical progression through a few generation, and see if it does not lead to quite as "striking consequences" as Mr. Malthus' peopling of the solar systems.[16]

It is customary to point to Ireland as a confirmation of Malthus' theory. There the poverty of the people is said to be due to their excessive number. Writing in 1838, when the population of Ireland was about eight millions, M'Culloch said that only half that number could be maintained there in a moderate state of comfort.[17] Mr. George asks what, then, ought to have been the condition of the people when the population numbered two millions only? Yet when the similar numbers inhabited Ireland, the poverty was so great, the surplus population so large, that Swift penned his scathing satire about eating the redundant babies.[18] "Had Ireland been by nature a grove of bananas and breadfruit, had her coasts been lined by the guano deposits of the Chinchas, and the sun of lower latitudes warmed into more abundant life her moist soil, the social condi-

tions that have prevailed ere would still have brought forth poverty and starvation."[19] He concludes his examination of established facts by denying that,
in any instance on record, increase in numbers has decreased the relative production of food, and by expressing his conviction that the famines in India, China,
and Ireland are as little due to overpopulation as the famines in sparsely peopled
Brazil.

He finds the inferences from analogy as weak as the facts adduced in support
of the Malthusian theory. No valid comparison can be drawn between the
strength of the reproductive force in the animal and vegetable kingdoms and
that of the human species. Animals increase at the expense of their food, while
the increase of man involves the increase of his food. When a seal eats a salmon,
there is a salmon the less; but man by placing salmon spawn under favorable
conditions can provide more salmon than he eats. Had bears instead of men
been shipped from Europe to North America, the number of bears there would
depend upon the amount of bear food obtainable, the probability being that the
bear immigration would lessen the total number of bears, owing to the supply
of their food not being equal to the sudden demand.

There are now upwards of 50,000,000 of people in North America instead
of the few hundred thousand who inhabited that Continent when the Pilgrim
Fathers landed there, yet the supply of food for each of the 50,000,000 is
infinitely in excess of that for the few hundred thousand. The supply of food
is larger, simply because the population is greater. "In short, while all through
the vegetable and animal kingdoms the limit of subsistence is independent of
the thing subsisted, with man the limit of his subsistence is, within the final
limits of earth, air, water, and sunshine, dependent upon man himself."

Nor is there any force in the allegation that land diminishes in productive-
ness through cultivation. Though true of a particular piece of land, this is not
true of the area of the earth's surface. Man neither creates nor annihilates; what
he takes from the land he restores to it again; life does not use up the forces
that maintain it; therefore, the limit to the population of the globe can only be
the limit of space. While asserting this, Mr. George regards the limitation of
space, the danger of the human race not finding elbow room, as "so far off as
to have for us no more practical interest than the recurrence of the glacial period
or the final extinguishment of the sun."

Mr. George goes the length of directly contradicting the theory that the
increase of population tends to reduce wages and deepen poverty in the way
stated by Malthus, and reaffirmed by John Stuart Mill. He asserts that if poverty
increases along with population, it is because social arrangements are unjust
and not because Nature is niggardly. He also asserts that new mouths require
no more food than the old ones, "while the hands they bring with them can in

the natural order of things produce more." The question which he would submit to the test of fact is: "in what stage of production is there exhibited the greatest power of producing wealth? For the power of producing wealth in any form is the power of producing subsistence – and the consumption of wealth in any form, or of wealth-producing power, is equivalent to the consumption of subsistence?"[20]

He instances California as a region where the increase of wealth has been in proportion to the addition to the population. Wages there are far lower than they were in 1849; but, then, the increase in the power of the human factor, owing to the greater efficiency of labor through the roads, railways, and machinery now available, have more than compensated for the decline in the power of the natural factor. Facts show that in California the consumption of wealth is now much greater, compared with the number of laborers, than it was formerly:

> Instead of a population composed almost exclusively of men in the prime of life, a large proportion of women and children are now supported, and other nonproducers have increased in much greater ratio than the population; luxury has grown far more than wages have fallen; where the best houses were cloth and paper shanties, are now mansions whose magnificence rivals European palaces; there are liveried carriages in the streets of San Francisco and pleasure yachts in her bay; the class who can live sumptuously on their incomes has steadily grown; there are rich men beside whom the richest of the earlier years would seem little better than paupers – in short, there are on every hand the most striking and conclusive evidences that the production and consumption of wealth have increased with even greater rapidity than the increase of population, and that if any class obtain less it is solely because of the greater inequality of distribution.[21]

He argues that what is true of California is true of other regions also, and that the richest countries "are not those where Nature is most prolific, but those where labor is most efficient – not Mexico, but Massachusetts; not Brazil, but England." He contends that, in all the cases hitherto seen, an increase of population "does not mean a reduction, but an increase in the average production of wealth." He says that twenty men working together where Nature is niggardly will produce twenty times the wealth that one man can where Nature is most bountiful, adding

> The denser the population, the more minute becomes the subdivision of labor, the greater the economies of production and distribution, and hence, the very reverse of the Malthusian doctrine is true; and, within the limits which we have any reason to suppose increase would still go on, in any given state of civilization, a greater number of people can produce a larger proportionate amount of wealth, and more fully supply their wants than can a smaller number.[22]

Mr. George repeats that the enigma which perplexes the civilized world, the appearance of want where productive power is greatest and the production of wealth largest, is not solved by the Malthusian theory, according to which want is caused by decrease in the productive power.

According to Mr. George, the real law of population is that the tendency of the human species to multiply is not always uniform, but "is strong where a greater population would give increased comfort and where the perpetuity of the race is threatened by the mortality induced by adverse conditions; but weakens just as the higher development of the individual becomes possible and the perpetuity of the race is assured." Having satisfied himself that it is the unequal distribution of wealth; and not the superabundance of mouths to fill which condemns men to want in the midst of wealth, he puts forth his remedy for the evil which he deplores.

He first reviews the laws according to which property is distributed and finds that the three factors of production are land, labor, and capital; the term "land" includes all natural opportunities for forces, "labor" all human exertion, and "capital" all wealth used to create more wealth. The whole produce is distributed in returns to these three factors: "that part which goes to the landowners as payment for the use of natural opportunities is called rent; that part which constitutes the reward of human exertion is called wages; and that part which constitutes the return for the use of capital is called interest."

He considers rent to be "the price of monopoly, arising from the reduction to individual ownership of natural elements which human exertion can neither produce nor increase." He holds also that wages and interest do not depend upon the produce of capital, but upon what is left after rent had been deducted. Now, as three things unite in causing production, labor, capital, and land, three parties share the produce, the laborer, the capitalist, and the landowner. "If, with an increase of production, the laborer gets no more and the capitalist no more, it is a necessary inference that the landowner reaps the whole gain." As, then, the value of land depends upon the power of its owner to appropriate wealth created by labor, the increase in the value of land is at the expense of labor, hence the reason why "the increase in productive power does not increase wages is because it does not increase the value of land. Rent swallows up the whole gain and pauperism accompanies progress." Mr. George thinks everyone must have observed that the contrast between wealth and want appears as land rises in value and that, where its value is highest, civilization exhibits the greatest luxury side-by-side with the most piteous destitution. In proof of this, he adds: "To see human beings in the most abject, the most helpless and hopeless condition, you must go, not to the unfenced prairies and the log cabins on new clearings in the backwoods, where man singlehanded is commencing the struggle with Nature, and land is yet worth nothing, but to the great cities, where the ownership of a little patch of ground is a fortune."[23]

Mr. George's remedy, then, for the ills of modern society is "to make land common property," to have no other landlord than the State. The rent received by the State would be a substitute for all taxes and the increasing rent would enrich the State as a whole instead of private persons. The surplus, after paying the expenses of government, would be distributed among the community in proportion to individual needs. As an alternative to making the State, the sole landlord, he would leave landlords in possession and levy a tax on land equivalent to its value in rent. This, he thinks, would open new opportunities, as, under such a system, "no one would care to hold land unless to use it, and land now withheld from use would everywhere be thrown open to improvement. The selling price of land would fall; land speculation would receive its deathblow; land monopolization would no longer pay."

We have stated Mr. George's views as much as possible in his own words. Though startling, they are not always so novel as may be supposed. In the writings of Mr. Herbert Spencer some of Mr. George's strictures on Malthus and some of his views about the nationalization of land are foreshadowed, if not formally anticipated. It is noteworthy that the necessity for the drastic reform which he proposes is admitted by him to be as pressing in the United Sates as anywhere else. He holds that his own country is quite as full of the elements of disturbance as any country in the Old World, owing to inequality in the distribution of wealth.

Indeed, the calm and thoughtful observer of the marvelous advance made by the United States must be struck with the increasing tendency there towards the division of the people into two classes, the smaller one being wholly composed of persons rich beyond the dreams of avarice and adding daily to their enormous wealth, the larger one being largely composed of persons who daily find it harder to live and who are sinking into the condition of paupers.

Mr. George takes a gloomy view of the condition and prospects of his own country, where he thinks that the transformation of popular government into a vile despotism has already begun. He considers it an evidence of political decline that legislative bodies are deteriorating in standard, the men of ability and higher character eschew politics, that the arts of the jobber are more influential than the reputation of the statesman, that voting is done recklessly, and that the use of money to control politics is increasing. He points out that the ruling class in great American cities is composed of "gamblers, saloonkeepers, pugilists, or worse, who have made a trade of controlling votes and of buying and selling offices and official acts." He holds that, though his countrymen are intense democrats in theory, there is growing up among them a class having the power without the virtues of an aristocracy. "We have simple citizens who control thousands of miles of railroad, millions of acres of land, the means of livelihood of great numbers of men; who name the Governors of sovereign states as they name their clerks, choose Senators as they choose attorneys, and whose will is as supreme with Legislatures as that of a French king sitting in [the] bed of justice."[24]

Still worse than this, in the opinion of Mr. George, is the prevalence of corruption and the excessive power of the purse, while the most ominous political sign in the United States "is the growth of sentiment which either doubts the existence of an honest man in the public office or looks on him as a fool for not seizing his opportunities. That is to say, the people themselves are becoming corrupted. Thus, in the United States today is Republican government running the course it must inevitably follow under conditions which cause the unequal distribution of wealth."[25]

If Mr. George's remedy for the evils which, in his judgment, are working mischief alike in the Old and New World were certain to prove efficacious, he would deserve to be hailed as a benefactor of his species. But his scheme has drawbacks, which he overlooks or underestimates. Indeed, we must regard all these schemes for regenerating mankind much in the same way as Coleridge did ghosts, having seen too many to believe implicitly in any single one.[26] From the days of St. Thomas More to those of Auguste Comte, from the gathering together of transcendentalists at Brook Farm to the formation of the Oneida Community, innumerable plans have been proposed for rendering earth a paradise;[27] yet the earth still continues much the same as it was after the expulsion of Adam and Eve from Eden.

Every proposal for restoring the world to its pristine perfection appears exceedingly attractive and quite feasible on paper; but, when put to the test, unforeseen obstacles mar its success and it collapses as quickly and completely as the many-hued bubbles which look so lovely, and are so evanescent. Mr. George's ideal will long be found in his book only; nevertheless, *Progress and Poverty* well merits perusal. It contains many shrewd suggestions and some criticism of economic doctrines which future writers on political economy must either refute or accept. Mr. George's reading has evidently been wide; he has reflected deeply; he is an acute reasoner, and he is the master of an excellent style. The readers of his book may dissent from his statements and conclusions without regretting the time they have spent over it, and, if conversant with economic doctrines and interested in the problems of social science, they will find in its pages much to ponder with care and much that is highly suggestive.

MR. HENRY GEORGE ON THE PROGRESS
OF HIS PRINCIPLES

(Pall Mall Gazette, January 7, 1884).[28]

Mr. Henry George is back in England in first-rate spirits and ready for the battle. His campaign against the landlords is to begin in St. James' Hall next Wednesday, and he is engaged to lecture at a number of the large provincial towns. His visit to England will last for about three months. With regard to the object of that visit and to the political and social situation generally Mr. George has just made some remarks to a member of our staff which may be interesting to the public. It is only fair to preface them with the observation that in a hasty conversation ranging over a great variety of subjects it would have been unreasonable to expect from Mr. George any such clear and connected statement of his theories and policy as he will doubtless put before his hearers in the lectures he is about to deliver. Mr. George, though his business is not to apply opinion but to make it, nevertheless keeps a sharp eye upon current political events. We have been moving fast in this country, he thinks, since he was last over here.

> There is a wonderful growth of opinion. One indication is the progress made by Mr. Chamberlain. He seems to be a man to go as far as public opinion will follow him, one who will stick at nothing, but who will not run his head against a stone wall. He is not a visionary, but a politician in the true sense of the word. I do not think there is another instance in your country of a man bringing so much business ability and experience to the highest walks of politics.

With regard to the prospects of his propaganda in America Mr. George was exceedingly sanguine. Hopefulness and cheeriness indeed are the characteristics which most strike one about Mr. George just now. Though he still has a wild glint in his eye at times, he is less "dour" by far than he seemed when he was last over here. Perhaps the immense success of his book, and the recognition which he has received in so many quarters, sufficiently account for his good humor. As for the future of land nationalization in the United States he entirely scouts the idea that its success is excluded by the number of small properties. The land question in America is not different from the land question in England, but only in an earlier stage.

> We have a large number of small proprietors, I know; but so had you only a couple of centuries back. But they are beginning to disappear with us, as they disappeared with you, and from the same causes. Only with us things will go faster. As material progress goes on the value of land rises. It is the most secure possession and a visible sign of wealth. The rich can afford to give more for it than the poor, and the poor man finds he can do better by selling his land and investing the proceeds in some other way. That is the story

of concentration all the world over. It is going on in America now at a great rate. There are nominally some 4,000,000['s] of farmers owning their own lands, but the figures are delusive. Not as much as one-half of these are really owners. A great number of them are heavily mortgaged, and the mortgagee is the real owner. The mortgage is the prevalent American form of tenancy.

But while convinced that time was fighting on his side, and that the course of economic development in America would of itself convert men to his doctrines, Mr. George confessed that they had not yet much hold upon politicians. "It is not time yet for the politicians. The Socialistic movement in America is growing rapidly, but it is still below the surface." "But with all your talk about Socialism," it was objected, "you have not been able to put a single spoke in the wheel of your Goulds and Vanderbilts."[29] "Jay Gould and Vanderbilt!" exclaimed Mr. George, "I wouldn't stop them for the world; they are the best propagandists we've got. And so are your Englishmen who come over and buy millions of acres of land in our country. I wish we could have all your House of Lords over in America to do the same." The idea that the spectacle of men owning whole counties will quicken revolt against private ownership in land altogether, is very strong in Mr. George's mind. For the rest he is convinced that the new movement, though not yet in the political stage, will soon reach it. The fight will begin about the tariff question. That question only came up in the last Presidential election, but the next will turn on it.[30]

"And what about the railways?"

I am in favor of making them State property – worked by the State – as soon as ever we can. I do not fear the jobbery which would result from such a course. Civil Service reform would be compelled, and under public management the railroads would cause less political corruption than at present. The jobbery in our postal system is as to out-of-the-way places; in the main it is as well managed as yours. The people will not stand jobbery in any public service which nearly touches their daily lives. And no jobbery could be worse than what goes on at present.

On the latter subject, indeed, Mr. George is very eloquent, and not only with regard to railways. He has a story to tell, for instance, about the Western Union Telegraph Company, and how it ruined a democratic paper of which he knew something, and that by a direct breach of contract. There might have been redress, no doubt, he thought, if the case had been taken before a court of law, but to do so, and to fight it over and over again in all the costly appeals which the wealthy corporation was certain to resort to, would have required a fortune.

It will be seen that Mr. George occasionally spoke as if he included himself among the Socialists. In the strict sense of the term, however, he denies that he is a Socialist, though he believes, of course, that he is in possession of the better way to realize the Socialist ideals. The English Socialist, however (Mr.

Hyndman for instance), are inclined, he thinks, to emphasize their differences with him, and one of them even asserted that, as against Mr. George, he would "shoulder his musket in defense of the landowners." Mr. George's doctrines are commonly reported to have been described by Karl Marx or one of his chief disciples as the last ditch of the capitalist class.[31] It will be interesting to observe the attitude adopted by the Democratic Federation to their ambiguous ally during his present lecturing trip. Mr. George himself evidently expects considerable results from it.

Landlordism, he thinks, is coming down, and the sooner the better, for it is a bad thing. Our reporter, though sympathizing to a great extent with Mr. George, happens to be strongly impressed by the immense strength of the established order in this country, and has his own opinion about the likelihood of landlordism, or capitalism, or anything else coming down with a run. But Mr. George could not be shaken in his faith in the instability of the existing social fabric in England. "Of the times and seasons," he thinks, "knoweth no man." An institution may stand for a long time after it is ready to fall, but you can never tell when seeming accident may overturn it. "Your Monarchy, for instance, may last almost any time yet; but something might happen to upset it in a few months." It is strange that a people so enlightened as the English should have so long retained antiquated institutions. They have little traditional reverence left in them. They are, in fact, in essential characteristics very like the Americans, the differences between the two people being greatly exaggerated in the popular conception.

Once convince the body of Englishmen, as he is sure they are rapidly being convinced, of the cause of their present economic evils, and a social revolution, to which that lately witnessed in Ireland would be a trifle, may begin here any day. "The Irish burn like chips. The fire spreads quickly among them and makes a great blaze in a little time. But the English burn like coal. They are hard to ignite, and may smoulder underneath a long time before you notice it, but when the flame does break forth it lasts." But Mr. George has no doubts about the beneficial effects of setting fire to the English on the land question.

THE MORNING PAPERS: EPITOME OF OPINION; GEORGE THE FIFTH'S UTOPIA
(*Pall Mall Gazette*, January 11, 1884).[32]

The *Daily Telegraph* says: "Our whole system is complex. It is sustained by personal aspiration, and industrial ambition; if we make robbery a State law we undermine all. We throw back the world thousands of years in order that Mr. Henry George may see his dreams realized. Were the Utopia dream of this

American innovator realized, the England of the future would be much more like the wild wood of the outlaw than the realm of order, progress, and government which we see today."

The *Morning Post* says: "It is quite clear that Mr. Henry George is calculated to do admirable service to the cause of order and property by the lucidity with which he demonstrates that the Socialist torrent once let loose cannot turn aside to spare this kind of institution rather than that."

The *Leeds Mercury* (L.) says: "Mr. George invites us, in order to encourage industry, enterprise, and ensure universal prosperity, to revert to the common tenure of barbarous times, and to turn the streams of industry into a very flooding of corruption."

The *Yorkshire Post* (C.) thinks that Radicals may discover in Mr. George's case, as in Mr. Parnell's, when they have called up Frankenstein, that it is not easy to keep him as fully employed as it is desirable to do even in their own interests.

The *Sheffield Telegraph* (C.) says: "This apparition of American Socialism puts its left-handed allies in the position of naughty boys, who have 'with wand and circle, book and skull,' raised the very fiend, and who in their terror at his appearance begin to cross themselves, and to try to exorcise the awful visitor with pious aphorisms and godly conjurations."

The *Birmingham Daily Gazette* (C.) says: "Mr. George is opposed to landowning; therefore Mr. George is opposed to civilization."

The *Birmingham Daily Post* (L.) remarks: "Nationalization of land, to be accepted as a question for consideration in England, must be propounded with the condition that existing owners are to be honestly paid for it. This is a *sine qua non*."

The *Manchester Examiner* (L.) says: "There are the greatest possible obstacles in the way of such a revolution. To mention no more, we like to possess at least a garden and potato plot of our own, and we are not irreclaimably dishonest."

The *Manchester Guardian* (L.) says: "Mr. George's proposition is one which all right-minded people must regard either with dissent or horror. If the right to property is to be abolished no one will have any motive to save, and the accumulation of wealth will cease, for it is difficult to see how property in land can be separated from all other kinds of property."

MR. HENRY GEORGE AT OXFORD: DISORDERLY MEETING
(Originally from *Jackson's Oxford Journal*, March 15, 1884).[33]

Mr. Henry George lectured on Friday evening at the Clarendon Hotel, on *Progress and Poverty*. F. York Powell, Esq., Christ Church, presided, and there was a very large attendance, the building being crowded sometime before the hour at which the lecture was announced to commence. The audience consisted chiefly of members of the University . . . The lecture was fairly listened to, but subsequently, when questions were put, the meeting assumed a very disorderly character, and was brought to rather an abrupt conclusion about a quarter to eleven o'clock.

Mr. Henry George, who was received with applause and a little hissing, said he was sorry that the hall was so small; (a voice in a crowded part of the room: "So am I") nevertheless, he was glad to appear before such a brightlooking and intelligent audience. ("Hear, hear.") The men he saw present were the very men he would like to talk to; they were to be the men of power in the future, and they were the men who, taking all things together, seemed to have about the best places in England. If any audience in England had reason to be satisfied with things as they were, it was such an audience as he saw that evening. (A voice: "So we are.") He was glad they were. ("Hear, hear.") He wanted to talk to them that evening about those who were not. ("Hear, hear.")

Of all the cities he had ever seen, this city of theirs seemed to him to be the flower and crown of their civilization ("Hear, hear.") They had there everything to make men satisfied with things as they were; a beautiful country, libraries, institutions that could be offered for the physical and intellectual man; but let them look over this England of theirs, and how many people were there could occupy such a position. ("Oh, oh," and uproar.)

When they got quiet he would talk. ("Hear, hear.") If people with human nature were utterly selfish, he would feel that he was wasting his time in talking to them, but that he did not believe. He believed there was in every man something greater than mere regard for his own comfort, than the mere selfish wish to enjoy himself; it was to the strongest as it was to the highest motive that he would like to appeal.

He was told that it took an average of about £250 a year to maintain one of them at the University. (Cries of "oh" and "no.") How many men were there in England who, by hard-straining work, can make that much in a year? The great mass of the people of this country were condemned to a life of hard-straining toil for a bare living. What the majority of those present might enjoy the mass of the people could never hope to gain. It was for them he would ask them to do something, not for themselves. ("Hear, hear.") All over this country of theirs – the richest country in the civilized world – there were families crowded into a single room, there were that evening women prowling the streets like beasts in order to get bread to take home to their children; there were little children growing up not merely without moral or intellectual conditions that would give them a full and healthy development, but without anything to eat, stunted, and deteriorated, even in body.

He was in one of their great libraries the previous day, and he saw there an illuminated manuscript, the picture being the massacre of the children by Herod; very quietly and without any expression at all on his face a man was cutting the throat of a little child, and people stood by and looked on with the utmost complacity. Did they know that there were that night in England very many children whom it would be a kindness to put out of the world in that way? Children who were growing up under conditions in which nothing but a miracle could keep them pure, children that were growing up under conditons which doomed them inevitably to the penitentiary or the brothel.

But it was said that things were getting better, and Mr. Giffen, an eminent statistician, wrote a pamphlet in which he marshaled figures to prove that the condition of the working classes was improving. ("Hear, hear.") Mr. Gladstone (loud cheers and slight hissing) wrote a letter in which he stated that was the best answer that could be made to Mr. George. ("Hear, hear," and a voice: "By George.") Well if that was the best answer he did not know what the worst could be. It was a very easy thing to prove anything whatever upon statistics. ("No, no.") Yes it was; let them give him the figures, and he would prove almost anything. Let them look at the facts which they read in their daily papers; hardly a day passed without the[ir] read[ing] some tale of destitution, some tale of degradation that would appeal to a savage.

All over this country human beings were living as among no tribe of savages, in any normal time, had human beings lived. The condition of the masses improving! What could they have been before when without this improvement men and women actually died of starvation. Professor Wallace had made an answer to that statement, in which he showed that as a matter of fact and by figures pauperism was actually increasing in this country. ("No, no.") Yes, yes. ("No, no.") They might go anywhere almost through this country and see with their own eyes human beings who had gained nothing whatever by the advance of civilization.

Go up into the north and see the cotters; there they would see people were working as their fathers thousands of years ago worked, cultivating the ground with the same rude instruments, digging with a wooden spade, threshing with a flail; there they were living . . . poorer and poorer than their fathers had been, crowded down and driven off good land onto poor land. Their crops had been diminishing, and they could not keep their cattle; the women were used as beasts of burden to do work that horses ought to do. If they went into their great cities they would see men and women living in dens in which no decent man would keep a dog. Then look at the mode of advertising; they made a man turn himself into a placard; the cheapest thing that there was today in this rich England of theirs was human labor.

Talk to him about improvement; improvement that they had to look at with a microscope in order to see. And if they took his country (America), they could see clearly and plainly all the advances they had been making in this century were only partial in their benefits; that while they elevated one class and gave them more, there was another class they crowded down. In their newer cities as material progress went on there came the tramp and the pauper. In the early conditions there were none rich and very poor; but as material progress went on, as their cities moved forward in the ways of progress, then came the almshouse and the penitentiary.

Say what they might about improvement, here was the fact that there was today all over the civilized world suffering and degradation that called on every man with a heart in him to do his best to remedy it. (Applause.) He neither did his duty to himself, or to his God, or his neighbor, who simply shrugged his shoulder and let this thing go on without some attempt at least to improve it. ("Hear, hear.") What could be worse than the doctrine virtually preached in their churches that these things existed by virtue of the dispensation of God. ("Oh, oh.") Was God a botch? (A voice: "Dry up.") If any man had a world to make, would he make one in which three-fourths of the people were condemned to a hard struggle to merely live, in which one-fourth were crowded down to the verge of starvation, and in which the few could develop their faculties and enjoy the pleasures of life?

He thought virtually that they had some responsibility for this. He thought there were but few men who, looking round them and seeing the misery and vice and degradation that existed, could rest content without doing something; their charitable societies, with the enormous sums that were spent in efforts to alleviate the condition of the poor proved that. ("Hear, hear.") But what was accomplished? A man might have the wealth of a Rothschild,[34] and go through this country and spend it all, and leave hardly a trace behind him; something more was needed than charity, and that something was Justice; (applause) and that was the highest call that could be made to any man.

Why, if they were passing through a desert, and saw a starving woman, they would stop and share with her; if one of the great Atlantic steamers flying across the ocean were to pass a vessel on which there were some shipwrecked mariners struggling for life, it would be counted shame, disgrace, and a crime if they did not stop and hazard property and life in the attempt to make a rescue. (Applause.) Yet in these cities of ours, in the very heart and center of civilization, they passed day-by-day men and women and little children who were struggling for life just as truly as they would be under these conditions. And they passed them by doing nothing save here and there to dole alms. Something more was the duty of everyone of them; it was the first and the highest duty that they should address themselves to this question, that they should ask what was the cause, and having found the cause should allow nothing to stand in their way in order to secure a cure. (Applause.)

Now what was the cause of this? If they were to find a great piece of machinery working ill in all its parts, the first thing they would do would be to go right down to the first wheel that gave motion to all the rest, and so let them go at once to this; what is the primary relation of man to the soil? Man is a land animal. His very existence depended on land; all that he called wealth came from land; his very body was drawn from the soil. Now given these facts, and having the soil on which and from which men must live monopolized by a few of their number, what could they have else than poverty among the rest? ("Hear, hear.") Whichever way they examined they would come to that conclusion. Prove deduction by induction? Commence at the other end, and ask why it is that wages are so low? Trace it up, and they would come to the fact that men could not employ themselves upon the natural source and opportunity of employment without having to pay a large portion of the produce of their labor for permission.

He should only take a little while out of their time by the lecture, because he understood that some questions were to be asked him; and there was one satisfaction in talking to an audience of that kind; he did not have to talk at length, but simply to drop a hint here and there.

Now, if this was the cause – and he believed the more they examined the more clearly they would see it was the cause – how could it be relieved? They could not relieve it by halfway measures; they must go to the root. As it was by land and from land all men must live, therefore, to give a firm and true base to the social edifice, they must give to every man that which was rightfully his, the produce of his labor; they must secure the equal rights of land.

Now, how could that be done? It could be done by cutting the land in equal pieces and giving every man his share; that would be an utter impossibility ("hear, hear") and if equality could for a moment be secured in that way, it would not continue. People would sell or give away shares of their land; the change of population would change the value of land, and they could only make such a division to have inequality come again. But if they could not divide land they could divide the revenue that came from land, and that was all that was necessary to secure equality. How this should be done was a matter of detail. ("Hear, hear," and laughter.) He believed the easy way – at least the easy way to begin with – was to go back to the old plan, and impose the weight of taxation on the value of land. ("Hear, hear.")

In the United States, although their system of taxation was, in some respects, worse than ours, in others it was better, so that system would enable him to explain what he meant. Their local taxes were by the assessment on the value of all property; once every year all property was assessed, or supposed to be assessed. The value of land was assessed separately, then the value of the buildings and other things. What he proposed was simply that they should levy their taxes on the value of land, and exempt all buildings and improvements. The tax on the value of land – as they knew, who knew anything at all about political economy – was certainly the best of taxes, inasmuch as it was a tax which could be collected with less expense, with less danger of corruption; it was a tax which bore less upon production, which, in fact, did not bear on production at all ("oh, oh") which, in fact, was a tax that stimulated production, for one of the reasons that kept production back was the holding of land by people who did not want to use it; those who prevented others from using it until they could get a very high price for it.

This was what he advocated, this was what he believed in, and he made his remarks short, as he understood questions were to be asked him, and he would reply to them. (Applause.)

Mr. Marshall (Balliol College) said, as members of the University, they prided themselves on not shutting their ears to any doctrine. They were most delighted to hear a man like Mr. George, and were prepared to give him a hearty reception. ("Hear, hear.") At the same time what they gave to him they claimed themselves – liberty to speak straight. What was it that separated Mr. George in his desire to promote the well-being of the poor from Lord Shaftesbury, Miss Octavia Hill, John Stuart Mill, and Mr. Toynbee?[35] (A voice: "Lord Salisbury.")

Mr. George said: "If you want to get rich, take land," and he was far from saying if they wanted to get well-off they should work well and be thrifty; that was hardly noticed in his book. Mr. George had not attempted to prove his proofs. ("Hear, hear.") He has stated that the only way to remedy poverty was to divide up land, and he had not given a shred or fraction of a proof of it. ("Hear, hear.") He would tell Mr. George what had gone on lately. It happened to be his (Mr. Marshall's) duty to lecture on political economy, and he had challenged repeatedly, over and over again, any person to show him one single economic doctrine in Mr. George's book which was new and true. But no one had come forward. ("Hear, hear.")

He might say he thought Mr. George in his books had not in any single case really understood the author whom he had undertaken to criticize. But he did not find fault with him for getting wrong on economic subjects without the special training that was required for understanding them. In doing that he was in very good company indeed; (laughter) a great many others had done it, and with them the world, as a rule, had no quarrel. It was because Mr. George proposed, as a person who wished to do so much good to the working classes, and had given them just that advice which, if acted upon, would prevent them from rising from their low condition ("hear, hear"); that they could not accept him as cordially as they should had he confined himself to addressing them – the well-to-do class.

He thought some fancied that the opposition to Mr. George arose partly from the belief that he was inspiring the well-to-do class. He did not share that feeling one atom; the more the case was investigated the clearer it became that land and other kinds of property were intimately bound up together. What they blamed Mr. George for was this, that he had used the magnificent talents he had, that singular and almost unexampled power of catching the ear of people, and he had used this power to instill poison in their minds. ("Hear, hear, no, no," and uproar.) He had not gathered from any of Mr. George's speeches that he had the smallest notion of the responsibility that he undertook when he said many of the things that he did. (Applause.)

Mr. George said he was perfectly willing to answer any questions, but he submitted that Mr. Marshall was piling them a little too thick. Mr. Marshall said he had already refuted his (Mr. George's) doctrines. ("Hear, hear" and "no, no.") Well, he was a good deal like their English General, he did not know when he was beaten, and he thought there were a great many other people in the same position. (Laughter and applause.) He was willing to answer the questions one at a time; his head was small and his mind was tired, and he could not remember so many questions when they were put together.

Mr. Marshall said he should like to ask Mr. George why in his book *Progress and Poverty* there was only one chapter on thrift, and in that chapter he showed only working men how they could not benefit their position by thrift and industry. ("Hear, hear.")

Mr. George said he submitted that he was not there that evening to answer questions on *Progress and Poverty*. ("Oh, oh.") It was a good while since he had the pleasure of reading that book (laughter) and his memory might be a little rusty about it. As to why he only gave one chapter or a part of one to thrift, and a number of chapters to something else, he did not think it was worthwhile answering.

But he would tell the gentleman if he wanted to know why thrift would not improve the condition of the working classes. Let one man save and he would get ahead of his fellows, but let the whole class save, let them reduce the expenses of living, and by an inevitable law so long as land is private property wages must fall proportionately. ("Hear, hear.") If the working classes of England were today to agree to live on rice like a Chinaman, how long would it be before wages would come down to a rice-eating level? They stood merely on the verge of starvation, and the only thing that kept wages above a certain point was that below that point men, with the habits of Englishmen, could not live.

The reason was simply this, when the man who owned the land could command all that came out of the land, he saved enough to introduce labor to produce that wealth. Take it that they, all of them there that evening, were on an island and he owned the land, and the men were fools enough to acknowledge his right. (A voice: "There's no chance of that.") They would only live on his wishes. ("No, no.") Yes, he would be as truly their master as if he had bought their bodies or their souls. They would have to come and beg for the opportunity of work, and he could give it only on his terms, and the only limit of his terms would be just what the men could live on. Then suppose they voluntarily reduced their scale of living he could simply increase the price of his rent. ("Oh, oh.") And supposing that improvements were made, that labor-saving machinery or [a] labor-saving discovery went to a point that he could do without their labor, he would simply have no use at all for them, and if he was charitable he might keep some of them in an almshouse or emigrate them.

Man was by nature an expensive animal. ("Oh, oh.") He was an animal that could make a microscope and reveal things in the water; that could make a telescope and bring far-off stars near; who could weigh and measure the sun, and yet he was just to keep himself above starvation. The thing was not to consume less, but what they had to do was to make more. (Applause.) How few there were that had profited by all the advance of civilization; instead of human labor being the cheapest thing in the world it ought to be the dearest, because on human labor guided by a human intellect all the powers of Nature wait. (Applause.)

Mr. Marshall said he did not mean to contrast thrift with mere energetic work; he meant both. (Applause.)

Mr. George: I did not hear anything about energetic work. (Applause and uproar.)

The Chairman said they were wasting time by making so much noise. Let the gentlemen on each side have a fair hearing.

Mr. George said let them keep quiet. Mr. Marshall had now coupled energetic work with thrift. The human muscle was one of the tiniest of forces; it was by his brain that man produced. What was the consequence of the present state of affairs? They turned men with brains and intellects into mere machines. Give every man an equal chance to develop his mental as well as his bodily powers, and they would have more than enough for all – Nature was no niggard – they might then have not merely all the necessities of life, but all the luxuries, and all that they wanted to secure was Justice. (Applause.)

Mr. Marshall said admitting that if the island was owned by one person he would have everybody in his power, but he wanted to take them to an island owned by thousands, not acting in combination. They had a rent for which Mr. George spoke, but it did not amount to much more than a shilling in the pound; he believed that landlords had not been able to take away more than that.

Mr. George said the gentleman raised too many issues in the same breath. ("Oh, oh.") He should like to talk about the shilling in the pound by-and-by. Mr. Marshall admitted what he said would be right if the island belonged to one man, but he denied that it would be the case if it was the property of many. He contended that it would be very nearly the same. (A voice: "Prove it.") He would prove it. Let one man own the soil on which and from which another man lived and he was that man's master. Let him own the soil on which the people must live, and he was their god on earth. ("Oh, oh," and uproar.) Let a class own it, and that class ruled the people who ruled the men of England. Who were the men to whom they applied the same title that they applied to the Deity, your Lord? (Great uproar and cries of "shame.") It was not the title that gave the power; it was the ownership of land that gave the title. There were on this island today men who trembled in the presence of their landlords almost as slaves. (Cries of "question.")

As Adam Smith[36] said a hundred years ago . . . (A voice: "Oh, the question.") He was answering the question. ("No, no.") Fair play was a jewel. ("Hear, hear," and "go on.") He had sat there and listened to their exponent, and he simply wanted to be heard on the other side. ("Hear, hear" and "no, no.") If the majority of them want to howl he would bow to their will and sit down while they howled. As Adam Smith said a hundred years ago they were almost under a tacit combination, and so they were. ("No, no.") If they wanted a correct instance he would give it them in this island. (A voice: "Let us have it.") He said he did not want to say anything personal, but if they wanted it they could have it. ("Hear, hear.") In Scotland, then, he knew when a farmer offended a landlord he could not get another farm, and a man was turned out of his farm for voting for Mr. Gladstone. (Cries of "shame" and "name.") The gentleman's name was Hope, and the landlord's name was the Duke of Buccleuch. (Great uproar.)

An undergraduate rose in the body of the room and said he wished to call attention to the fact that ladies were present, which many there seemed completely to have forgotten. ("Hear, hear.")

The Chairman said he had appealed once for order, and he hoped to call attention to do so again. There were many people there who wanted to hear both sides, and he hoped they would be more silent.

Mr. George said the landlords could hold out for the highest price for the land and they could wait; the man who must eat could not wait, and the man who could not wait must give way in the bargain to the man who could. (Applause.)

Mr. Marshall said he had only to ask the question over again. He wanted Mr. George to prove in an island owned by many, who were not acting in combination but in competition, it would be possible for the landlord to screw the people down to the verge of subsistence.

Mr. George said he had only to appeal to facts. It was the competition for the land in Ireland which forced up the rent. The English farmer was intermediary; his true place was that of capitalist. The man who cultivated the soil was the laborer, and how much did he get above the mere living? It was utterly impossible for him, by a life of the hardest toil, to save enough to keep him in his declining days. (A voice: "Quite true.") On this island today there were men who were paying for the privilege of living on the land, paying more than by any human possibility they could get out of it.

Mr. Marshall said there was a great doubt as to the wages of the agricultural laborer. There was, no doubt, a rise in 1870. Could Mr. George show them the proof that the people of England were in the power of the landlords. The landlords could only get as much as competition allowed them, and he maintained that was 1s. in the pound.

Mr. George said he did not say that one landlord was the same as many. He admitted there was a difference. He understood the point was whether wages could be forced down to starvation point if there were many landlords instead of few. That was a different thing, and he told them that wages were so today. There was no use going to a theoretical island, here was the island. ("No, no.") What was the use of men talking like that; they could go out into the cities and country, and hire men for almost anything.

Mr. Marshall said he submitted that Mr. George had not answered his question.

Mr. George said Mr. Marshall ended as he began, in mere assertion.

Mr. Marshall said he would leave it to the audience.

An Undergraduate said Mr. George had stated that the agricultural laborer only received wages on which he could just live. He had had twenty-five years experience in agricultural districts, and he could give that statement the lie. He knew in Staffordshire in time of harvest, and through most of the year, a throughly competent laborer got as much as 20s. a week or more. There was no comparison between the cost of living in town and in the country. He knew the case of a man who, on 13s. and 14s. a week, brought up a wife and twelve children respectably, and lived to a green old age (loud laughter) and he died with £200 in the bank; that was thrift. (Cries of "question, name," and "order.")

Mr. George said the man referred to ought to be placed in a glass case. A man had brought up twelve children on 14s. a week! He should like to ask the gentleman how much it costs to keep an average pauper in the workhouse?

The Rev. A. H. Johnstone said he should be very obliged to Mr. George if, instead of continuing to insist on the fact that there was great destitution in this country – which all of them knew and deeply deplored, he would address himself to the question as to whether really the cause of it was the monopoly of the land by a few. He wanted a simple answer to this problem without any sentiment whatever. If the land was nationalized, and there was an overwhelming population, would not the competition for wages ensue, and wages be reduced to [the] starvation point?

Mr. George said in a natural state of things they would never have an overwhelming population. It is not that kind of world.

Mr. Johstone said he wished to state as Mr. Marshall had done, that Mr. George had not answered his question. He asked of him not for sentiment, but to address himself to a theoretical problem, and he would not do it. (Uproar.)

Mr. George: Will the gentleman please state his theoretical problem; and in case his memory should fail him, will he put it on paper. (Great uproar.)

Mr. Johnstone: I will do what the Chairman suggests. (A voice: "Sit down.") My problem is given in the land nationalized, and an overwhelming population; would not a competition for wages at once commence, and would not wages fall nearly to [the] starvation point?

Mr. George: Get a pint pot and pour into it a gallon, and what would happen? If "ifs" and "ands" were pots and pans. That is an insult to the intelligence of this audience. (Uproar.)

Mr. Robinson (New College) said he understood that Mr. George proposed to sweep away the whole of the taxation of this country and to have a magnificent surplus by which all sorts of good objects might be promoted. The taxation of the country at the present time, including Imperial and local, amounted to about one hundred millions per annum. Further, the economic and ground rent of this country, which Mr. George proposed to apply, according to the very best estimate they could get was but sixty millions per annum. He wanted Mr. George to tell them how these two figures were to be squared?

Mr. George said let them suppose he had been too sanguine; the principle was the same. They would also gain economy of administration. They would not have to keep a cordon of custom officers round their shores, and very many other expenses might be saved.

Mr. Hugh Hall said he did not know whether Mr. George was a believer in the Ten Commandments or not. ("Oh, oh.") He wanted to know how it could possibly be fair or reasonable to take away property from a man who had acquired it by the sweat of his brow, and invested it in land, and give it to the nation. Take the case of a father of a family who by his labor had saved a few hundreds or thousands, and he bought one son a farm and another a shop. He understood that Mr. George would take away the property from one son and leave the other in possession. He maintained that was a breach of the Commandment.

Mr. George said he believed in the Eighth Commandment: "Thou shalt not steal;" it was attested by every fact of Nature, and the indictment that he brought against the present state of things was that it ignored that Commandment. Today, all over the civilized world the men who labored were robbed. If there was to be any compensation it ought to be to the people who had been disinherited. (Uproar.) Did the fact that robbery had gone on with impunity in the past give a man the right to continue it in the future; not a bit of it. ("Hear, hear," and uproar.)

From this point the meeting was of the most disorderly character.

Mr. Conybeare (University) described this nostrum to confiscate the land as scandalously immoral. ("Hear, hear" and "no, no.")

Mr. George: Are you a member of the University? (Uproar.)

Mr. Conybeare: Yes.

Mr. George: Then I take back all I have said of them after such insulting terms. (Great uproar.)

Mr. Conybeare: I said this proposal was, as I believed, scandalously immoral.

Mr. George: You stigmatized it as a nostrum.

Mr. Conybeare: I should like to ask Mr. George if he likes people to be sincere with him?

Mr. George: I do, but I like people to be gentlemen with me if they can. (Uproar.)

Mr. Conybeare: I consider, sir, I have been so with you. (Great uproar, which lasted several minutes.)

An Undergraduate said he thought Mr. George should withdraw his imputations on Mr. Conybeare.

The Chairman: I do not think Mr. George wished to make any imputation on Mr. Conybeare.

Mr. George: No, I will not withdraw anything. Mr. Conybeare says it was gentlemanly. All I have to say is that he was raised in one school and I in another.

Mr. Lodge proposed that the meeting decline any longer to listen to Mr. George. ("Hear, hear" and "no, no.")

The Chairman said he thought it would be extremely regrettable if a meeting like this broke up on a question of mere personal conflict. He was sorry it had arisen, but he thought it was over.

Mr. Conybeare said he was the best of friends with Mr. George. He did not mean any harm in what he said, and he thought he was justified in using the word "immoral" with regard to the scheme. He did not mean to signify that Mr. George was an immoral character. He was simply frank in what he said.

Mr. George: One can be frank and not rude.

Mr. Conybeare said he was not criticizing any men but ideas. He wished to know if Mr. George wished to take the land without compensation?

Mr. George said he proposed to take it away without a bit of compensation. He did not propose to take away from the landlords anything that belonged to them, but he proposed to give them their equal share.

Several other questions were put amidst great noise.

Mr. George said this was a University town, and it was the most disorderly meeting he had ever addressed. He would not answer any more questions.

Mr. Marshall proposed a vote of thanks to Mr. George.

The proposal was received with cheers for Mr. Marshall.

Mr. George said he undertook to thank them for the cheers which had been given for Mr. Marshall, and he moved a vote of thanks to the Chairman, who had presided with great dignity over a most disorderly meeting.

Mr. Faulkner seconded the proposal.

Mr. Abbey protested against the behavior of the Undergraduates to Mr. George. The meeting then broke up with groans for "Land Nationalization" and "Land Robbery."

MR. GEORGE AT OXFORD
(*Pall Mall Gazette*, March 8, 1884).[37]

Mr. George addressed a large audience at the Clarendon Hotel last night upon the subject of *Progress and Poverty*, those present consisting almost entirely of undergraduates. The lecturer commenced by remarking that he believed it cost each member of the university £250 a year to be at Oxford, while there were thousands of strong men in the country who worked from morning to night for a bare subsistence. There were thousands who were living in a state in which no decent man would keep a dog. Justice was needed more than charity. Man had his subsistence from the soil, and given its possession by a few, what could they expect but poverty for the rest. To secure equality the division of the land would not be the right way, but the proper way would be to divide the revenue upon the land, and he would suggest that taxes should be levied upon land alone, and not upon what was in it in the shape of buildings. Mr. Marshall, Balliol College, blamed Mr. George for instilling poison into the minds of people.

A SNIPPET FROM THE *PALL MALL GAZETTE*
(January 14, 1885).[38]

Mr. Henry George, whose views upon things in general and his own crusade against landed property in particular are fully reported in another column, called at our office this morning. He is full of enthusiasm concerning the uprising of the masses, and for the sake of the agrarian millennium which is dawning in the United Kingdom before his enraptured vision he could almost wish he were an Englishman. Apart from the natural exultation of an agitator on discovering that he has convinced a Cabinet Minister[39] that it is worthwhile to trim sails to the Socialistic breeze, Mr. George's chief ground of satisfaction is to be found in the support he receives from the Financial Reform Association, on whose old fad – a 4s. in the pound land tax – he declares that the battle for the land will be fought and won. Mr. George leaves for his native land on Monday.

HENRY GEORGE IN ENGLAND: PLAN OF HIS SINGLE-TAX CAMPAIGN TO HELP THE IRISH
(*The Standard*, March 16, 1889).[40]

Henry George arrived at Southampton today on the steamer *Ems*, and his friends took him aboard a tender amid the cheers of those who had voyaged with him. He said the passage had been a pleasant one, and he was full of enthusiasm for his work in England. In an interview with the *World* correspondent this evening at the Westminster Palace Hotel, London, he said:

I am here with the idea that England is the most vulnerable point in the world for us. Politics here are in a peculiar position. In the first place, there has been a tremendous revolution in favor of the Irish, which culminates in the exposure of the Tories. When I first came to England in 1882 the intensity of the people against everything Irish was something that could hardly be stated. Nothing was too bad to believe of Ireland. The result of the Gladstone fight, putting the Liberal Party with all its machinery upon the side of the Irish party, has been to enormously educate and enlighten the English people. Now that party and its machinery has been virtually put against the Tories on the purchase scheme. That is about the last refuge of English landlordism. The feeling is very strong. My friends insisted on my coming over here in view of the fact that, although the parliamentary election might not come off for three years, it was liable to come off in three months, and they wanted, as far as possible, to infuse radical ideas into the minds of the voters, and to so bring them into the election that the successful members would have to pledge themselves. Recent elections to the County Council have shown the depth of the radical sentiment. For instance, the London joint committee for the taxation of ground rents, which I addressed when I was in London last, and of which association I am a member, was formed only a year ago. Thirty-eight of the members who formed it have been elected to this new county council, and there is no question that a clear majority of the members were in favor of taxing the landlords, and a majority of that majority, I should say, were in favor of taxing them as I would, so as to utterly abolish them.

In this body and in other similar bodies the question will come up, not for decision, but at least for presentation to Parliament, and at the next parliamentary election it is the confident expectation of our friends that a majority of the new members can be secured.

What, in your opinion, will be the effect of the outcome of the Parnell commission?

The breakdown of the *Times'* case is a crushing disaster to the Tory government and may so far precipitate things as to bring the election on at any time. The smash up aids the radical movement tremendously. Probably nothing has happened within the present century that will have such an effect. When I was last in London the *Times* was evidently only fighting for delay and to drain the Irish funds.[41] Outside of the bigoted Tories there was an expectation that the *Times'* case would finally collapse, but nobody seemed to anticipate such a terrific collapse as has come. The government is so thoroughly mixed up with the prosecution of the case that it affects it as disastrously as it does the *Times*.

What of the progress of your movement in America, Mr. George?

It is very strong and gaining every day. The people everywhere are beginning to discuss the subject of taxation. Among tariff reformers radical free trade ideas are coming to the front, as was shown in the Chicago convention, and the timid movement for the reduction of the tariff is passing into a demand for the abolition of all tariffs.[42]

Mr. George says he is going to make a three months' campaign of it through England, Ireland, Scotland, and Wales. He is jubilant over the prospect and more confident of a satisfactory outcome even than when he came so near to being elected mayor of New York a couple of years ago.

HENRY GEORGE IN ENGLAND: THE ENGLISH CAMPAIGN
(*The Standard*, March 16, 1889).[43]

There seems now no reasonable doubt of the success of the campaign upon which Mr. Henry George is to enter within a few days after these lines are in the hands of our readers. The times are ripe for such a campaign. On all hands we receive the most gratifying assurances of the progress which is everywhere being made by the views with which Mr. George's name is identified. The Birmingham program, the great uprising of London in its first municipal election, the great and determined meeting at St. James' Hall, followed by the introduction of a bill into the House of Commons to tax land values in order to provide better houses for those whom landlordism has impoverished – these things, and such as these, are the outward and visible signs of a great awakening all over the country. The lecturers of the Land Restoration League find audiences, larger and more enthusiastic than ever before, eager to settle down to the detailed discussion of practical proposals for the abolition of landlordism. Everywhere single-tax ideas are to the front. If our friends do their duty during the next few months the slowly but surely rising tide of popular indignation against landlordism will be converted into a torrent. . . .

The invitations already received have come from the most various sources. It is hopeful that so many of them should come from religious leaders. One of the first, if not the very first, of Mr. George's meetings will be held in an important South London chapel, at the invitation of a young men's society connected therewith; another is to take place in connection with Westminster Chapel; the City Temple will probably be available for a third, and other arrangements of a similar character are under discussion. . . .

In London, meetings will be held east, west, north, and south in large halls, where the whole of the local arrangements will be made and carried on by working men's clubs combining together for this special purpose. . . .[44]

HENRY GEORGE IN ENGLAND: THE SINGLE-TAX CAMPAIGN THERE ATTRACTING GREAT ATTENTION
(*The Standard*, March 23, 1889).[45]

The certainty that the land theories of Henry George are attracting widespread attention in England just now and are being studied to an extent hitherto . . . belief is abundantly manifested by the fact that the gentleman is being courted, feted, and consulted to a degree that would turn the head of a man of less perfect mental balance. Since his arrival here Mr. George has been sought out by men who a few years ago would have felt ashamed to incur the suspicion that they had ever wasted sufficient time to read his book, not to speak of giving serious consideration to his ideas, and there is a well-founded suspicion abroad that several of the leading members of the new London county council have a decided leaning toward the practical application of his theories to the future disposition of municipal lands.

Another unmistakable proof of the hold which Mr. George has secured upon the popular mind is furnished by the fact that the *Times* has deemed it necessary to denounce two members of Parliament who dared to signify by their presence at a reception to Mr. George by the radical clubs Saturday evening, at least their appreciation of his exceptional abilities.

It will be remembered that when Mr. George's *Progress and Poverty* was published in its first edition, the *Times* complimented the author by reviewing the work to the extent of a page. Upon this occasion the paper remarked that although it was forced to dissent from the doctrines laid down by the writer, it was compelled to admit that the work was one of the most important of the century, adding that upon some future occasion Mr. George would have to be "reckoned with."[46]

Manifestly the *Times* has reached the belief that the time has arrived when Mr. George should be "reckoned with," and the suspicion is abroad that the paper's present attitude in the matter indicates that the Conservatives are becoming alarmed at the growth of the single-tax apostle's popularity in England, and are preparing to counteract it by disciplining such public men as exhibit the courage of their convictions by extending the hand of welcome to men who prefer seeing things as they are to looking at them through the refracted lens of Toryism.

The Two M.P.'s the Times Names

The *Times* attacks Messrs. Cremer and Clark, radical members of Parliament, for joining with the radical clubs in the welcome to Henry George on Saturday. The *Times* says: "Mr. George's doctrines come as near spoliation as anything that has been proposed by a man with a character to lose since the French Revolution. Men like Mr. Cremer wish to apply the operations of the Irish leaguers to England, Scotland, and Wales."

In Rev. Stopford Brooke's Pulpit

Mr. Henry George has been making land-tax speeches all the week, principally under the auspices of the London preachers. Tomorrow (Sunday) he occupies the pulpit of the Rev. Stopford Brooke, which has never before been filled by an itinerant layman.

A Big Meeting in Camberwell

Henry George delivered a lecture at Camberwell tonight before a large and extremely enthusiastic audience. His hearers plied him with questions, all of which were manifestly answered satisfactorily. It is noticeable that each of

Mr. George's successive visits to England is attended by a large increase of interest in his theories.

WHAT THE NEWSPAPERS SAY
(*The Standard*, March 23, 1889).[47]

Henry George is safely on the other side once more. It would be a thoroughly agreeable thing to the American people if arrangements could be made for him to stay there. His land heresies are working incalculable injuries. (Newport, RI, *Observer*).

Henry George is in big luck in England. The London *Times* is jumping on him with all four of its feet. (*New York Press*).

Henry George seems to be making a great stir in England. Henry should beware. The kinship between stirabout and soup is perilously close.[48] (*New York World*).

Henry George is in England again, whooping things up for his single tax on land values, and seems to be creating a great deal of enthusiasm. If he could manage to make the Irish landlords tremble along with their English brethren, perhaps they would give Balfour a hint to let up on the poor peasantry. (*Boston Globe*).

Henry George is in London. He took his single-tax idea with him and expects to place large blocks of it in English craniums. If there is any people on the globe's face that has cause to accept Mr. George's plan it is the British. The progress of his work on the other side will be watched with considerable interest by a large number of Americans upon whom the single-tax theory seems to have taken a firm hold. (*Chicago Mail*).

INTERVIEW WITH HENRY GEORGE:
Progress of the Single-Tax Movement in Various Parts of the World; Answers to Questions about the Proposed System; The Irrepressible Widow and Orphan
(*The Standard*, May 25, 1889).[49]

Yesterday one of our representatives had an interview with Mr. Henry George in his hotel. Mr. George, who is a good specimen of the frank, easygoing American, received our representative with courtesy, and when it was made known to him that an interview was wished in regard to his propaganda, Mr. George readily agreed. He was asked:

What progress is the cause you espouse making in Great Britain?

Very great progress. This is perceptible to me wherever I go. However, it has been a progress, as it were, beneath the surface – a progress in thought and education, and its full effects are not yet seen in politics, but will be so as soon as the land question comes fairly up.

Have you as many followers proportionately in Scotland as in England?

In proportion to the population I am inclined to think more.

I suppose you have made and are making a large number of "converts" in America?

We have been all these years steadily moving forward in America, and during the last year, and especially since the election, our ideas have been progressing with great rapidity. This is largely owing to the discussion on the tariff question, between protectionists on one side and tariff reformers, who are half protectionists, on the other: we single-tax men, as we are called, are the only ones who precisely know what we want. We are out-and-out free traders, and on this question we have magnificent opportunities to bring our views under discussion. We supported President Cleveland in the last election tooth and nail, but were perfectly satisfied with his defeat.[50] I not only felt this myself, but on going home after visiting this country I found that the feeling was shared by all our friends throughout the states. If Mr. Cleveland had been elected president a Republican Senate would have blocked the way to any radical action, while the protectionists who yet remained in the democratic body would have claimed the victory as due to their moderation. As conditions are now, the free trade element must get full possession of the Democratic Party by the next presidential campaign, or else split it in two, and the struggle will be a square one. Education on the tariff question is going rapidly forward in the United States, and the late victory of the advocates of protection only means their more thorough defeat in the future.

Have you any information of the progress of your ideas on the continent of Europe and Australia?

They are making some progress on the Continent. They have started a paper in Germany. Michael Flürscheim, a large manufacturer in Baden-Baden, has left his business and now devotes himself to the propaganda.[51] In Holland I am also assured that the cause is making steady way, and although I can only speak English I have accepted an invitation from our friends [in] Amsterdam to go over there and address them. In France we have some friends; I cannot tell how many. And so in other countries in Europe. *Progress and Poverty* is now being translated into Turkish. In Canada the progress we are making is almost as rapid as in the United States. In Australia and New Zealand we have also been making great strides in advance, and in South Australia a law has already been passed, giving to localities the power to place their taxation on land values alone. I have not been to Australia, though I have been for a couple of years most strenuously urged to go out there.[52] I have learned from South Australians

whom I have met here that they believe that that colony will be the first to
adopt the single tax. Mr. William Webster, now of Aberdeen, was the first, I
believe, to propagate our ideas in Australia.

What is your opinion regarding the claim of the eight thousand families in
Western Australia to be put in possession of the entire colony?

Well, I presume what they want to do is to establish a state as we call it in
America. I don't see why they should not. As an American and follower of
Thomas Jeffferson, I believe most strenuously in local self-government, or, as
you call it here, home rule. I believe the people of every locality should have
the right to manage their own affairs, and that general affairs should be managed
by the representatives of the various communities embraced in the federation.
The federation principle, which I hope some day will be adopted on this side
of the Atlantic, is indeed the only principle which can unite flexibility and
strength over an extensive area.

I understand your proposal is to take the unearned increment of land from
the present owners and to give it to those who really create it – that is, the
whole people?

Yes; that is to the community.

Where would you draw the line in applying that rule? Suppose a sum of
£2,000 or £3,000 belonging to a widow with a family of three or four children
had been invested within the last ten or twenty years in feus within the boundary
of a city, would you take from a family in such a position the unearned increment without compensation?

Yes; I would make no distinction.

Are you not rather cruel in dealing with people in the circumstances I have
mentioned?

No; I don't think so. That argument was wont to be used against us in
America. The injustice is in the continuance of the present state of things. If
widows and children are to be hardly used by the proposed change they could
be taken care of. Such widows are not one-hundredth part of those widows
who suffer under the present iniquitous system.

Would you apply the same doctrine to houses where, in consequence of
improvements effected by the community, rents were doubled or trebled?

The improvements made by a community never increase house rents or
the value of houses. What improvements increase is the value of the land.
Where the rent paid both for houses and ground is increased in a growing
community the increase is always in the value of the land, never in the value
of the houses.

We will shortly have a case in Dundee which, if your ideas were carried into
effect, would be easily settled when the time arrives. Almost in the center of

the city we have a fine breathing space – it has been called one of the lungs of the town – locally known as the Barrack Park. The park was for a number of years used as a military parade ground, but in 1854 it was leased to the town for nearly thirty-six years at an annual rate of £25. The lease terminates in 1890, and it is now proposed to feu the park, and it is expected that it will bring at least 15s. per pole.[53] It belongs to an earl, of whom the people generally know less than of the shah of Persia. Still, he is the proprietor of the ground, his forefathers having received the park in the eleventh century as a free gift. In recent years the park has risen very much in value, simply because of the growth of the city. How would you deal with a case of that kind?

Under the plan I propose his resumption of the ownership in the park would compel him to pay a tax so heavy that he could not afford to retain the land in his possession and hold it lying idle, and would take from him what he could get by feuing it for buildings. Under such conditions he would be willing to get rid of the ground for little or nothing. The great advantage in making municipal improvements under the plan I propose is obvious. If the city authorities wished to widen a street or to obtain a site on which to erect public buildings, or to turn ground built on into a public park, it would only be necessary to pay a little more than the value of the buildings to be demolished.

At what figure do you estimate the economic rent of Great Britain?

I have not made an estimate myself, but undoubtedly it is enough to pay all imperial and local taxation and leave a surplus.

You do not propose to interfere with improvements?

No; improvements should be absolutely free of taxes or burdens. I would tax nothing that would make the nation richer.

Are there not some improvements which merge into value of land so insensibly that you cannot distinguish them?

Yes; there may be, for instance, in the case of a hill leveled by the Romans, or a marsh filled up, we would not now be able to distinguish from the operations of Nature, or is there any reason for our doing so.

WHAT THE PAPERS IN GREAT BRITAIN SAY
(*The Standard*, May 25, 1889).[54]

The Ayrshire Post: There is more danger in underrating than in overestimating the importance of the movement in favor of land nationalization. It has reached a stage at which it can be no longer blinked out of existence. Its opponents may not like it, but they are better to look at it and to understand that the crusaders are getting more numerous every day, and that wherever Mr. Henry George raises his standard they come literally rushing in. Two or three years ago Mr. George was not regarded at all as a dangerous man. His opinions were dangerous, not himself. He dropped upon the unruffled surface and there was a splash, but the wiseacres, and, indeed, the country at large thought the circles created by the contact with the waters were not extending, and were not likely to extend. What society wished to take place, it thought would take place. But society was mistaken, as it generally is.

Wherever Mr. George goes today he is met by crowds of sympathizing admirers; his views are readily assimilated, and when he leaves he does not take them away again with him. They remain; they keep on spreading, and where a few years ago it was very rare to find a man who talked of land nationalization or of taxing the landlords off the soil, supporters of these views are now so common as hardly to be considered extremists. And the views themselves are, in their intent, very "taking." They gripe. People see for themselves what a terrible gulf there is between the many and the few. Wealth is in few hands, land is in fewer; and the masses get poorer and poorer. It is no matter for surprise that men, working and slaving for a bare pittance should welcome anyone or anything that holds out the prospect of relief to them. The landlords keep rigid grasp of all they have on all they can get. In the country districts the populations keep steadily going down. In the cities the East Ends are becoming more and more crowded. The colliery owner pays men a beggarly pittance for their toil; the sweater makes life miserable by his parsimoniousness and his grinding exactions.[55] The land is hedged in and walled about so that the masses can no more than look at it; and they are expected to rest satisfied with their public parks and the amenities of the country roads. Glens and straths that once carried large populations now carry large droves of sheep or deer.[56] The emigrant ships are loaded with those who cannot get room to live at home. Discontent is being fostered all round, and, though there are brighter spots here and there in the darkness, the darkness itself is abounding.

Into this unhealthy moral condition Mr. George drops his balm. He has not to manufacture a case. It is at hand, and his audience know it and feel it. They can tell what unremunerative labor and grinding poverty mean, and they turn naturally to the man who says he can heal their afflictions. And without doubt his proposals are, in their way, both simple and enchanting. The scheme is simplicity itself, and if it were only feasible and honest no fault could be found with it. We could with pleasure see the land nationalized; but to simply tell the landlords that they must go, seems to us to involve a wrench of common fairness in dealing, and of principle, for which there is no justification. We do not take this view from any sneaking sympathy with the landlords.

Many of them have inherited land which their forefathers stole, without waiting to consider anything about the morality of their transactions; and those in that position, who hold up their hands because the people propose to treat them as their forefathers treated others, are not worth wasting sympathy on. And of the many who have come honestly by their estates, there are comparatively few who do not use their rights in the most selfish conceivable fashion. They call their lands their own. They claim absolute right to do what they like with them; and the poor man who dares to walk their acres is rudely ordered on the highway. The landlords as a class have shirked their duties and their responsibilities. They were bound at one time to maintain the army, for instance; but the taxpayer knows who has to pay for the armaments of the kingdom. The position of the lairds bristles with anomalies and with monopolies; and if they are doomed to fall they will have brought down their doom on their own heads. They retain every privilege they can get hold of; and if they could help it, the masses would have nothing save the beggarly allowance they now possess, and the right to live. Hardly even that.

But retaliation in kind is not honest; and of the two alternatives – to take their land without paying for it, and to tax them out of existence – it is hard to say which of the two is the more reprehensible. Better a sadden than a lingering death; and better on the whole to bid them go than to multiply their taxation until their domains are riven from their grasp by the inexorable tax-gatherer. Yet – and there is no good trying to blink it – it is one or other of these alternatives that they may have to face unless they make haste to agree with their adversary quickly while they are yet in the way. The country has no desire to rend them separate from their inheritance, without compensation; but if they keep on in their present attitude and defiantly insist on the "rights" which they have given themselves – stolen in most cases – the fault will be their own if they land themselves in destruction.

For socialism is spreading, and changes come rapidly on the country. The more quickly things grow in great political and social matters, the more accelerated is the tendency to progress; and therefore, if the present *non possumus*[57] attitude is maintained, there can as yet be no foretelling what the end may be. Of one thing, however, we are certain – they cannot be worse than they are. It is a most gratifying sign of the times that the working classes are taking their self-amelioration into their own hands. It is they alone who can work it out. They can expect nothing, look for nothing, hope for nothing, from the landowners as a class. The farmers know that, in bitter experience. If this amelioration is to come at all it will not be without a drastic change in the existing conditions; the sooner the few look to their position the better.

Reynold's Weekly: The English Land Restoration League, be it noted, is preparing extensively to petition Parliament to take the full economic rent of land in town and country in lieu of all rates and taxes whatever. The Single Tax on Land Values is the true antidote to Budget jugglery and mystification.

It is the potent germ of the cooperative commonwealth that is to be. More power to the Land Restorers and the Prophet of San Francisco.

Dumfries and Glasgow Saturday Herald: Mr. George is the only political economist who has managed to make that dismal science fascinating. The volume on which his reputation rests – *Progress and Poverty* – is as interesting as any romance.

Labor Tribune (a London miners' paper): Mr. Henry George has been addressing some meetings of miners in Ayrshire this week, and has been opening the eyes of the men to some grand truths. His new gospel is very popular here, and only wants someone to head the movement, as Bright and Cobden headed the Anti-Corn law agitation – to place it in the front of the burning political questions of the day.

Greenock Telegraph (Scotland): The visit to Greenock of Henry George, the famous American writer and land-law reformer, has been a marked success, so far as the attraction of popular interest is concerned. Notwithstanding the important circumstance that admission to the town hall was by price tickets, that spacious meeting place was yesterday evening – keeping off the gallery – quite filled. A glance at the area audience showed that it was composed of from twelve to fifteen hundred of the cream of the workingmen of the town and district. That such a number of working people should have been attracted to the hall, and should have persuaded themselves to pay for their seats, is a striking proof of the advance of the land question in public opinion as a matter of practical politics, and also a remarkable testimony to the power and influence of the author of *Progress and Poverty*. The address, a report of which will be found in another column, was a clever piece of platform work – logical, incisive, and pawky.[58] As a gentleman in the audience last night remarked: "His talk is even more racy than his writing." There is considerable truth in the observation.

HENRY GEORGE IN MANCHESTER:
The Financial Reform Association's Great Meeting in Free Trade Hall; A. D. Provend, M.P., in the Chair and J. Hampden Jackson, Editor of the Financial Reform Almanac, Presents the Resolutions: Urging a Further Reduction of the British Tariff and the Substitution of a Tax on Land Values; Henry George's Speech in Support of the Resolutions; Much Enthusiasm Evoked; Answering Questions
(*The Standard*, June 15, 1889).[59]

Henry George, who was received with great applause said: I am a believer in Home Rule.[60] I believe it is the right of every community, be it small or large, to manage all things that relates to itself, without let or hindrance from anyone else. Nevertheless, I am glad, and I feel honored that you, Mr. President, and you, the gentlemen of the Liverpool Financial Reform Association, have given to me as an American the privilege of supporting this resolution. I do it most heartily, and I do it as an American and as a representative of a great and daily increasing body of men, the free traders of the United States. (Applause.) And I do it, not in reference to your own affairs alone, but in reference to the affairs of the world. The question of free trade is more than a local question. As Richard Cobden said, free trade is, indeed, the international law of God. If we ever see it, we shall see that it is indeed the peacemaker; we shall see that it has power to knit the nations together into fraternal bonds; that it has power to disband the standing army, to still the war drum, and to bring on the era of peace and prosperity and brotherly love of which the poets have always sung, and the seers have always told. (Applause.)

As an American and an American free trader, I feel it to be a privilege to stand on the platform of the Free Trade Hall at the request of an association that is the true representative today of the great Anti-Corn Law League that a generation ago did so much – to stand here in the presence of men who like my good friend Mr. Thomas Briggs,[61] stood by Richard Cobden in support of that great movement. (Applause.) After all our advances, after all our conquests, there are two things of which we, the great kindred people of the English-speaking race on both sides of the Atlantic have most reason to be proud as having occurred during the century. One of these on our side of the water was the movement that struck the chains from the limbs of chattel slaves (Applause.) And the corresponding movement on this side of the water was that great movement which began in the city of Manchester, and was led by Richard Cobden, and which took the first step in the direction of free trade. (Applause.) I stand here tonight to support the resolution with all my might as the first effort towards another great step – towards the final victory of the principle for which the greatest Englishman of our time strove.

The Program Of The Financial Reform Association
If I only saw in that resolution the abolition of the duty on tea, the duty on coffee, the duty on gold and silver plate, I would not waste much time upon it. It is good to have plenty of tea, plenty of coffee, and plenty of gold and silver plate. But I see in that resolution more than the abolition of the duty upon these things. (Applause.) I see in the resolution what Richard Cobden saw in the abolition of protective duties, a clear and decided step in the advance toward a state of things in which poverty shall cease to exist. (Applause.) And to me the appeal in its support is the appeal that Richard Cobden made to John

Bright: "There are tonight in England women and children dying of hunger – of hunger made by the laws. Come with me, and we will never rest till we repeal these laws." (Applause.)

This is the Free Trade Hall
Some Englishmen think that England is a free-trade country. I don't! Nor measured by the standard of Richard Cobden is it. When the principle for which he fought is fully incorporated in the laws and institutions of this country, there will be no more starvation in England. The men who began the work have ceased from their labors: it is ours to carry it on still further. (Applause.)

You have no more secured free trade than we in the United States have abolished slavery. What we did was simply to abolish one crude form of slavery which consists in making property of the man himself. Slavery still exists in the United States – aye, and with our advances slavery is broadening and deepening. It is no longer chattel slavery: it is industrial slavery. (Applause.) That form of slavery is more widespread, more insidious, and in some of its results more revolting than the system that makes property of the man himself. It is the slavery which comes from the system that makes property of the element on which man must live, if he lives, at all. ("Hear, hear.") They denounce you, Mr. Jackson; they denounce the Financial Reform Association, and say it is radiant, and perhaps worse. Why, that is the very sign and symbol of your mission, the very token by which men may know that your association is indeed

The True Successor Of The Anti-Corn Law League
Did not they denounce the Anti-Corn Law League? ("Hear, hear.") Were not the few men who fifty years ago made memorable, made illustrious, this city of Manchester by meeting here and forming that association – were not they denounced in far worse terms? Why, what were they told by cabinet ministers? To repeal the corn laws! That it would bring destruction upon industry (laughter) – that it would sweep away the very foundations of Church and State. (Laughter.) Lord Melbourne – he was a very intelligent lord, as lords went in his day (laughter) – did not he get up in the House of Commons and did not he say: "I have heard many mad things in my life, but before God the maddest thing I ever heard is this mad proposal to repeal the corn laws." (Laughter.) When a deputation went to another member of the government, Sir James Graham – was not that his name ("yes") – the speakers presented to him the deplorable condition of the laborers, and did not he turn round to them and indignantly say: "Gentlemen, it seems to me you are Levellers.[62] Am I to understand that you consider that the laborers of this country have any interests in the estates of the landowners?" When they presented reports and statistics

showing the great distress in the commercial centers, were they not met just as the men who do such things today are met? Sir Robert Peel[63] and others told them: "This cannot be. Why, the consumption of cotton is increasing as statistics show. This cannot be. Why, statistics show there is more food consumed in England than is consumed per head in Prussia." And did not the Duke of Wellington[64] say, and thank God for it, that this was the only country in the world in which a man was willing to work and to save could get a comfortable existence. (Laughter.)

Denounced? Why, without exception, the whole of the newspaper press of the country denounced this movement. The early free traders had to buy any place they got in the newspapers for the publication of their reports. Men like my friend Mr. Briggs know those things. This generation has to a large extent forgotten them. ("Hear, hear.") Aye, but truth is mightier than all things else. That commercial traveler, that calico printer of Manchester, whom no one had ever heard of before outside of his little commercial circle, set himself against Parliament, in both houses, exclusively, or almost exclusively composed of the landowning class. He set himself against the press, against the Church, against all organs of power and influence. Everything was against the movement but truth, but Justice. They were on its side. (Applause.) And they were stronger than all. ("Hear, hear.")

So do not bate your breath, Mr. Jackson, and men of the Financial Reform Association. You have behind you, if you will but avail yourselves of it, the power that is stronger than anything else – the appeal to men's consciences, the appeal to men's hearts, the appeal to men's sense of right, the appeal to the first perceptions of men, and the harmony of the universal laws. What carried Richard Cobden into the agitation, whose first step was such a quick and brilliant success? No selfish motive. It was because he saw around him lives debased; because he saw around him men willing to work and unable to get employment; because he saw women and little children suffering, starving, dying for what the Creator had provided in abundance. It was that that nerved him to the fight, and that was the spirit of his struggle. That is the spirit that will carry this movement on. (Applause.)

How To Get Rid Of Our Slums

That is what the ticket says I am to speak of tonight. Aye, and consider your slums all over this country, from your great capital, the great metropolis of the world, where human beings are crowded together, hundreds of thousands of them, under conditions of which a savage would be ashamed, up to the north, where your farm laborers, men and women, are housed, I should rather say herded, in bothies.[65] Consider such facts as that 125,000 of the population of

Glasgow are living whole families in a single room ("hear, hear") that most of the miners of Scotland, as the secretary of the association tells me, live whole families or more in one room. Consider that one of your papers, the *Sunday Chronicle* (applause) is publishing. How to get rid of slums? There is a sovereign recipe. It consists of two words – free trade. Carry out the principle of free trade and your slums will disappear.

England And Full Free Trade
Last autumn, when I was stumping my own country for Mr. Cleveland and free trade (applause) I never posed before any audience as a tariff reformer.[66] I would reform the tariff just as the man cut off the tail of the mad dog – right behind the ears. (Laughter.) The protectionists would ask me: "If what you say of free trade is true, how is it there is pauperism in England – how is it that wages are so low when England is a free trade-country?" My answer was: "That is because England is not a free-trade country. ("Hear, hear.") When England is a free trade country all that will have disappeared." England is not a free trade country, although it has started on the road to free trade. But, thank God, the people of England are awakening for another stride forward. (Applause.)

How to get rid of your slums! There was not bread enough in the country. Clearly, the taxation on bread was a bad thing. ("Hear, hear.") There are not houses in the country. Just as clearly, is not the tax on houses a bad thing? ("Hear, hear.") Free trade – what does it mean? It means the removal of all restraint, of all fines on the production of wealth. (Applause.) Free trade is good simply because trade is an important mode of producing wealth, and just as free trade requires not merely the abolition of protective duties, but the abolition of every tax that interferes with foreign trade; so, also, to carry out this principle it is requisite to abolish every tax on domestic trade, or upon any form of domestic production. (Applause.)

And when this is carried out to the full extent, you come to the point where it is but one step further to abolish the primary wrong, the essential injustice, that is today the bottom cause of all social and political difficulties – to secure to every man and to every woman and to every child born in a country, an equal right to the use of the land in that country. (Applause.) To abolish all the taxes that rest upon the products of human industry, to abolish all the taxes that seek to fine and punish men for being industrious or getting rich, what is it but to come of necessity to the taxing of land values for the support of government, and from that there is but one step further – the taking of the whole of the land values for the benefit of the people. (Applause.) I believe there is a finality in taxation, and I believe that neither our chairman nor the Financial Reform Association will be disposed to dispute my assertion, that although in form we propose to substitute one tax for other taxes, yet it is merely in form, and that in reality what we propose to do is to abolish all taxation ("hear, hear") because the imposition of

A Tax On Land Values Would Only Be In Form A Tax

In its nature it is but taking for the use of the community the rent which is due, not to the individual landowners, but to the community. (Applause.) There is another thing in which I would differ. I would not put the objection to taxes that are now levied – the objection to the income tax, to the tax on tea, to the death duty that falls on personal property, and takes away from the children a large portion of what their father may leave them – I would not object to those taxes on matters of expediency, on matters of detail, on matters [of] relative injustice. I believe the true objection to them is on a matter of right.

I believe that when the state takes from the individual what properly belongs to that individual, the state robs the individual! I believe that if a man increases his income by his own exertions, he is entitled to the whole of that income. (Applause.) I believe that if a man imports a pound of tea by sending to the countries where tea is grown some of his own productions, the whole of the tea belongs to him, and that neither a half, a third, nor a tenth of it ought to be confiscated by the community. (Applause.) And I believe that if a man builds a house that house belongs to him entirely. What the individual adds to the wealth of the country and to the wealth of the world, ought to be that individual's against the whole world. And any community can afford to leave it to the individual, if it will only take what belongs to the community itself.

Why, these land values which now escape taxation, to whom do they belong? To the people. And for what reason? Because the people created them. (Applause.) Why do such enormous values attach to the land of this island today? Because the landowners are here? Why, let everybody else emigrate, and leave only the landowners, and how much will the land be worth? (Applause.) It will be worth mighty little, I can tell you. Why are the land values so great in Manchester? Because the whole people of Manchester are here. Why are the land values so great in London? Because of its great population, and the fact that it is a great center of exchanges. Every family that comes and settles here, every public improvement that is made, adds to the value of land.

How Public Improvements Increase Land Values

You of Manchester are engaged in a very commendable enterprise in bringing water from the lake country, and have to pay pretty dearly for it. You have to pay for the privilege of taking the water, for the rain which falls from heaven and collects in the lake country is not supposed to fall for the benefit of the whole people; it belongs to the landowner. (Laughter.) Then you have had to buy a big hillside for the purpose of collecting the rain, and you have also had to pay for the privilege of laying the pipes from the lake down here to the city. This water supply will make Manchester a better place to live in, but who will reap the money value of that? The owners of the land of Manchester?

You people of Manchester have another great enterprise on hand. You are making a canal which you hope will bring up here ships that cross the Atlantic, and make Manchester a seaport. What will be the result? You will do just what Mr. Gladstone told the people of London they had been doing by expending money on the Thames embankment – adding to the fortunes of the landowners adjoining the embankment. Do what you please, improve as you may, you won't find that cloth is of any more value in Manchester, you won't find that horses increase in value, or that iron or houses will sell for more. On the contrary, the tendency is always to a decrease in the value of these things as progress diminishes the cost of production.

The one thing you will find going up in value is land. The effect of your forward step in abolishing the system of protection has been to enormously increase your wealth and your commerce, but the lion's share of this increase the landowners have already got. And although many men besides the land-lords have become rich, it is only by intercepting for a little while what the landlords must ultimately get. You may see the whole process. In the city of London there are large blocks of buildings in business parts of the city which have fallen in to the Dukes of Portland and Bedford. What has been the result? Rents have gone up eight or tenfold. Ultimately the landlord, the man who owns the indispensable element, can squeeze all other owners; and although a manufacturer here and a manufacturer there, if he be sheltered a little from competition by some sort of a monopoly, may make money for awhile, ulti-mately the landlord will get the whole increased value.

Now to end that; to give every man an opportunity of exerting his powers, and of profiting by that exertion; to remove the cause that brings biting poverty into the very centers of wealth, the cause that in the midst of enlightenment is produc-ing ignorance, the cause that is making the average life of a man of the working class only twenty-five years while the average life in the idle rich class is fifty-five years – to remove this you must carry free trade to its ultimate limit; leaving to the individual all that individual industry, or energy, or forethought, or thrift can pro-duce or accumulate: taking for the community that value that attaches to land by reason of the growth of the community. When that is done, then, indeed, will England lead the world. When that is done your prosperity will be so great, the condition of your people so good, that the whole world will look upon you with envy and admiration and the whole world will follow you. (Cheers.)

And if you do not do it first, we on the other side of the Atlantic will. The great struggle has commenced there, and there can be but one result. And so I most heartily support this motion.[67] (Cheers.)

The resolution was then put to the meeting and carried unanimously.

The Chairman: A number of questions have been sent up in writing. I will read out these questions, and no doubt Mr. George will be willing to answer them.

Answering Questions

Question: If free trade will remove poverty, how do you explain the Savior's declaration that "the poor ye shall have with you always?"[68]

Answer: Because he was talking to the scribes and Pharisees and hypocrites. (Laughter and applause.)

Q.: Considering that the land and the wealth it contains are essentially necessary for the happiness and prosperity of a people, are the people justified in adopting any means to regain possession of it?

A.: Justified in adopting any means? Yes. But there is only one means, and that is education. If in any country in the world the people are oppressed and robbed it is simply because they are ignorant. (Applause.) Forcible means can never really accomplish anything, for the reason that force does not educate.

Q.: Do you consider that industrial slavery is caused by individual ownership of land?

A.: I do.

Q.: Do you consider that industrial slavery is caused by the aggregations of capital?

A.: No; the aggregation of capital cannot cause industrial slavery. You can aggregate capital as much as you please, so long as you leave to labor the raw material.

Q.: Please tell us how you mean the people to get hold of the land?

A.: I am glad a gentleman in the audience said some time ago "no politics." The Financial Reform Association ought to do nothing to keep off any Tories who want to help on these reforms. But the only way to get reforms carried through in a country like this is by making them political questions. And the way to get the land back is not to take actual possession of it. Let the landowners keep their land. Do you simply put taxes on land values. (Laughter and applause.)

Q.: Suppose the landowners were made to pay twenty shillings in the pound as taxes, would such taxation prevent the labor market from being overstocked with unemployed laborers, or in any way prevent capital from exploiting the wealth [of] producers?

A.: Unquestionably it would. If you were to tax land values twenty shillings, in the pound, no one would want to be a landowner. (Laughter.) You would see how soon he would want to sell out. (Laughter.) But land will be just as valuable as ever, nay, much more valuable to land *users*, and when there are no landowners compelling natural opportunities to go unutilized, when land is only profitable to the land user, and only so long as he uses it, you will find plenty of land in Great Britain and Ireland for labor to use, and then your capitalist cannot exploit labor. Labor, so long as its feet rest on land, cannot be crushed. What does labor need?; simply one thing – Justice.

All that labor need ask is free play, and that was what Richard Cobden saw. What he saw by the repeal of the corn laws was the beginning of a movement for the repeal of all laws that imposed restrictions. What he saw was the natural harmony of God's laws; what he saw was that the Creator had put here enough for all, and that the only thing that produced poverty and want in this nineteenth century were the restrictions placed on labor. Sweep them away – remove all restrictions. Give to natural forces their free play, let men gratify their innocent desires, and you will have wealth, not in the hands of a few, but of all.

Make no mistake – there is no conflict between labor and capital. (Cries of "oh.") The real conflict is between labor and monopoly (applause), and of all monopolies the monopoly of land which is and must be the standing place, the workshop, the reservoir of all men, is the most fundamental and the most important. Break that up, and then it will be easy work to deal with other monopolies. Don't be led away. Strike at the root of the evil tree; do not go fooling with its branches. Strike for the land, for your rights in the soil. (Applause.)

As for capital, what is capital? Why, it is merely a derivative factor of production. What is capital? It is merely the product of labor exerted on land. Give labor land, and every laborer will become a capitalist. (Applause.) This unnatural divorce system between men who have the capital, men who have the land, and men who have only their labor will cease to be. God never made a landless man. He made the land before He brought men upon it, and He made the land for all men. (Applause.) Give each man his birthright, and then every man who chooses to work and chooses to save a little will become a capitalist. (Applause.)

Q.: To tax land unless the ownership was vested in the state would be to increase the rent, and the balance remaining to the landlord would remain the same. Is that not so?

A.: Oh, no, it is not so.

Q.: If the change you advocate is accomplished, will not the land be cultivated for the benefit of the capitalists, they being able to pay most rent for its use; and if so, how will the worker be better off, seeing that the capitalist will still be his master?

A.: If the capitalist rents the land under that system, if he pays the taxes upon its value, that value will be fixed by the demand of the people, and that value will go to them. If capitalists do all the cultivation and use all the land, why, all that the capitalists are compelled to pay to the land will be shared by the laborers. Put a tax to the extent of its full value on land, and who can afford to hold the land idle? The man who holds it must use it, and to use it he must employ labor or he will have to sell out or give up to somebody who will. (Applause.)

If men then choose to hold great tracts of the country for deer parks, and so forth, as they are doing in Scotland today, they will have to pay the full rate that the crofters would give to cultivate it. (Applause.) If men around your cities choose to hold land idle, as they are doing today, for building rents, they would under that system, have to pay as much as would be paid if the buildings were already on it. (Applause.)

And they cannot shift that tax. The tax on land values, as any economist of repute will tell you, is a tax that must be borne by the owner, who cannot shift it onto the tenant. Abolish the tax that is at present imposed on the tenants of the houses, and the rents of the houses will be lowered. Put that on the landlord, and how will that reduce the supply of land or tend to reduce it? You can tax land values all you please, and there won't be a square inch less land in the kingdom. Instead of making it more difficult to supply the demand you will force land now held idle into the market. If the capitalist is today master of the laborer, it is because the monopoly of land makes labor helpless by depriving it of the element without which labor cannot be exerted. Break that up and labor will be its own master.

The Chairman after a few words explanatory of a point in the resolutions which had been adopted earlier in the evening, moved a vote of thanks to Mr. George which was responded to with general applause.

Mr. Edward Wainscott then took the floor and said: I rise for the purpose of seconding the resolution moved by our chairman. Mr. George is called an American prophet. I only wish that Mr. George had been an Englishman. I believe he is the greatest writer on political questions that this or any other century we read of has ever had. I believe he is reforming the political economy of the world, and that before very long the system known as "George's" will be taught in the universities throughout the civilized world. I have the greatest pleasure in seconding the vote of thanks to Mr. George for his grand lecture this evening. (Applause and interruption.)

Mr. George: There is a knot of men in the middle of the hall who seem to wish to ask a question. If they do, I am ready to reply to it.

A man in the middle of the hall: I have sent up six questions to the Chairman, and the Chairman has asked Mr. George only one. I want Mr. George to reply to the others.

The Chairman: I did not put them to Mr. George, because he had already in effect answered them.

The man in the middle of the hall proceeded to put his questions:

Q.: If the landowners have no right to monopolize the earth's surface, what right have the people to monopolize what is beneath?

A.: The raw material beneath the earth is land.

Q.: If the landowners have no right to claim rent from the agricultural laborers for cultivating the soil, what right have capital monopolists to claim profits out of the commodities produced by the workers, when the materials used come from beneath the land which you say ought to be nationalized?

A.: No one has any right to make a charge upon God's bounty. Rent, that bonus which attaches to the use of land, ought to go to the whole people, and not to any individual. But my advice to you is to find out the distinction between land and capital.

Q.: Why do you not advocate the destruction of all monopolies, so as to give the laboring community, to the fullest extent, the product of their labor?

A.: So I do. I advocate the abolition of all monopoly, and I would begin with the greatest and most important of all – the land.

Mr. George: I have now to move a vote of thanks to our Chairman. I don't think he and I agree in all things, but I hope the day is not far distant when a majority of the members of the House of Commons will be ready to take the step indicated in the resolution. I see that Mr. Stanhope, the representative of a district in the Black country, made a speech last night in which he declared in favor of putting taxes on land values. To such a man give your support. Now, I have the pleasure of moving a hearty vote of thanks to the Chairman. (Applause.). . . .[69]

GEORGE IN ENGLAND
(*The Standard*, June 29, 1889).[70]

Henry George is back in London from his great provincial tour. He is off to Paris in a day or two to attend a conference of land nationalizers, and when he returns he will address a few meetings in Ireland.[71] He is looking ruddier, stronger, and fatter than when he was first with us, and he is full of spirits. A soundhearted, soundheaded man is Henry George; never loses faith in his principles, and is always looking steadily forward to the triumph that is to be. He called at the *Star* office, and plunged at once into the story of his tour.

My speaking trip through this country was ended on Friday night by a speech at Maidstone, Kent. It has been a most gratifying and successful one to me, and has shown clearly the great advance that public sentiment has made in the social question since I was over the country in 1884–[8]5.

How did you notice this Mr. George?

In the audiences, which have been large, in spite of the fact that at nearly all my meetings a charge has been made for admission; and still more in the character of the audiences.

In the enthusiasm, Mr. George, or in the status?

The men who have managed the meetings, and who have filled the platforms, who have generally acted as chairmen, and who have everywhere been present in large numbers, are the active men of the radical wing of the Liberal Party, the men whose sentiments and ideas show what is coming in politics. The number of clergymen taking part, and who have invited me to speak in their churches, is also a very significant indication. This has been specially noticeable among the Congregational body, who represent today the old "Independents," whose present position in English politics is not without a suggestion of what it was two and half centuries ago.[72]

Did you take part in any chapel services about the country?

Yes; for instance, at the great meeting at the city hall in Glasgow, where the religious service was conducted by the Rev. Mr. Cruikshank; in the church of the Rev. Donald Macrae in Dundee, and in the Congregational Church at Newcastle. And I have had a great many invitations I have been obliged to refuse, as my ordinary engagements have been generally for six nights during the week.

You spoke of a change in the tenor of your audiences, I think, Mr. George.

Yes, they seemed very much better informed, and very much more sympathetic than when I was here before. The truth of the matter is that our ideas have been advancing not merely by direct propagation and conscious acceptance, but by diffusion. They are, as the phrase goes, "in the air;" and, so far from our ideas of the single tax seeming now to be strange, it strikes the ordinary man among large classes as something obviously just and expedient. A curious illustration of this change in public sentiment is the idea that I heard expressed in many places that I had changed my position since I was in Great Britain last. The truth is that I stand in the same place, and these people themselves have changed their position.

What were your observations as to socialism?

Socialism is such an indefinite term, and is used so indefinitely in England, that it is hard to answer that in a word. Men who see the necessity of social improvement frequently call themselves socialists, and are called socialists, but as to the true meaning of the term "the state socialist" I found them stronger in London than elsewhere, and not at all strong, even in London. Socialism in this sense must yield to the single-tax idea, which assigns an adequate cause to social injustice and advocates a definite and simple remedy.

Would you be willing in conjunction with the single tax to advocate such social palliatives as the taking over of the tramways and the reduction of the hours of labor, and other reforms in which London is interested?

On the contrary, I advocate the running of tramways by the municipal authority and at municipal expense, as one of the proper functions of the government. We draw the line wherever competition ceases to act. Every business which is in its nature a monopoly is in our opinion a proper subject for governmental control. We are in reality antimonopoly men. We do not believe that there is any really necessary conflict between labor and capital, but that the real conflict is between labor and monopoly; and we would abolish all monopolies and all special privileges, putting all citizens on an equal plane of opportunity, and giving to all fair play. But the most fundamentally important of all monopolies in our view is the monopoly of the land, the indispensable element to all labor and to all life.

As for the reduction in the hours of labor, we regard any action which can tend in that direction as a good sign, as promoting increased intelligence; but we despair of accomplishing any large and general reduction by arbitrary means. In our opinion men do not overwork themselves because they want to, but because they are forced to; and the relief which would come by opening land to labor and giving productive forces fair play, would so increase the opportunities of employment, and so raise the rate of wages, that it would be in a little while impossible to get men to devote the greatest part of their waking life to a mere effort to maintain life.

What of your talked of debate with Mr. Hyndman?

Whether it is to come off or not I do not know. I am informed by the committee in whose hands the matter was left, that the Social Democratic Federation have up to this time refused or neglected to comply with the condition which they at first proposed, that both parties should put up half of the preliminary expenses. I seek no controversy with socialists, but I am, as I said when asked the question, perfectly willing to meet any one of their representatives men under proper conditions.

What is your opinion of the course of English politics in regard to the Irish question, and its bearing on your special subject?

I think the truth that the just and proper relations between Great Britain and Ireland are those that exist between our American states has made enormous progress and is steadily gaining ground. There is among the great body of the liberal body not merely the desire to give to Ireland her just rights, but a warm sympathy with the oppression to which her people have been subjected. And there seems to me on the other side to also be a mergence in Great Britain of the Irish movement into the great democratic movement, and the Irish party in England seems to be taking its proper place in the great English democratic party now forming.[73]

One of the great agencies to this end has, I think, been the *Star*. The establishment in England of a popular paper, edited by a prominent member of the Irish parliamentary party, and that is taking a leading part in advocating the reforms that are desired as earnestly by the English democracy as by the Irish, is at once an indication and a most powerful agency in promoting this change. I have all along believed that this was the true course of the Irish leaders, and I am confident that in this they have at last struck the right track.

What do you think of the commission, Mr. George?

So far as the Irish people are concerned it seems to me that the game's not, and never has been, worth the candle. The effect on public sentiment has been to clear the leaders of charges which everyone who knew them knew to be groundless; but the terrible expense to which they are subjected must exhaust their resources. If Michael Davitt and two or three others of them had discarded counsel and gone into the court for themselves it seems to me that they might have accomplished as good a result at much less cost. But, at any rate, it is a striking commentary upon the manner in which the machinery of the law can be made to give substantial advantage to those who have the longest purses. There is a good deal in this country as well as in America, to make one think there was some method in the madness of Peter the Great, who, after sojourning in London, said there was only one lawyer in Russia, and when he went back he intended to hang him.[74]

Next year, Mr. George, the government will probably complete their scheme of land purchase. What do you think of the situation which will then be developed?

I look to it with a good deal of hope. I think the effect of the debate on the five millions appropriation of last winter was to firmly set the masses of the Liberal Party throughout Great Britain against land purchase on any terms. Mr. Gladstone, Mr. Parnell, and the other leaders, hampered by their previous positions, only ventured to fight the appropriation on matters of detail, but in the radical clubs and liberal associations the effect was precisely the same as if they had fought it on a matter of principle. I have been curious to inquire on this subject from well-informed men in the localities I have visited, and their universal opinion is that even Mr. Gladstone, powerful as he is, could not bring the masses of the party into any acquiescence in a scheme of land purchase.

The present coalition of Tories and Unionists is of course strong enough to pass any bill they like, and the effect of the adoption of such a scheme as is proposed will be to delay the settlement of the Irish land question, and perhaps to relegate Ireland for sometime to the rear in the general advance; but as educational effects on the British people can, I think, hardly be overestimated.

Do you think the Liberal Party in Parliament will oppose it?

I think they must oppose it if from no higher motive than political necessity. To accept such a measure from the Tories would virtually be political suicide. And if any section of the present Liberal leaders acquiesces in it the effect will, it seems to me, be like the Unionist split, to further purge the opposition of elements which at present retard its advance.

A WONDERFUL IMPRESSION ON THE BRITISHERS[75]

Henry George has made a wonderful impression on the Britishers. The London and provincial press agree in saying that he has swayed the masses more powerfully than any man who has appealed to them from the platform in a generation.

HENRY GEORGE AND SAMUEL SMITH, M.P.:
They Meet in Debate Before the National Liberal Club in London; The Hall Crowded and the Liveliest Enthusiasm Prevailed; Loud and Long Cheers at the End
(The Standard, July 13, 1889).[76]

Henry George and Samuel Smith, M.P., did not quite manage to settle the land question between them last night, but they made to its consideration a contribution of the first importance. The conference room at the National Liberal Club was filled with clearheaded radicals – among them a sprinkling of ladies – and the volleys of applause which wavered now one way now the other, showed that a close and impartial hearing was being given to the two champions. Mr. Halley Stewart, M.P., who presided, had to exercise his authority more than once to restrain some of the more enthusiastic listeners, who wanted to go in and help the debate along.

Of the two hours' solid speaking it is obviously impossible to give more than the bare outline of the arguments on each side. Under this treatment which involves the omission of much descriptive and illustrative matter, Mr. George's speeches are especially sure to suffer. But Mr. George has this advantage – that the omissions may be filled in from his own works, which are now at least superficially known to everyone who pretends to be instructed in political affairs.

Mr. George opened the debate with some definitions. By a single tax he meant a tax levied on land values only, a single basis, in fact, of taxation. By land values he meant the value attaching to land irrespective of improvement or use made of it, the value which grows by the increase of population and general improvement. He founded his contentions first on the ground of Justice, arguing that that which man provides by his exertions belongs to him entirely and the community ought not to take any part of it except under pressing necessity. It was, he contended, inexpedient to tax wealth – using the term in the strict sense – in any form. "It is stupid to tax the things which we all want. Tax capital in any of its forms, and you tend to drive it out of the country." But that cannot be so with land values. In the next place the single tax would enormously diminish the complexity of government and reduce the number of officials and the temptations to fraud and evasion. It would be a tax only in form. It would take from no individual that which was due to his exertions or his savings, but simply what belongs to the community.

Mr. Smith has a very weak voice, which puts out of the question any attempt to rival Mr. George oratorically. But he is a man of business, and he marshaled his arguments in business-like fashion. He dissented from his opponents scheme firstly because he believed it would not accomplish what was desired, and secondly because he could only regard the project as immoral. Appropriation of rent by taxation really meant confiscation of rent. Now, the rent of agricultural lands in the United Kingdom amounted to about £60,000,000. He was informed by one of the largest land agents in the country that of the total rentals an average of one-third was spendable personally by the landlords.

But Mr. George proposed to confiscate the whole. Further consols[77] and lending on land were the chief directions in which trust moneys were invested. So they were asked to confiscate the provisions that have been made for hundreds of thousands of widows and orphans. "The fact is," quoted the honorable member, "land is the platform built tier after tier on the credit of the country; and to pull out the lower story and say that you hurt no one but the owners is lunacy." Who knew but that another Mr. George, going a little further – that is, Mr. Hyndman – would propose to confiscate capital. State ownership of land, he further contended, always meant poor cultivation.

Mr. George pooh-poohed the terrible forecast of the destruction of banks and commercial concerns generally. That was, he said, much what was prophesied as to the Channel tunnel, what the Tories said would follow the abolition of the Corn Laws, and what in the United States was predicted would follow the abolition of chattel slavery. He agreed with Mr. Smith that there had been progress in an age which saw the introduction of the steam engine, the electric telegraph, the abolition of the protective system, and so on. But had the advance been commensurate, and had the gains been fairly shared? "Only some hundreds starved," said Mr. Smith. "But there are hundreds and thousands who die of diseases really caused by starvation." Mr. George hammered this nail well home amid the cheers of his audience.

Mr. Smith failed to see that the proposed remedy would give them more access to the land. He also contended that to the working classes relief from other taxation by the adoption of a single tax would be trivial. The burden of revenue was at present borne by taxes on alcohol, tobacco, and tea. "Does anyone think that to remit these taxes would turn the poor into rich, the drunken into sober, and give work to the workers."

Real property – that is, land and houses – was taxed for imperial purposes to the extent of about 40%, but personal property was only taxed to about 2% of its capital value. There was no form of property that was more properly subject to taxation than personal property, such as that of the great moneyed capitalists. But Mr. George would come down on peasant owners of land. Of all property owners the landowner had the smallest income for his own use.

Mr. George said that the only way of taxing the capitalist was to go for the property – which was ultimately the land. In taxing capital they did not tax the capitalist, for the tax ultimately rested on the consumer. Of course, the landowner would object to the new proposals, for everywhere he would have to give room for the land user. That was the great advantage. Emigration was indeed, not merely to get land, but to get the reward of labor. It would only be where a community had arisen and given value to land that the tax would operate. It was not a question of agricultural land values alone, but of town and mineral-bearing districts.

Mr. Smith, in closing the debate, returned to his muttons. He was sure that the single tax spelt confiscation. But further he contended that the difficulties over the assessment of what he called state rents would cause to grow up a great bureaucracy. If there were not a government valuation and land were put up to competition, rents would be much higher, and hundreds of farmers would be evicted. Finally, he argued that his opponent was upsetting the first principle of finance – that taxation and representation should go together – by removing all taxes from the greater part of the people.

There were loud cheers as the debate came to a close, and votes of thanks were accorded to the two principals.

THE WARNING OF THE ENGLISH STRIKES
(North American Review, October, 1889).[78]

How it may come that the New Zealander shall yet sit and meditate on the broken arch of London Bridge, the strike of the London dock laborers gives something like a suggestion.

War is the great destroyer. Of all wars, civil war is most destructive. Of all civil wars, that which rages, not between different sections of the same country, but between different classes in the same territory. And most destructive of all civil wars of this kind is that waged in great cities.

Such strikes as that in London are in reality incipient civil war of this kind. Passions that make man the most destructive of animals are aroused. They find expression, it is true, only in passive, not in active form – in refraining rather than in doing. But it is like superheated water, which, so long as confined, retains its liquid form. Once let the pressure be relieved, and with explosive force it flashes into steam.

Consider the spectacle that the banks of the lower Thames have presented for some weeks past. The wheels of industry blocked, commerce paralyzed, perishable cargoes rotting, ships unable to go to sea, trade driven away, enormous losses going on, ordered armies of tens and scores of thousands parading, great bodies of men fed by public rations – for the time it could hardly have been worse if the Channel fleet had been annihilated and Continental squadrons blocked the river's mouth.

Yet law and order have reigned throughout. The forces that keep the peace are strong in London. The police are numerous, of splendid physique, and well-disciplined, and large bodies of the flower of the regular army are constantly in reserve. And the leaders of the strikers have used their influence to prevent violence. But what is going in London is nevertheless war – a contest in which both parties have been trying to inflict loss and injury on each other, counting for success on doing so. Only passive war, it is true; a contest of endurance, not of physical force. But the war spirit was there – the sense of injury and feeling of animosity; the passion that leads to the taking of life and destruction of property. And the danger was there. It needs but an accident to convert passive into active energy. [John] Burns said two weeks ago that his influence alone had prevented the firing of London in several places.[79]

Greatest of great cities, the world's metropolis, Babylon of Babylons, yet steadily growing with accelerated rapidity, London is today what New York and Chicago and St. Louis and San Francisco promise to become; the type towards which all great cities tend. It is in London of all places on earth that one may see and feel the strength and weakness of modern civilization, its glory and its shame, the high possibilities it is opening, and the explosive forces it is generating. There is London and London, it is true – the London of society, of science, of politics, of religion, of philanthropy, of business, of amusement. And he who has password and key may see one London and be hardly conscious of the existence of any other. But there is also the London of hard strain and bitter pinching, of want and misery, of vice and degradation.

And if one strives to realize what the whole great city is and what it holds, the streets of London are the place where it may most clearly be seen that the tendencies of modern civilization are towards catastrophe; and that, as Macaulay saw, the Huns and Vandals that may be yet to come, will come not from without, but from within.

A few weeks ago I stood [on] Bond Street before a painting in a dealer's window. It was by a noted painter, Moscheles, but of an everyday scene.[80] Not only an everyday scene in London, but in New York and other great cities. In the background fine shops, with the passers and loiterers one sees in fashionable retail streets; at the curb a private carriage, the liveried coachman erect on his box and a fur-caped footman springing to open the door for a richly-dressed lady coming out of a jeweler's establishment. In the foreground, three sandwich men tramping along.

I noted for a few minutes those who stopped: They glanced at this picture a moment, and then, as though it were too familiar for attention, turned to [a] picture of [a] mountain or a seacoast, of fruits or flowers, or the graceful female form. A man with high-lettered hat and long oilcloth coat all printed over glanced in a moment from his handing out of circulars. I asked him what the picture was. It was a picture advertising a theater, he said.

It was more. Into this picture of familiar things the painter, with the subtle power that is in his art, had put the problem of modern civilization. Underneath it, a piece of cardboard bore the legend: "In the year of our Lord, 1889!"

Almost twenty centuries; and in the greatest and richest of Christian cities, whence missionaries are sent to Asia, to Africa, to the islands of the sea, human labor is cheapest of commodities, and man, "the roof and crown of things," is turned into a signpost! It is this paradox and problem that this London strike brings out.

This strike of dock laborers is, in many respects, the most remarkable of the industrial conflicts which in recent years have been so many; remarkable because of the class of men embraced, the endurance manifested, and the sympathy excited, and for the growing ideas and new influences it discloses.

The London dock laborers consist of a small class who have something like steady employment, and a larger class of casuals who are taken on by the hour as needed. The work being unskilled labor, the London docks have been the last chance of unemployed men physically able to do such work. It is at their gates that the pressure for employment has most strikingly shown itself. Around them have gathered every morning thousands of men – men with hungry wives and children, men who had walked the streets all night, or got what rest they could in alleyways, beneath railway arches, in doorways, or behind boxes – waiting the appearance of the contractors or dock foremen to pick out those to be set to work, and then pressing, scrambling, fighting for the chance like hungry dogs for a bone. Journals like the *Pall Mall Gazette* have made their readers familiar with these scenes, not merely by pen-pictures of onlookers, but by accounts from men who have gone amid the crowds and struggled for work; and sympathy has been excited which has shown itself substantially during this strike.

But for this deep and wide sympathy the dock laborers would have been unable to maintain a strike for more than a few days, if even beyond a single day. Besides the contributions that have flowed in through newspapers and directly to the committees, every parade and public meeting has been made a means of collecting money from the crowds, some £400, mostly in pennies, having been taken in at the Hyde Park demonstration; and the streets in the East End have been scoured by strikers' collectors, who accosted every passer and boarded every omnibus. Besides the contributions of money – in which far-off Australia bore the palm – the shopkeepers in the vicinity of the docks made liberal donations in kind, even the pawnbrokers having reduced or waived interest on articles pawned by strikers or their wives. The Salvation Army soup houses reduced their prices to strikers and their families one-half at the first, and afterward, I believe, dealt out food on ticket without charge.

All the charitable and religious societies working in the east of London seem, in the same spirit, to have done what they could to assist and support the strikers. Pressure has been brought to bear on the dock directors by city businessmen to induce them to yield, while the religious elements working to the same end included such extremes as General Booth, of the Salvation Army, the skirmishers, so to speak, of the low wing of the Established Church, and the Bishop of London, its official local head; Albert Spicer, the leading Congregational layman; and Cardinal Manning, the foremost representative of English Catholicism.[81]

Perhaps even more striking than all this – at least even more ominous to "things as they are" – is the fact that the policemen detailed to march with the first procession of strikers subscribed among themselves to pay for a band, and that the guards in their barracks at Birdcage Walk cheered the Hyde Park procession as it passed. It was when the French guards sided with the mob that the Bastille fell. And when the day comes that policemen refuse to club and soldiers to shoot men to whom they are bound not merely by human but by class sympathy, the guarantees of the existing order, on which all over Europe the House of Have so confidently leans, are gone.

It would be a mistake to suppose that all this sympathy which has enabled penniless, unskilled men to hold out for a month is merely sympathy with poverty and suffering, such as might go out to victims of flood or fire. It is something more. It has in it the feeling – ranging from uneasy suspicion to passionate conviction – that the dock laborers are victims of social injustice. It has in it, in large degree, a desire to do more than to help the dock laborers – the desire to raise the spirit and promote social discontent among the most downtrodden of the English people. And it is an evidence of the growth of such discontent that this strike is most significant. It is in this respect far more significant than any of the strikes of the skilled trades.

An English politician of the first rank, then a cabinet minister, said to me some years ago:

> In spite of its shocking contrasts, the existing order of things is secure in England. Go to the entrance of the park of an afternoon in the season. There you will see in long procession the utmost extravagance of luxury, the very ostentation of wealth. Look at the faces of the poorer people crowded together to watch the show. You will not find in them the expression of envy or hatred, the consciousness that their robbery provides this luxury, but of pure admiration. This is really the feeling of one extreme towards the other. The poor admire the rich. The man who cannot find work does not feel bitterly towards the great landlord or capitalist, but towards the man whose competition he thinks is depriving him of employment. The man who has three shillings a day envies only the man who has three and sixpence. And so through all gradations of society, each class is more than content to see others above it, because of the conscious superiority with which it looks down on those beneath it.

This is indeed, the strength of the existing order. But this keen observer made, I think, the same mistake as one who, from the temper of the crowds that watched the carriages rolling from Paris to Versailles something over a century ago, had argued the permanence of the ancient regime. And though he is doubt-less now of the same opinion as when he thus spoke, yet during the seven years that have passed strong influences have been at work beneath the surface of British politics and society. On the one hand, a recognition of the fundamental injustice which denies to the great body of the British people their natural, equal, and unalienable right to the use of British soil, and makes the element on which and from which all must live the private property of some, has spread widely and deeply. On the other hand, Socialism has been making way. The two things are widely different and in some respects antagonistic, but both foster social discontent.

And beneath all this is the effect of compulsory education, of the extension of the franchise, of cheap newspapers and cheap books, of the efforts that have been making to improve the intelligence, the morality, the physical well-being of the poor – from the teetotal movement, the Salvation Army, and the People's Palace, to the flower missions and the taking of children from the slums for a month, for a week, or a day in the country, there giving them glimpses that forever after make them dissatisfied with what before seemed natural because it was all they knew.

None of these agencies have yet completed their work; they are only beginning it. The board schools have not yet been in operation for twenty years. The only really radical halfpenny paper is but two years old. The system of registration which disfranchises great bodies of workingmen, and the property suffrage which gives to richer men two, four, and six votes apiece, still prevents democratic strength from fairly showing itself. And the generation is not yet fairly on the stage in which other ameliorative influences will tell.

But the steady, ofttimes cruel, work of school boards and truant officers in driving even hungry children to school is beginning to show in the decadence of local dialects and the disappearance of illiteracy. The *Star* has a circulation of over two-hundred thousand. John Burns,[82] the leader of this strike, is a member of the County Council, on which peers of the realm are glad to sit, while candidates for office find it to their interest to show sympathy and send money to the strikers.

Let it be granted, as some contend, with an air of thus settling all social problems, the condition of the masses is, on the whole, growing better. Man is not an ox for whom any standard of contentment can be fixed; who, given so much food and drink, will fill his stomach and chew the cud. His desires grow by what they feed on; are aroused by glimpses of new gratifications. According to De Tocqueville,[83] it is when things are growing better, not when they are growing worse, that revolutions come. This at least is certain: that hope is an essential element in the social discontent that shatters institutions. American slaveholders were right when, in the interests of the "peculiar institution," they made it a crime to teach a slave to read, and sought as far as possible to prevent his seeing or hearing of a free black.

Each year as it passes is making English thought and English conscience more restive under existing social conditions; is making more certain either peaceful readjustment or blind and forcible revolt. Between these two everyone of any influence must take his choice. If he will not aid the one, he is helping on the other.

As I write, pulsing flashes in the cable mirror have told the Western world that the strike is ended and London breathes free again. After a month's strife the Cardinal whom men know only to reverence and love, and whose strength in meekness has been abated neither by years nor by Rome's purple, has effected a compromise.

Think of it. So has our civilization soared that what happened in London when the sun was sinking is told in New York ere the shadows have more than begun to lengthen. Think of what advances in the arts of production this suggests. Then think of what this London strike so forcibly brings out – that in the distribution of wealth we are in reality no further advanced than when barbarian fought barbarian.

We girdle the earth; we weigh the stars; we rule scales to the hundred-thousandth of an inch; we make instruments so delicate that they record and give back again the finest inflections of human speech. Yet when it comes to dividing the product of their joint exertion between labor and capital, we have nothing better than:

"The good old plan,
That they should get who have the power, and they should keep who can."[84]

The London strike is over! There was war. There is a truce. And with the next quarrel war will begin again.

The lesson of this London strike. What is it but the lesson of the strikes and lockouts in the Illinois coalfields; of the New York freight-handlers' strike; of the Chicago strike, out of which grew the explosion of the Haymarket bomb and the hanging of the Anarchists; of the Southwestern Railroad strike; of the Pittsburg[h] riots[85] – the lesson of all strikes, coming sharper and clearer as the years go on? It is the lesson that the social problem cannot be ignored; that unless the moral advance of our civilization is commensurate with its intellectual and material advances, civilization itself is doomed. A civilization in which the arts of production advance by leaps and bounds, and distribution is left to war, though but passive war, is like an Eiffel Tower standing on one leg.[86] The higher we carry it the more certain the final crash. That is what we are doing here in the United States, as there in Great Britain. And every year it becomes more dangerous.

For every year society becomes integrated, industry more complex and interdependent, and the stoppage of one function more likely to involve and paralyze others. You may cut a worm in two, and both pieces will live. But a bodkin's touch in a vital place, and your man is dead.[87]

The loss caused by the London strike is estimated at from two to three million pounds. Had it continued in its highest intensity during the whole time, the loss would have been much greater. Had the stevedores and the wharfmen and the coal-handlers stayed out as long as the dock laborers, every factory in London and its neighborhood might have been closed. As it was, coal rose to forty-five shillings a ton, mail steamers were delayed for days, excursion steamers with full passenger lists had to give up engagements, while colliers, turning back, literally carried coal to Newcastle. The gas-stokers refused to come out when asked, for they had just won a concession on a threatened strike of their own, and sent a donation instead. Supposing they had not won the concession, and had come out, and left London in darkness but for a night – London, where no man knows his neighbor; houses certain in such cases to be sacked, and their spigots set flowing ere the peal of twelve! The manifesto of the strike leaders calling on all workmen to stop work fell flat. But supposing it had been followed in only a few of the more vital occupations, as at a time when the strike spirit was more rife it might have been, what would have been the loss? and the danger?

As it is, the strike has cost from ten to fifteen millions of our money [dollars], to speak of no more than money cost. What has it settled? There is a gain in some ways, but nothing is settled. The dock companies will hesitate before refusing the next demand that involves a strike, and other employers will be warned by their loss. But on the other hand, the substantial success of this strike will prompt . . . others. The question of division between the employer and employed is still left to force; the labor question, the question of questions for civilization, is untouched.

See precisely what has been the gain. Fresh spirit has been infused into a downtrodden body of workers, and the social question again forced on public attention for the warning of those who have eyes to see and ears to hear; the Socialists have made a gain, and their propaganda will go on with more ardor and zeal than ever; and a considerable body of the strikers will, for a while at least, have better wages and more permanent work. But the dock companies will employ fewer casuals. So far from any vent being opened for the mass of unskilled unemployed labor constantly congregating in London, such poor vent as the docks afforded will be narrowed, perhaps closed, for some of the dock companies do not employ casuals.

This is the hopelessness of trades unions and strikes, so far as any settlement of the labor question is concerned. It is like leveeing a river subject to flood. Every levee that is raised requires constant watching, and every new levee increases the pressure on all. Nay, the illustration is hardly strong enough. For the rise of the water also increases the swiftness of the current that carries it off. But the restrictions with which trades unions keep each their own little territory from inundation by unemployed labor do not add to the facility with which that labor finds employment.

Am I, then, opposed to strikes? My answer would be: No! A strike is the necessary weapon of the trades union, and without it the trades union would be of as little effect as a prohibitory act without a penal clause. I believe, as I have never neglected an opportunity of telling workingmen, that trades unions can accomplish nothing large and permanent, and that the method of raising wages by strikes is the method of main strength and stupidity. But I also believe that trades unions and strikes, and especially among such a class as the London dock laborers, may so raise the spirit of men, so, temporarily at least, improve their condition, as to enable them to act in a more promising line. There is no hope from the very poor. They are as dangerous as the very rich.

The organization of the dock laborers in Glasgow and Belfast,[88] which preceded the London strike and won considerable concessions, was initiated by Scottish single-tax men, and the money which enabled the Glasgow strikers to hold out until they brought the employers to terms was advanced by an English single-tax man. These men look on labor organizations and their necessary weapon, the strike, merely as a means for infusing heart and hope into a down-trodden class, and so improving their condition that they may be able to see in the monopoly of the natural element of all production the real cause of the unnatural competition in the labor market, and in the restoration of equal rights in the soil, the simple remedy to be applied through the ballot. The London organization and strike seems to have been initiated and managed by Socialists, though supported by all who have sympathy with the condition of the laborer, and brought to a final close by the offices of Cardinal Manning.

Nor do the thoroughgoing Socialists hope by mere trades unionism and sporadic strikes to accomplish more than preparatory work. In fact, they are opposed to trades unionism as it has developed in Great Britain and in the United States, as forming an aristocracy of labor as will enable labor to appro-priate all the tools and means of production, to control all industry, abolish competition, and do away with the wage system by putting everyone on the payroll. It must have been a glad Saturday afternoon to them when the mani-festo calling for a general strike in London was issued. The hope proved delusive, but for a few hours at least they must have thought that the time long waited for had come.

Now that I have answered the question, some readers would like to ask: Let me put one. I would like to ask the intelligent, well-to-do-people, of whom the readers of the *North American Review* on both sides of the Atlantic are so largely composed – the professors, clergymen, doctors, and lawyers, the bankers, merchants, manufacturers, and capitalists – whether they are in favor of strikes. I do this because on both sides of the Atlantic the influence of this class is, in the main, passively or actively exerted in favor of strikes.

I know that there is a good deal of gush about profit sharing as preventing strikes, or arbitration as taking their place, talked by, or rather to, this class – for they, and not workingmen, are its consumers. But this amounts to nothing, unless it be to the admission that organization among workingmen, with some method of enforcing their "reasonable" demands, is, in the nature of things, necessary. The profit sharers assume that all employers have profits, and that wages called by another name would cease to be wages. The believers in arbi-tration assume that men who have the power of enforcing their demands will consent to submit them to arbitration, or they vaguely contemplate some kind of courts which will have power to compel employers to pay wages they do not wish to pay and workingmen to work when and where they do not want to. This is the only sort of arbitration that would prevent strikes, for all human law rests ultimately on force, and though a league of nations might put an end to international war, it could only be on the condition that the power of the league should be turned against the nation that refused to obey the mandate of its tribunal. The profit sharers and the arbitrationists ought to go to the Socialists, where they belong, for the State Socialists, with their organization of all industry and fixing of all prices, are the thoroughgoing profit sharers and arbitrationists. Indeed, they are there already, for they are, in fact, but rosewater Socialists.

There are also those who condemn strikes with the confidence and vigor of men who, from premises in which essential facts are suppressed, argue, by logical methods, to false conclusions. They say that it is the right of every employer to employ, and of every workman to work for, whom he pleases and on what terms he pleases, and that while workmen have an unquestioned right to stop work individually, they have no right to combine to force others to stop. These men have shut eyes for boycotting by employers, but are alive to the wickedness of boycotting by workmen. Against combinations of capitalists to freeze out business rivals, to blackmail and rob under forms of law, they have little or nothing to say, but are bold as lions in inveighing against the viola-tion of personal liberty by labor combinations. Though there is an implied falsehood in their premises, the vigor with which they push to their conclu-sions is better than the weak tea of the shilly-shally school. But their day both in Great Britain and in the United States is gone, and what influence they may exert on workingmen is the opposite of what they wish.

But the position of the main body of the class I speak of, as shown in press, pulpit, and university teaching, has, in certain fashion, advanced beyond that of the ultras who would put down strikes and substitute nothing. Generally it regards strikes as all right if not carried too far, or, at least, as a necessary evil. Not merely is the influence of this most influential class mainly exerted to prevent the spread of ideas that aim at something better, but it acquiesces in and fosters ideas that look to restriction, regulation, and interference as the only way of doing anything for workingmen.

I have not space to point out how in England the real strength of Socialism comes from the upper rather than from the lower classes, but in some respects this is obvious. A committee of peers, the chairman being Lord Dunraven, has been for sometime taking testimony in regard to the sweating system. The character of the measures they will propose is clear in advance. To cure evils caused by restriction they will propose more restrictions; the enforcement by law of such prohibitions and regulations as trades unions try to enforce by combinations and strikes. Michael Davitt raised in Ireland the cry of the land for the people. How have the ruling class striven to head off the agitation thus begun? By gross interferences with what they declare to be the rights of property, by stepping in between man and man and fixing prices. And that not sufficing, by furnishing one particular class with capital to buy farms at the cost and risk of the whole body of taxpayers. What is the difference in principle between supplying Irish tenants with money to buy farms and supplying English operatives with money to buy factories, or London costermongers with money to buy donkeys? And since the purse of government is only the purse of taxpayers, since governments produce nothing, but can merely give with one hand with what they take with the other, what is the difference, save as a matter of adjustment, between furnishing money to buy these things and the simpler plan of taking them from one set of men and handing them over to another? The difference between thoroughgoing State Socialists like Mr. Hyndman and the majority of Parliament is not of direction, but of degree. And in this country the same tendency may be seen. What is our protectionism but a form of Socialism?

The great loss of the London strike falls on noncombatants. This is the case with all such strikes, and increasingly. And designedly. Just as towns are bombarded to make garrisons capitulate, so it is the true policy of strikers to make the general public feel injuriously their opponents' obstinacy.

As to strikes being all right so long as they do not involve force, why, all strikes in occupations which unemployed labor can enter must involve force of some kind. This London strike – thanks largely to Mr. Burns, who is really a superior man, and of whom I remember saying to my companion when some five years ago, I first met him, that if he lived he would be heard of, and thanks to such men as Cardinal Manning, to say nothing of the police – has been, by all accounts, a most decorous and well-bred strike. Nevertheless, there was something that kept men who would have been glad to take the places of the strikers from applying for work, and I read in a London paper of a train of meatwagons, which the strikers mistakenly thought had come from the docks, being compelled to turn back. Did the drivers turn back merely because they feared the pickets would feel bad if they passed on?

Are strikers necessary? Under present condition in which the opportunity of employment is a privilege, in which men talk and think of those who "furnish work" as benefactors, they are inevitable, and must increase in magnitude and intensity. When wealth is concentrating in great blocks, when capital is combining in all directions, the growing intelligence and increasing aspirations of the laboring masses will not permit them to be crowded to the wall without struggling. Under these conditions strikes can only be prevented by laws which will destroy liberty and put aside the rights of property. But is there no alternative?

What is the real justification of the strike, both in the minds of the men who engage in it and in the minds of those who support it? Is it not that the men who thus try to force their employers have no power to employ themselves?

What more obvious, stated nakedly, than that, while all men have a right to work or not to work as they see fit, no one has a right to prevent anyone else from working, and that no one has a right in any way to force another to employ him or to compel him to assent to the terms of such employment. Yet declare this, as applicable under present conditions to mere laborers in London or New York, and you but mock them. For to say nothing of the minor restrictions and taxations which prevent men from working, our treatment of the natural basis of all work is one which prevents men from working.

Go up in imagination, as it were in a balloon, above London or New York, or any city in which unemployed men are struggling for work, or preventing others from working in order to compel some poor little advance in wages. Look down, as it were, from a height.

What is man, the animal who builds cities, and excavates docks, and lays wires under the ocean and drives ships over it? Is he not a land animal, whose very body is composed of land? What is his production but the bringing forth on land of materials drawn from land, by moving, combining, separating them so as to satisfy his desires? Look, and in every direction see land half-used or not used at all. Why should there be any scarcity of work; why should men willing to work suffer and strain for want of the things that work produces, while land, the natural source and means of all production, is so abundant? There is no reason in the nature of things. The reason is simply that the natural element on which all men must live and work, if they are to live and work at all, is by human law made the exclusive property of some men, who thus can and do prevent other men from working, and rob them of the produce of their work.

Here is the root of the social problem, of all the paradoxes of our modern civilization.

The lesson of the London strike – its seems to me to be that modern society has but the choice between the single tax and Socialism, between Justice and war.

NOTES

1. This first editorial on George in the London *Times* (on September 14, 1882) was written by John Milne: PDM. The following lines appeared as a footnote: "Progress and Poverty: An Inquiry into the Cause of Industrial Depressions and of the Increase of Want with Increase of Wealth. The Remedy." By Henry George. New York: D. Appleton & Co. 1880. "The Irish Land Question: What it [I]nvolves and [H]ow [A]lone it [C]an be [S]ettled. An Appeal to the Land Leagues." London: William Reeves.

2. A reference to the latter-named work in the footnote above.

3. George, "The Land Question," *The Land Question*, pp. 9–10.

4. Gaius Marius (ca.155–86, BC) was a Roman statesman and general, who was popular with the people. Between him and Lucius Cornelius Sulla (138–78, B.C.) another Roman general from the ranks of the aristocracy, a bloody civil war broke out that the latter won, becoming dictator in 82 B.C.

5. George, "The Land Question," *The Land Question*, 13.

6. A reference to the depression that began in 1873 and ended in 1876.

7. Thomas Babington Macaulay (1800–1859) was an English historian, author, poet, and Whig parliamentarian. His most famous work is *The History of England from the Accession of James the Second*. Edward Gibbon (1737–1794) wrote the monumental *Decline and Fall of the Roman Empire*, which appeared in six volumes between 1776 and 1783.

8. A reference either to the American economist Mathew Carey (1760–1839), who advocated protective tariffs, or more probably to his son Henry Charles Carey (1783–1879), who was also an economist instrumental in the development of economic nationalism.

9. See Henry George, *Progress and Poverty*, pp. 29–30.

10. Ibid., 86.

11. Thomas Malthus (1766–1834), an English economist and forerunner of modern demographic studies, the author of the 1798 work *An Essay on the Principle of Population*.

12. George, *Progress and Poverty*, 104–105. Jonathan Swift (1667–1745) was the author of *Gulliver's Travels* which was published in 1726.

13. George, *Progress and Poverty*, p. 110. Annie (Wood) Besant (1847–1933), an English social reformer, was unsuccessfully tried on the grounds of immorality for her advocacy of birth control. She also championed theosophy.

14. Confucius (551–479 B.C.) was the Chinese proponent of a system of ethical teaching.

15. Possibly a reference to the K'ang Hsi reign of Sheng Tsu (1654–1722), a Manchu emperor who ruled from 1662.

16. George, *Progress and Poverty*, pp. 112–113.

17. John Ramsay McCulloch (1789–1864) was an English economist.

18. A reference to Swift's 1726 work *Gulliver's Travels*. See footnote no. 12.

19. George, *Progress and Poverty*, p. 127.

20. Ibid., p. 142.

21. Ibid., p. 146.

22. Ibid., p. 150.

23. Ibid., p. 224.

24. Ibid., p. 534.

25. Ibid., p. 537.

26. Samuel Taylor Coleridge (1772–1834) was a leading English romantic poet, critic, and philosopher noted for the "Rime of the Ancient Mariner" and "Kubla Khan."

27. Sir Thomas More (1478–1535) was a famous English statesman and humanist who wrote *Utopia*. (See pages 1–2). August Comte (1798–1857) was the founder of sociology and the school of positivism. Brook Farm, an experimental community founded in Massachusetts in 1841 by a number of transcendentalists, was disbanded six years later. The Oneida Community was established in 1848 in New York as a communal religious refuge and dissolved in 1880.

28. "Mr. Henry George on the Progress of His Principles," *Pall Mall Gazette* (Jan. 7, 1884): PDM.

29. Jay Gould (1836–1892), one of the more famous railroad robber barons, schemed with others to corner the gold market which precipitated the financial panic in 1869 known as Black Friday. Cornelius Vanderbilt (1794–1877) was another well-known robber baron.

30. A reference to the 1880 election between the Republican winner James A. Garfield (1831–1881), the Democrat Winfield S. Hancock (1824–1886), and the Greenback-Labor candidate James B. Weaver (1833–1912).

31. Karl Marx (1818–1883), founded his socialism of economic and social development in German historicism and metaphysics, has given birth to a multiplicity of interpretations.

32. "The Morning Papers: Epitome of Opinion; George the Fifth's Utopia," *Pall Mall Gazette* (Jan. 11, 1884): PDM.

33. This article, "Mr. Henry George at Oxford: Disorderly Meeting," is taken from the Appendix to George J. Stigler, "Alfred Marshall's Lectures on Progress and Poverty," *The Journal of Law and Economics* 12 (1969): pp. 217–226. Reprinted from *Jackson's Oxford Journal* (Saturday (Mar. 15, 1884). The chairman's introductory remarks are omitted. Gratitude is extended to Fred Harrison for obtaining these proceedings.

34. The Rothschilds were a prominent international family of bankers. Their name has become synonymous with wealth.

35. Anthony A. Cooper, the seventh earl of Shaftesbury (1801–1885), was a noted social reformer who promoted legislation favorable to the working classes and the poor. For John Stuart Mill see footnote no. 73, page 101. Arnold Toynbee (1852–1883) was an English reformer and pioneer economic historian.

36. The Scottish economist Adam Smith (1723–1790) is best known for the 1776 work *An Inquiry into the Nature of the Wealth of Nations.*

37. "Mr. George at Oxford," *Pall Mall Gazette* (Mar. 8, 1884): PDM.

38. This article is taken from the *Pall Mall Gazette* (Jan. 14, 1885): PDM.

39. For Chamberlain see footnote no. 83, page 102.

40. "Henry George in England: Plan of His Single Tax Campaign to Help the Irish," *The Standard* (Mar. 16, 1889); it originally appeared in the *New York World* (Mar. 10, 1889): GRD.

41. Many Irish who had emigrated to the United States donated sizeable amounts of money for the cause of freedom for their homeland.

42. Democratic presidential candidate Grover Cleveland (1837–1908) lost to the Republican Benjamin Harrison (1833–1901) in 1888, beating him four years later. George's support of Cleveland on the issue of tariff reduction split the ranks of the single taxers, some of whom opted for a separate uncompromised movement.

43. "Henry George in England: The English Campaign" is taken from *The Standard* (Mar. 16, 1889); it originally appeared in the *London Democrat* (Mar. 1, 1889): GRD.

44. A final line deleted by the editor reads: "We append a list of all the dates definitely fixed at the time of going to press." Then a list of eighteen places in England, Wales, and Scotland are mentioned with dates of visitation ranging from March 13 to May 19.

45. "Henry George in England: The Single-Tax Campaign Campaign There Attracting Great Attention," *The Standard* (Mar. 23, 1889); it originally appeared in the *New York Telegram* (Mar. 18, 1889): GRD. The following sentence was deleted by the editor: "The following cablegrams received by the daily press [from London] speak for themselves."

46. See pages 213–229.

47. "What the Newspapers Say," *The Standard* (Mar. 23, 1889): GRD.

48. Stirabout is a porridge made of cornmeal or oatmeal boiled in milk or water and stirred.

49. Henry George, "Interview with Henry George," *The Standard* (May 25, 1889); reprinted from the *Dundee Advertiser* (May 7, 1889): GRD.

50. See footnote no. 42.

51. Michael Flürscheim (1844–1912) was an active German supporter of Henry George's single tax. See chapter 11 by Michael Silagi in Wenzer, *An Anthology of Single Land Tax Thought*, pp. 341–375.

52. George visited Australia in 1890.

53. A pole can be a unit of length equal to sixteen and a half feet or a unit of area equal to one square rod.

54. "What the Papers in Great Britain Say," *The Standard* (May 25, 1889): GRD.

55. A colliery is a mine. A sweater manipulated or forced people to produce a great deal of a product or service in return for little pay.

56. A glen is a valley and a strath is a wide flat river valley or the low-lying grass-land along it.

57. The literal meaning of the Latin term *non possumus* is "we cannot," but it denotes an inability to move or act in a particular matter.

58. The term pawky means canny or artfully shrewd.

59. Henry George, "Henry George in Manchester," *The Standard* (June 15, 1889); reprinted from the June issue of the *Liverpool Financial Reformer*: GRD.

60. A list of eighty-three names preceeded this section, including that of Henry George, and then speeches by A. D. Provend and J. Hampden Jackson were deleted by the author. This meeting was held on May 21, 1889.

61. Thomas Briggs of London was a friend and supporter of George.

62. The Levellers were a group of radical extremists who flourished between 1647 and 1649 during the Puritan Revolution and were put down by Oliver Cromwell. They advocated political and social reforms far ahead of their time, such as a democratiza-tion of government, inalienability of individual rights, and popular sovereignty; by extension the term was used to describe anyone who sought the removal of inequali-ties. (See page 3.)

63. Sir Robert Peel (1788–1850) was an English statesman noted for his furtherance of reforms. He was prime minister in 1834 and from 1841 to 1846.

64. Under the command of Arthur Wellesley, first duke of Wellington (1769–1852), allied troops defeated Napoleon at Waterloo in 1815. He was prime minister from 1828 to 1830.

65. A bothy is a small lodging or hut used by farmers or workmen.

66. See footnote no. 42.

67. The resolution passed was for the abolition of all tariffs on goods and food and the adoption of a tax on land values apart from improvements.

68. From the Christian Scriptures, Matthew 26:11.

69. The final paragraph has been deleted by the editor. It reads: "The Rev. Dr. Macfayden – I have much pleasure in seconding this resolution. Our chairman said at the beginning of the meeting that we disagreed on many points, but I think everybody will be of one mind that he deserves our thanks for the fairness with which he has conducted the meeting. (Applause.) This brought the proceedings to a close."

70. Henry George, "George in England," *The Standard* (June 29, 1889); reprinted from the *London Star* (June 6, 1889): GRD.

71. This international conference of land reformers, in which George was elected honorary president, took place in 1889.

72. Independents was a term that was first used in the 1640s for people of various Protestant denominations who supported Cromwell; later many of them were absorbed by the Congregationalists and lost much of their Puritanical leanings.

73. Possibly a reference to the split from the Liberal Party and the formation of the Liberal Unionists, including Joseph Chamberlain. They were against the repeal of Union with Ireland and were moving to the right politically with the Conservatives, though some Unionists were progressive on social and economic issues. For Chamberlain see footnote no. 83, page 102.

74. Peter I, known as "the Great" (1672–1725), ruled Russia from 1682 until his death.

75. "A Wonderful Impression on the Britishers," *The Standard* (June 29, 1889); reprinted from the *Boston Globe*: GRD.

76. Henry George and Samuel Smith, "Henry George and Samuel Smith, M.P.," *The Standard* (July 13, 1889); reprinted from the *London Star* (June 27, 1889): GRD.

77. A consol is the shortened form for consolidated annuities, an interest-bearing British government bond without a maturity date and redeemable on call; it was first issued in 1751.

78. Henry George, "The Warning of the English Strikes," *North American Review* 149, 395 (Oct. 1889): pp. 385–398: PDM.

79. John Burns (1858–1943), an English labor leader elected to Parliament in 1892, became in 1905 the first workingman to obtain a cabinet post.

80. A reference to either Ignatz Moscheles (1794–1870) or Felix Moscheles (1833–1917).

81. Henry Edward Manning (1808–1892), an English cardinal of the Catholic Church, advocated social reforms and workers' rights.

82. See footnote no. 79.

83. Alexis de Tocqueville (1805–1859) was the French politician and writer of the 1835 classic *Democracy in America*.

84. These lines are from William Wordsworth's (1770–1850) 1803 work *Rob Roy's Grave*.

85. Labor violence punctuated the American scene in the 1880s during the hard economic times.

86. The Eiffel Tower, erected in 1889, was designed by Alexandre G. Eiffel (1832–1923).

87. A bodkin can be a small pointed instrument used for making holes in cloth; a blunt needle-like tool for for drawing cord; or a pin-shaped instrument used to fasten woman's hair. The first meaning is probably intented here.

88. One of them being a system of pay ticket [sic], cashable in certain public houses, at the price of taking a drink. I have been struck in England and Scotland by the number of ardent teetotalers who are working for social reform, [John] Burns, by-the-by, being a strict temperance man. The temperance people who imagine that could drinking be abolished poverty would cease, in large degree mistake cause for effect; but that every advance in temperate habits among the working classes does increase both the disposition and the power to overthrow the conditions that produce poverty is clear (H.G.).

SECTION IV: NATIONALIZATION, COMPENSATION, AND SOCIALISM

NATIONALIZATION: NOT A NEW PLAN FOR THE SETTLEMENT OF THE LAND QUESTION; J. Finton Lalor in '48: "No Man Can Plead Any Right to Private Property in the Soil;" Bishop Nulty on the Same Platform

(*Irish World*, August 5, 1882).[1]

To The Editor of the *Nation*.

Sir: I have to thank you for your report of my lecture in the Rotunda on Saturday, June 10th, though, of course, in the compression of a lecture of that length into a couple of columns, many points are necessarily omitted.[2] But a couple of quotations were made by me in that lecture which I would like to ask you to republish, inasmuch as there is an endeavor in certain quarters to create the impression that Mr. Davitt's proposition for the nationalization of the land of Ireland is the importation of novel ideas.

In the lecture to which I refer I contended that the Land Question could not be justly nor satisfactorily settled by any scheme for enabling the agricultural tenants to buy out the agricultural landlords, as those most interested in the defense of Landlordism now desire; but that, on the contrary, the only way to settle the Land Question justly and permanently, so as to secure their full rights, not merely to the tenant-farmers, but to the laborers, artisans, and all classes as well, is by nationalization of the land – that is to say, the assumption of the whole soil of Ireland as the common property of the whole people of Ireland by the appropriation of all ground rents to common purposes.

While giving Michael Davitt all honor for his able and consistent advocacy of this principle (always excepting his proposition for the compensation of landlords, in which I do not believe), I asserted that this was no new idea or novel principle; but that it was, on the contrary, the commonsense principle that underlaid the ancient laws and customs of the Irish people before their enslavement (as it has those of every other people); and as showing how other eminent Irishmen had appreciated the same truth and enunciated the same principle, I made a few quotations which, as they are not so well known as they deserve to be, I hope you may find space to republish.

The first is from a remarkable series of articles published in the *Irish Felon* in July, 1848, by James Finton Lalor.[3] He says:

> The principle I state and mean to stand upon is this – that the entire ownership of Ireland moral and material, up to the sun and down to the center, is vested of right in the people of Ireland; that they, and none but they, are the landowners and lawmakers of this island; that all laws are null and void not made by them, and all titles to land invalid not conferred or confirmed by them; and that this full right of ownership may and ought to be asserted

and enforced by any and all means which God has put in the power of man. In other if not plainer words, I hold and maintain that the entire soil of a country belongs of right to the entire people of that country, and is the rightful property not of any one class but of the nation at large, in full effective possession, to let to whom they will, on whatsoever tenures, terms, rents, services, and conditions they will; one condition being, however, unavoidable and essential – the condition that the tenant shall bear full, true, and undivided fealty and allegiance to the nation, and the laws of the nation, whose lands he holds, and owe no allegiance whatsoever to any other prince, power, or people, or any obligation or respect their will, orders, or laws. I hold, further, and firmly believe that the enjoyment by the people of this right of first ownership in the soil is essential to the vigor and vitality of all other rights; to their validity, efficacy, and value; to their secure possession and safe exertion.

No man can legitimately claim possession or occupancy of any portion of land, or any right of property therein, except by grant from the people, at the will of the people, as tenant to the people, and on terms and conditions made or sanctioned by the people; and every right, except the right so created and vesting by grant from the people, is nothing more or better than the right of the robber who holds forcible possession of what does not lawfully belong to him.

No length of time or possession can sanction claims acquired by robbery or convert them into valid rights. The people are still valid owners, though not in possession. To my plain understanding the right of private property is very simple. It is the right of man to possess, enjoy, and transfer the substance and use of whatever he has himself created. This title is good against the world, and it is the sole and only title, by which a valid right of absolute private property can possibly rest. But no man can plead any such title to a right of property in the substance of the soil. . . .

The people, on the grounds of policy and economy, ought to decide (as a general rule, admitting of reservations) that these rents shall be paid to themselves, the people, for public purposes, and for behoof and benefit of them, the entire general people.

I commend these extracts from the writings of an Irish patriot of '48 to those who, in the endeavor to bring the Irish revolution to such a lame and impotent conclusion as will really strengthen Landlordism, are endeavoring to create the impression that Michael Davitt derived his ideas from "an American Communist." And in order to show whether there is anything "Communistic" in these ideas let me ask you to reprint another extract which I read at the Rotunda. It is from the letter addressed in April of last year to the clergy and laity of the diocese by [the] Most Rev. Dr. Nulty, Bishop of Meath, one of the most eminent of Irish prelates and theologians:

The human race cannot now any longer live on earth if they refuse to submit to the inevitable law of labor. No man can fairly emancipate himself from that universal decree which has made it a necessity for everyone "to earn his bread in the sweat of his brow." Now, the land of every country is to the people of that country, or nation, what the earth is to the whole human race. That is to say, the land of every country is the gift of its Creator to the people of that country; it is the patrimony and inheritance bequeathed to them by their common Father, out of which they can, by continuous labor and toil, provide themselves with everything they require for their maintenance and support, for their material comfort and enjoyment. God was perfectly free in the act by which He created us; but, having created us, He bound Himself by that act to provide us with the means *necessary* for our subsistence. The land is the only means of this kind now known to us.

The land, therefore, of every country is the common property of the people of that country, because its real Owner, the Creator who made it, has transferred it as a voluntary gift to them. *Terram autem dedit filis hominum*.[4] Now, as every individual in that country is a creature and child of God, and as all His creatures are equal in his sight, any settlement of the land of a country that would exclude the humblest man in that country from his share of the common inheritance would be not only an injustice and a wrong to that man, but, moreover, would be an impious resistance to the benevolent intentions of his Creator.[5]

Dr. Nulty goes on to show that this fundamental religious truth is a fundamental social truth, and that human society may be conformed to the Divine Will in relations to the land by the simple expedient of devoting rent by taxation to public purposes. He says, after fortifying his position as to the common ownership of land by numerous citations:

I think, therefore, that I may fairly infer, on the strength of authority as well as reason, that the people are and always must be the real owners of the land of their country. This great social fact appears to me to be of incalculable importance, and it is fortunate indeed that on the strictest principles of justice it is not clouded even by a shadow of uncertainty or doubt. There is, moreover, a charm and a peculiar beauty in the clearness with which it reveals the wisdom and the benevolence of the designs of Providence in the admirable provision He has made for the wants and the necessities of that state of social existence of which He is the author, and in which the very instinct of Nature tell us we are to spend our lives. A vast public property, a great national fund, has been placed under the dominion and at the disposal of the nation to supply itself abundantly with resources necessary to liquidate the expenses of its government, the administration of its laws, and the education of its youth, and to enable it to provide for the suitable sustentation and support of its criminal and pauper population. One of the most interesting peculiarities of this property is that its value is never stationary; it is constantly progressive and increasing in a direct ratio to the growth of the population; and the very causes that increase and multiply the demands made on it increase proportionately its ability to meet them, as I shall clearly show further on.

And Dr. Nulty does show how the nationalization of the land, substantially as Davitt proposes, would not only be of the greatest benefit to tenant-farmers, but would raise the wages and improve the condition of all classes of laborers and artisans, and how the general prosperity thus ensuing, by greatly adding to the value of the land, would increase the common fund, which would not only pay all taxes now levied, but afford a great surplus for beneficial purposes.

But my purpose in writing to you is not to go over the argument for the nationalization of the land, but merely to show that Michael Davitt is anything but alone in advocating nationalization and that proposition is anything but a new one. Nor is it a new one so far as he himself is concerned as it is, I believe, well understood that he has been in favor of the nationalization of the land since at least the formation of the Land League, and that he, Mr. Brennan, and Mr. Ferguson desired to make that the declared platform. And I think Mr. Davitt's demand of "The Land for the People" has been understood in Ireland as well as in America, as meaning the land for the whole people and not merely any part of them.

Yours respectfully,

THE NATIONALIZATION OF LAND
(The London *Times*, September 6, 1882).[6]

Last evening, in the presence of a large audience, at the Memorial Hall, Farringdon Street, Mr. Henry George, the author of a book called *Progress and Poverty*, delivered a lecture, under the auspices of the Land Nationalisation Society, on the subject of "Land Nationalization." In the unavoidable absence of Professor F. W. Newman, the chair was occupied by Mr. A. R. Wallace.[7] The secretary (Mr. A. J. Parker) at the outset of the proceedings read the following letter from Professor [F. W.] Newman:

My Dear Sir:

You desire of me some expression of my sympathy with the expected lecture of Mr. Henry George, for which I had agreed to take the chair if it came off on August 25. I should have felt it an honor to appear publicly at his side, and it would have been in accordance with my old convictions. More than thirty years ago, in a book long out of print, I avowed several cardinal truths concerning land in substantial harmony with the doctrines which he eloquently expounds.

English law has never admitted that property in land is identical in character with property in movables; yet in the mercantile classes, probably at all times, and recently in the school of Cobden and Bright, an effort has been made to claim and establish an identity, with much success in blinding our public. The monstrous despotism under which Ireland, Scotland, and England groan has been too truthfully and awfully depicted by our president, Mr. A. R. Wallace, in his noble book on the nationalization of the land.[8]

Mr. Henry George is our very valuable coadjutor in the main question. The despotism has been built up in the course of six centuries by landlords, locally supreme, as well as irresistible in Parliament, from which they long excluded all but landlords; besides they had the aid of the subservient lawyers. Such malversation in power does not imply that the class of landlords is specially unjust. No single predominant class is capable of making just laws. The clergy, the medical faculty, or the capitalists, if for six centuries they had the decisive voice in legislation, would have inflicted upon us evils different in kind, but probably not smaller. Mr. George's native America has a vast advantage over us in the fact that with them neither House of Congress is hereditary.

Many try to persuade us that a law or institution is sacred because it is old; but when laws or institutions sanction injustice, antiquity cannot make them sacred. The presumption is to the contrary – "institutions made in an age of violence and comparative barbarism are likely to be violent and barbarous;" and such are our oldest institutions. Only one Norman king – emphatically our Conqueror – claimed to drive out human population in order to make a hunting forest; and his memory is infamous for it in our history.[9] Yet this very thing is done by modern landlords, and our Home Secretaries at most do but regret it as an extreme use of legal power. As well might a king or queen claim to drive us all into the sea. And these modern landlords keep the land empty of men and barren of crops simply in order to get more rent from over [sic] rich men, who pay a fancy price for the pleasure of cruel sport. This is still more despicable than to be oneself devoted to hunting.

There is another matter into which the public ought to inquire – that is, with what right modern lords of the manor claim to be owners of the wild land. To me its aspect is that of simple usurpation. It is certain that in feudal times the barons were remunerated for their personal political service by customary dues from the tenants. They had the duty of "keeping the king's peace" on the wild land, but no ownership of it was assigned to them. The baron might take timber from the forest, stone or lime from the hill, gravel from the seabeach; but so might many others, it was not with him an exclusive right. If tenants then paid anything to him when they served themselves (of which I know nothing), it can only have been a payment fixed by custom, as was all rent in those days.

To claim back for the State all the wild land appears to me very natural and reasonable, especially that which has been artificially made wild by vile despotism. But this is only one branch of a vast subject, to elucidate which I do not doubt Mr. George will eminently and fruitfully contribute.

The Chairman, in introducing the lecturer, said probably the meeting had mainly heard of Mr. George because he had been thought worthy of the special attention of the Irish police recently (laughter), and also because he had been deemed worthy by the masters of the Irish police of being instantly set at liberty.[10] In his book Mr. George had dealt with the problem of human well-being. The main question which he put was: "How is it that with increased power over Nature and the forces of Nature for which the present century is so remarkable we have not been able to diminish the mass of poverty, misery, and crime that abounds in this country?" The solution of these ills was, according to Mr. George, the abolition of private property in land, which inevitably resulted in a land monopoly.

Mr. George, who was well-received, said this fight for the nationalization of the land was a worldwide fight. Throughout the world there was a growing discontent as the great masses of the people were becoming more and more conscious of the cruel wrong under which they suffered, as they were awakening more and more to the great truth that the Creator made all men equal (cheers) and that they were entitled to demand equal rights. Equally on this, as on the other side of the Atlantic, there was an overstocked labor market.

What was the cause of this? Allusion had been made to the right of men to the soil which the Almighty had given them for the satisfaction of their desires, and the employment of their labor. Now the land, instead of being in the hands of the many, was the monopoly of the few. What was the land question? It was the great labor question. ("Hear, hear.") He believed that the majority of the people had never paused to form a clear idea of what land was. People of Scotland were tabooed from the land to make way for the deer of the Duke of Sutherland, just the same as the South Sea Islanders had to give way before the missionaries who went to convert them.[11]

Let them go to Wiltshire – a lovely country. He went down to that county accompanied by a gentleman who was born in it. He said children had no such playgrounds as he remembered when he was a boy. All the grounds seemed to be fenced in. He (Mr. George) asked why the children did not jump over the fences, and the gentleman of whom he asked the question replied that if the children did they would be imprisoned within twenty-four hours. Was it not time that this state of things should be put an end to?

Carlyle said there were so many people in England, mostly fools (laughter) and he thought he was right. At the same time he did not think that people who dwelt on his side of the Atlantic had any right to throw this insinuation at the British people. They had some enormous advantages over [the] English people. They had, as he thought, a freer and higher political system. Every citizen of the United States was absolutely equal before the law. Any child born in the Union could hope to be President; but for all that they had made land individual property.

How was it that men who worked the most got the least for their labor? When they considered the physical fact that all wealth was the production of labor, they were confronted by the strange anomaly that the men who labored most, either by their head or their hand, were not the men who lived in luxury or even sometimes in comfort. The comfortable places of the world were reserved for the men who did nothing. How was it, he again asked, that those who did the hardest work got the least for their labor? He replied that if they came to analyze the matter they would find that the cause of this was primarily the monopoly which a privileged class possessed in this country.

Let them look at the enormous incomes that were derived from the rents of the lands of London. Those incomes amounted to millions. Who produced those millions? Necessarily the working people. (Cheers.) He did not mean to say, however, that when they had accomplished the nationalization of the land – when they had secured to every human being in the country his full and equal right to the land of the country they would then have accomplished everything. Much would still remain to be done; but they would at least have laid a square and true foundation – and it was alone upon a true foundation that they could erect a stable edifice.

Unlike Michael Davitt, he was not prepared to give the landlords compensation in his scheme for the nationalization of the land. He thought Michael Davitt would progress beyond his idea. (Laughter.) Mr. George resumed his seat amid loud cheers.

On the motion of the Rev. S. Headlam, it was resolved: "That private property in land is the monopoly by a few of an element essential to human existence; that it had its origin to a large extent in force or fraud or economic ignorance; and its continuance is a national wrong and a danger to the stability of the community."

At the instance of Sir John Bennett, it was determined: "That this meeting is of the opinion that nationalization of the land is the only effectual remedy for the evils of the present land system. Among other priceless advantages it would carry with it a great progressive alleviation of our fiscal burdens, would stimulate cultivation, and largely increase the production of food; and, finally, would render every person free to share in the benefits which the possession of land for personal occupation is calculated to afford."[12]

LECTURE IN BIRMINGHAM, ENGLAND
(June 23, 1884).[13]

Mr. George, who on rising was warmly applauded, said: Ladies and gentlemen, citizens of Birmingham – I thank you for your cordial greeting. I am glad to meet a Birmingham audience. I like this town of yours. I believe in it. I believe in you. I well remember my first notions of Birmingham. They were very vague and indistinct. I had heard of it and thought of it as a place where they made gimcrack jewelery and little brass idols for the heathen. (Laughter.)

But I learned to know something more about you, and one Sunday a couple of years ago – one Saturday night – I landed in this town without knowing a soul; only one man by reputation and correspondence, my friend Mr. Walker.[14] (Applause.) And I went out, Sunday morning to take a walk through your town. I had been in a good many places in Ireland and in England and in Scotland, and I had got sick and tired of seeing everywhere statues to the butcher of mankind[15] ("hear, hear"); and I came into one of your public houses and there I saw a monument to a man who never killed another man in his life or was the cause of killing him (applause); to a great man and a good man and a man of whom I then knew something and of whom I now know more – George Dawson. (Reserved applause.)

And I honor you, too, men of Birmingham, because you have honored a man whom every truehearted American must delight to honor – John Bright. (Cheers.) When John Bright, standing up with the courage of his convictions, was unseated at Manchester, Birmingham took him up. I admire John Bright. He has worked well in the cause of human freedom. That he goes with me now I of course do not expect. There is given to each of us but a little truth to see, and the truth that we grasp, if it be truth, is greater than any man who grasps it knows; and it swells and widens and grows with the progress of mankind. Birmingham to me represents new England – the new spirit that is abroad in old England (applause); the life and the vigor and the strength of the Democratic idea. (Renewed applause.)

I come before you tonight to talk freely to you, to express my own convictions as man to man. Look on me, if you please, as a wolf among a lot o' sheep (laughter), or a sheep among a lot of wolves. (More laughter.) I bind nobody but myself, not even the men at whose instance I come here. And just now it seems as though all sections were only desirous of repudiating such [a] man as me. ("No, no.") The Conservatives are down on us; we are against religion, the Church people say; Mr. Bradlaugh is going to pulverize us (laughter); the Irish Parliamentary party have warned their leagues not to attend any of my meetings ("shame"); and the Socialists are generally down on us. (A laugh.)

I don't want to go into that question tonight but it is a curious thing that you find the most ardent defenders of the existing state of things occupying the same stand as the most Radical Socialists. On both sides it is said: "Why don't you let the landlords alone?" and "Why don't you go for the capitalists?" Well, by-and-by, when we get rid of the landlords (laughter and applause); then if it is necessary to go for capital. (Laughter.) For my part, I think capital hurts nobody. I believe that the sins that are attributed to capital are simply results of monopoly ("hear, hear"); that the thing which is responsible for the unjust and unequal division of wealth everywhere is simply the denial of equal rights, simply that some men are given privileges which do not belong to all.

I know a great many of you so-called Liberals are, in my opinion, a very Conservative sort to [deal] with ("hear, hear"); are rather Tories in disguise. (Applause and laughter.) I can understand how a real Liberal, with a Democratic feeling in his heart, may at the present time want to avoid touching such a question as this land question. The Liberal Party, as I understand it, or at least that active section that is called the Radical wing, has set before itself a great political task in the extension of the suffrage. ("Hear, hear.") I believe in universal suffrage. (Loud applause.) I believe in it, not as a boon, but as rather of natural right. (Renewed applause). The man who has no voice in the making of the laws of his country is not and ought not to be bound by the laws of that country. (Applause.) I believe in universal suffrage. I believe in it as a great educational measure. Once I doubted it. Once, when the question before us was whether the suffrage should be extended to the unfranchised blacks, slaves just emerged from slavery, men who had been kept in ignorance systematically, I was opposed to it. But the suffrage was extended, and I see that it was right ("Hear, hear.") Those men, who were then only "the damned niggers," are now "our colored fellow-citizens." (Applause.) Extend the franchise to every Englishman and you will at once see how the educational influences will extend to the very lowest ranks. ("Hear, hear.")

And I go further than that. I believe now, though I did not two or three years ago, in the doctrines enunciated by Helen Taylor last night; I believe in woman['s] suffrage. (Applause.) I think that our women are at least as important in public affairs as our men ("hear, hear"), and I don't think we will ever get the attention that we ought to have to these great subjects that must now be settled by the intelligence of the people unless we can interest our women. You are right to want to extend the suffrage.

But the fight for the rights of man in the soil will not hurt the fight for the suffrage. It will help it; it will strengthen it. Do you think that the class who will stand against it [does not] realize what it will mean? Don't you suppose that they well understand the power that is going [on] and what it can be used for. You cannot but meet with their opposition; but by showing men what the suffrage may do for them you will excite a strength that will overcome all opposition. (Applause.) The mere poking of a paper in a box – what use is it? What use are political rights unless man can be used to gain social rights. (Applause.) I care not what the form of Government may be – I care not how you treat of the rights of man – so long as the equal, inalienable right of men to gain a free fair living by their own labor is denied. ("Hear, hear.") Have the courage of your convictions. No halfhearted policy ever wins where the question is a great one.

That has been the fault of the Free Traders in the United States. Going for a halfhearted, halfway Free Trade, they have never rallied the strength that might be rallied for the international law of God. In the last presidential campaign I happened to be in New York.[16] I had come from California. I was not then very well-known in New York, but just before the election, when the tariff issue was sprung, the Democratic Central Committee sent for me, and they told me they had heard that I was the best man in all the country to talk to the working men on the question of the tariff. Well, I told them I didn't know about that, but I could talk to working men and I would like to talk to them about the tariff; and they asked me if I would go out and make some speeches. I said: "Certainly I will," and they made a great list of engagements for me close up to the day of the election, and I went out. Well, it seems what they were after was for somebody to tell the working men it didn't make any difference; the Democratic Party was as good as the Republican Party for the tariff.

I went to a crowded meeting. The gentleman who spoke before us made that kind of speech and then I was put on the platform. I told them that I had heard of a high-tariff Democrat, though I could not conceive how there could be such a thing; and I knew these were men who called themselves revenue-tariff Democrats, but there was also another kind of Democrat, and that was a no-tariff Democrat. And that what was wanted was to sweep away the custom houses and customhouse officers and have Free Trade. (Applause.) Well, you ought to have seen the man on the platform there (laughter); and I went off without a man to shake my hand. And I got that right, [and] as I was going to my other engagement, [I received] a telegraphic dispatch asking me to go by midnight train to New York. The chairman of the committee met me and begged me not to make any more speeches. (Laughter.) Well, General Hancock was defeated.[17] He could have been elected, had that party stood firm and stuck to its principles. I got all the opposition that could be got. It did not gather strength.

Don't be afraid of opposition: opposition is a good thing. (Applause.) Opposition! Why men who oppose a truth do more to forward it than they can who are its advocates. (Applause.)

Here is an instance. I come from the hotbed of protection – the State of Pennsylvania. When I was a little boy I remember talking with another boy about it and I said I couldn't for the life of me see how it was [a] good thing for the working men to have everything taxed. He told me that was a very deep question. He had a big brother in a lawyer's office who had studied the whole matter – knew all about it – and it was really a good thing. I subsided and accepted the authority (a laugh) and went out [to] California when a young man and was a thoroughgoing Protectionist – such a thoroughgoing Protectionist that when the *Alabama* used to sink our ships and the *Florida* used to play havoc among [ ] I thought that was a splendid thing for California – it used to help home industry.[18] (Laughter.) Only one night, when I was working at my trade, a fellow-workman said to me: "Let us go to a debating society," and I went; and I heard a very able long protection speech and I came out of that hall a thoroughgoing Free Trader. (Applause.)

I was in Cardiff the other day, and went into some Turkish baths there as soon as I got into town; and when we were sitting in the bath – you know how it is in those places – the men got talking about this American Henry George, who had come to talk to them – that he preached confiscation and robbery and thievery – and they said they hoped people would not go to hear him. The Americans were all a set of liars. (Laughter.) And as I chipped in; and I said:

"Yes, all excepting the Canadians" (renewed laughter), and then I just went and I stood up for the landholders and said there had been a lot of growling, and one thing and another. Well, in ten minutes they were howling confiscators. (A laugh and applause.)

I am not going to go over abstruse questions of political economy tonight. (A voice: "Why not?") Because I haven't time (applause); I am only going to talk for an hour, then you may choose to ask me questions. I am willing to argue the question from any standpoint with any individual – as a matter of political economy, as on the historical side or as a matter of expediency or any other way – and I am to speak, I believe, in this town hall of yours before I leave England (applause), and if there are any of those gentlemen who want to come up and argue that question with me I am very willing to meet them. (Applause.) It may be argued in a great many different ways, but I prefer the ground that its opponents have taken, the ground of Justice. ("Hear, hear.") I believe that Justice is the supreme law of the universe ("hear, hear"), and if any man will show me in anything I have advocated that which is unjust I will retract it there and then. (Applause.)

I have received a great many letters and a great deal of personal advice, urging me not to say anything about this matter of compensation. But it is the very heart of the question ("hear, hear"), and it cannot be avoided. Here is Mr. Fawcett, and his but an example of many others who say nationalization of the land is impossible, simply because compensation is impossible. They tell you, of course, you cannot think of nationalizing the land, of resuming your birthright without compensating the present owners, and that of course would saddle you with a debt two or three times as great as your public debt. There is a great deal of divergence of opinion upon this matter of compensation, and I attach no importance to this divergence. I am ready to argue with any man, no matter how far he goes.

There are some, like my friend Walker here, who are thoroughgoing compensators. Compensate the landlords, they say, after you have nationalized the land. There are others who say that the landlords ought to be compensated first and the land nationalized afterwards. Then there are others, like Miss Taylor, whom I heard last night lecturing for the Land Nationalisation League, that insist upon compensation; and Miss Taylor insisted very earnestly upon compensation – at least upon compensation to the men who could show that they had, by the earnings of their labor, paid somebody for the land. (Applause.) What Miss Taylor proposed in the ways of compensation was this: She proposed that the landholders of England should pay up, with interest and compound interest, all the back taxes at four [percent] on the pound from the time of Charles the Second (laughter and applause); and then she said that part of this money should be used to reimburse the people who had bought land with the fruits of their own labor. ("Hear, hear.")

Then there are other compensators. Why, a clergyman of the Established Church in the west of England the other day was telling me that he met another clergyman of the Established Church in a railroad train, and they got talking about this matter, and clergyman number two said that he was opposed to George because he did not advocate compensation and they got talking for awhile.[19] The clergyman said, compensation is Justice; but, come to find out, the men he wanted to compensate were the men who had been deprived all their lives of their rightful share in the land. (Laughter and applause.) I am concerned only with the principle, and it is a wide principle. It not only goes to the heart of the land question; it goes further and to other things – can we not abolish the wrong without paying those who have profited by the continuance of the wrong? ("Hear.")

Now, to give an example ... as it goes into the tariff question. They say, here all over the United States: "Manufacturing industries have been established under this tariff; they have invested their capital. It would be compensation to suddenly abolish the tariff. You must keep up a tariff. You are in honor bound to keep up a tariff. You are in honor bound to keep up a tariff for a long time – to let them slide quietly out." That I absolutely deny. ("Hear, hear.") I take my stand on the tariff question. As I have told audiences over there, so far as I am concerned I want it utterly abolished at ten o'clock tomorrow morning. (Laughter and applause.) When men will go to Congress and lobby and bribe and logroll to put on a tax upon the consumers of the nation by which they selfishly profit, I will not admit that they have any rights. ("Hear, hear.") And, further than that, I believe it would be the quickest and the easiest way to read just matters.

This long, lingering, dragging out of Justice is bad on all sides. If you are going to do a thing do it quickly and be done with it. Now the trouble about compensation is this: I really cannot see any way of compensating for an injustice without to a greater or less degree continuing that injustice. (A Voice: "Prove it to be an injustice.") If you propose to pay to the landholders all that their land is worth, what benefit is it going to be to the nation? We should die and our children would die before we see it. It is substituting one tax for another. ("Hear, hear.") And if you propose to take anything at all from the landlords why not go to the full extreme (laughter and applause): "It is as well to be hung for stealing a sheep as for stealing a lamb." (Laughter.)

This thing of compensation has been carried to an absurd length. Here, because Charles the Second or Charles the First or some other of your monarchs dead centuries ago (a laugh) settled a perpetual pension – upon some of his panders ("hear, hear") or some of his flunkeys, you have held here that people cannot stop paying that unless they buy 'em off. ("Oh.") Why, look at the rotten boroughs.[20] In Ireland when the Irish Parliament was abolished the men who made merchandise out of the lawmaking powers who hold their election year after year were compensated for it. ("Hear, hear.") You compensated too the West India slaveholders when you abolished slavery and instead of doing a good thing it seems it was that you did a bad thing – an evil thing that has strengthened slavery all over the world. (Hisses and applause.) A thing that led our people in the United States to believe that slavery would never be abolished without compensation and that kept the price of slaves up to the highest pitch until the [American Civil] War had actually commenced.

I do not believe in compensation because I believe in the sacred rights of property ("oh, oh") – because I do believe in the command "Thou shall not steal." (Applause.) I believe that the right of property lies at the very foundation of the social order and that no community can be prosperous or any state secure when the right of property is denied, and the heaviest indictment against the present state of things is that it is a denial of the right of property. ("Hear, hear.") What is the right of property? From what does it spring? Is it not, as Adam Smith says, that the first and most sacred rights of property is the right of a man to himself and to the produce of his own labor ("hear, hear"); and is not a system which takes from laborers the produce of their labor and puts it in the hands of men who do nothing whatever to earn it – is not that system a denial of the first and most sacred right of property. (Applause.)

Why, you have only to look around and see it. Here is this flourishing town of yours. You have a great landlord named Lord Calthorpe (hisses and applause), a man, I believe, who never comes to the town from year to year, yet who draws from it yearly an enormous sum. For what? For land that existed before he was, before any of you were, before our race was – for land that will exist long after he and we are dead and gone. Where does that wealth come from – that money, as we call it? It comes necessarily from the working men of Birmingham. (Applause.)

Here you have another family, I am told, who owns land in the center of the town, land that has gone up within some years past in some places 12,000%. What made it go up? It was not something that his family did, who don't live in Birmingham, but what you men of Birmingham did. The rents that men draw are so much taken from your property (applause), so much confiscation that goes on year-by-year. ("Hear, hear.")

Confiscation! Why, look at your city-improvement scheme; where you undertook to improve and beautify your city, and you had not merely to buy out the landlords, if I am correctly informed, but to pay blackmail to the landlords,

over and above the value of the property (applause), and today your improvements are stopped and you have awkward corners and narrow streets. Why?
Simply because these "dogs in the manger" sit there and say that you can't
improve your city without paying toll to them. Confiscation!

Why, look at the railroads of this country. According to Samuel Laing, M.P.,
in a recent article – and he ought to be a good authority – the railroad here
had to pay to landowners £50,000,000 over and above the market price of their
land, for the privilege of constructing railroads that added to the value of the
land that was left to the landowners over £150,000,000. (Applause and
"shame.") Where is the confiscation there! Today here, you men of Birmingham
are complaining of your railroad rates. How can you blame the railroads when
you suffer them to be robbed in that manner, when to ease off the tax upon
the landowners you impose a tax upon the persons who travel. ("Hear, hear.")
Confiscation! Why, I read in a west-of-England paper the other day of a place
where a local railroad had been absolutely abandoned for the time, after a great
deal of money had been spent, because two great landowners blocked the road.
Confiscation!

Why, a gentleman of this town told me that he had arranged to buy a piece
of property that was necessary for him, for £1,600, and when the landowners
found that it was he, and that he was anxious to have it, he stuck another thousand pounds on the price. (Laughter.) Confiscation!

Why, down in Cardiff they told me a case where a tenant and leaseholder
who had twenty years of his lease yet to run wanted to improve the house –
to add to it and to put it in better condition – and a term of the lease was that
at the expiration that house went to the landlord; but there was a clause in the
lease that he could make no changes without the consent of the landlord, and
the landlord refused to give this man permission to improve the house unless
he would take a new lease, give up the twenty years yet to run and raise the
rent from £5 to £50. Bah! To talk of compensation! (Laughter and applause.)

Well, there was another case a gentleman told me, in a Welsh town not far
from there, in which a tenant wanted to make some improvements in the house,
to better it and to paint it, and the landlord would not say either yes or no
about it, so the tenant got tired and went to work and did it. The landlord came
in, and looked over the house. He said: "It looks very nice; very fine; I think
you can pay a little more rent," and he did put it on and absolutely made him
pay it. (Laughter.) Confiscation! Why, that in Ireland has been done over and
over again. (Applause.) Confiscation! Why down here in Newport, near Cardiff,
there is a Wesleyan Methodist Chapel, on the land, I believe, of Lord Tenderness
– some such name as that. They have a lease that has yet twenty years to run.
They pay £8 a year ground rent. The congregation has outgrown the size of

the chapel and they wanted to make it larger; and they can't get permission to enlarge this place of worship unless they will give up their present lease and take out a new one at £100 a year. (Sensation.) Talk about confiscation! (Applause.)

Down there in Cardiff there is a company who wanted a little while ago to build a factory on a certain piece of ground. They would have employed hundreds of men and added much to the business of the town. But they could not get a lease where they wanted it; the only way they could get that ground to build on was at a notice of three months. Of course they refused to build. Why, here in the papers the other day, at some place near here, I think, a little manufacturing town, Lord Somebody-or-Other absolutely won't allow another cottage to be erected.

And look all over [ ] on and everywhere – in every one of your cities – how this monopoly of land, by the high prices that it compels, prevents the building of houses and forces families of English people to herd together as swine. (Applause.) According to Sir James Caird,[21] the increase in the rent of agricultural land – mind you, not the land of the cities and towns and mineral land, where there has been the great increase of rent, but the increase in the rest of the agricultural land – between 1857 and 1875, the increase in the value was £331,000,000, for which he estimated the landlords contributed about £60,000,000. Where did the rest come from? Created by the growth of the community. Created by you, common people of England, you common Englishmen and Englishwomen (laughter and applause), who have not a right to an inch of your own soil. An American or any other foreigner who comes over here and buys land, as one of my countrymen has in the islands of Scotland, cannot clear you out except at a preposterous price.[22]

Is it not about time that in thinking of these questions we came down to first principles? (Applause.) Is it not about time that we asked who made the land (renewed applause) and how does any man get a title to it? Why, go there to London – go down to Euston Station. You find the street blocked with a fence, an iron fence – I suppose there may be many people in this audience who have lost a train by having to go around that fence or wait while a great fat fellow comes out at a snail's pace and opens it – put up and maintained by his Grace the Duke of Bedford! Why, go a little further and look at those London squares – Lincoln's Inn Fields and all those squares – squares with grass and trees and sometimes, in the summertime flowers: great high fences, gates locked and barred, and not a living soul in them, while around in the bystreets within a stone's throw you find hundreds of little sickly squalid children playing in the gutter. ("Hear, hear.") Heavens and earth, men! Is it not time that some missionary should come from somewhere? (Loud and continued applause.)

Talk about the heathen! Talk about their little brass idols or any other idols (laughter) – where are there heathen so stupid as that? Only one example I know of, and that was down in the South Seas, where there existed the custom of the taboo and where you of England and we of America send missionaries to convert men. (A laugh.) These stupid ignorant subjects had such a veneration for their high chiefs that when a high chief tabooed a place not one of them dared to go on it. He might die with thirst or with hunger rather than drink at a tabooed spring or pluck the fruit of a tabooed grove. He would go round for miles rather than set his foot on a tabooed path. It was very stupid; but what have you here – all taboo. (Applause.)

They have land in abundance, land enough to give every family its house and its garden, while thousands and thousands of families are crowded – men, women and little children, into a squalid room. Thousands of acres you may see, as I saw when I was last in this country, lying vacant, absolutely vacant – rich land – while men all over the country are out of employment, only anxious for employment. ("Hear, hear.") What might labor have been doing on that land? What is that but the English form of the taboo? The truth is that vested interests ought not to stand and cannot stand before natural rights. ("Hear, hear.")

It is said that men bought the land. Well that is true, I think, of very little of the English land. (Laughter.) Whom did they buy it from? You know as a historical fact how it was that the land of England passed from the ownership of the people into the possession of the owners. Why, it is said that before William the Conqueror brought his gang of thieves and robbers over they sat down and they had got a map of England from somewhere (a laugh), and they just mapped it out between them and not a penny of compensation! (Applause and laughter.)

But, bad as that was it was not private property in land. Indeed, at that time nobody among our forefathers had thought of such an absurdity. This thing of absolute individual ownership of land is only a modern thing. It only dates fully in England for two centuries back. Then the land of England was merely divided up by William the Conqueror into knights' fees and the men who had the land were the men who had to do the fighting (hear, hear); and in all the wars that the England of our forefathers waged there was not a penny of debt until you came down to the time of William the Third.[23] I won't go over the whole story. You know it – how the Church lands, appropriated for purposes of worship and education, were divided among a crowd of profligate and greedy courtiers; how the Crown lands, that under the old system maintained the Civil List, were given away to profligate courtiers;[24] how the commons were enclosed ("hear, hear"), and are – yes, even today – being enclosed.[25] Why, a clergymen told me the other day that he went up to the estate of the Duke of Wellington and a Lord Somebody – I forget his name. He said to the gamekeeper: "Why, this is a new park." "Oh, yes," the man said, "this was a piece of land that was going to waste and nobody is doing anything to it, and so his grace took it in." (Applause.)

They told me at Cardiff that a great part of the most valuable estate of the Marquis of Bute was a moor. There are glassworks on it now, and docks, and it yields an immense rental, but it was a moor that a little while ago belonged to the town. Well, the only people who had any of the rights in that town were a body of freemen, just a handful of them, and the Marquis of Bute, the father of this one, used to give an annual dinner to these freemen, and he dined them and wined them (laughter), and finally they came to the conclusion that the land was not much use to anybody and about the best thing they could do with it would be to give it to the hospitable Marquis. And so they did. I don't know whether there was anything implied in a contract, but the understanding among them was that the dinners were to be kept up. (Laughter.) My informant tells me the dinners stopped. (Renewed laughter.) But the Marquis's son has the land now.

Some men did buy their land; unquestionably some men did. I don't think there are many. I am curious to know how many there are who have bought their land. Now, the reason you can't make the discrimination between these men who bought his land and the man who did not is simply this – that if you go to do this the men who have not bought their land will simply sell their land to somebody else. (Laughter and applause.) What is the difference? The principle is well-settled in law, and rightfully settled, that the purchases can get no better title than the man he purchased from has to give. Surely, if a man takes a thing that is not rightly his he does not convey a title when he sells it to somebody else.

Why, it is a piece of juggling. It reminds me of a story they used to tell when I was a boy about a Dutchman who used to keep a grocery store. A fellow came in to him and said: "I want to buy some crackers." The grocer weighed out the crackers and gave him the crackers. Then he said: "I have changed my mind; I would like to have some cheese. You take back the crackers and give me the cheese." He got it, and started off. The Dutchman said: "You haven't paid me." Paid you for what?" "Paid me for the cheese." "Yes, but I gave you the crackers for the cheese." "Well, but you haven't paid me for the crackers." "Well, but I didn't take the crackers." (Laughter.) Then the Dutchman said: "I know, I am out the cheese anyway; I haven't got anything for it. You will have to pay me."

In the same way may the people ask, since they are out the land, what are they going to get. (Laughter.) It is said that the present landholders have acquired title to their land, a valid title, by consent of the state. Who is the state in this case. (A voice: "That's the question," and applause.)

In the case of the lands of England is it the people of England or your upper House of Parliament, which is composed of men who sit there because they are landlords – your hereditary legislature? Your lower House of Parliament is largely composed of men of the same class. When since the Conquest [of 1066] have the real people of England had an opportunity to assert their right? (Applause.) If they have not done it, why? Because they have been impoverished . . . and kept in ignorance, and because it has been preached to them, day-by-day, and year-by-year, that this order of things was the natural order of things and if they wanted to avoid hell hereafter (applause) what they had to do was to obey the powers that be.

Why, if in a family to whom an estate was left, the older brothers were to simply seize it, and cut out the tongues, and put out the eyes of the younger ones, would it give them a right because these little ones made no complaint? What complaint can those poor people make from whom comes up that "bitter cry of outcast London?"[26] What can those agricultural laborers, men who must live a life of poverty and die the death of a pauper, what can they do? Why, is it not an absurd thing? Well may we ask with Herbert Spencer: "At what rate per annum does wrong become right?" Because a man takes my earnings today, and he had them the day before, does it give him the right to take them tomorrow? (Cheers.) Why, the thing is absurd; and observe, it is not proposed to take from the landlords anything save the power they now exercise of taking the fruits of the labor of others, without giving them any recompense. (Cheers.) No one I think proposes to deny to the landlords their equal share in the land of England; it is proposed to give them as much as anybody else. ("Hear, hear.") How can they complain of that? Now it is said that such a thing will injure the poor. Always so! Just as it was in the slavery fight when it was the poor widow that was put to the front, the poor widow who had only three or four slaves upon which she depended for a living – were we going to rob her? – it is the poor widow who is put to the front here. Speaking in St. James Hall the other night I spiked that gun[27] by proposing that the first thing should be to pay all the widows on a good round pension. Haven't heard anything of the widows since then. (Laughter and applause.)

But now it is the poor working man; it is the man who has worked and toiled, and bought himself a little home. Well, now, look at that case. Here around your towns is a lot of vacant land, land only used for agricultural purposes. What taxes are paid on it? Why, almost nothing. The moment that land is cut up into little plots and houses are built upon it, then down comes the rate-gatherer and wants one-third or one-fourth of the rent. Is that the way to encourage the building of houses? Why, take any man – let him be a free-holder, let him own his land out-and-out – say nothing of a tenant but let him be a freeholder – you would take taxes off his house, you would take taxes off all he uses and consumes, and make it up with a tax on the value of the bare land; and he will necessarily be a large gainer. ("Hear, hear.") As a mere matter of taxation it will be money in his pocket, to say nothing of the enormous stim-ulus that will be given to industry and trade. (Applause.) Justice hurts no one. ("Hear, hear.") Justice, I believe, is the eternal law. What we have to fear, as all history teaches, what we have to avoid, is not Justice, but injustice. (Applause.) That is the thing that brings the poverty.

I am not starting any paradox – I firmly believe it – when I say that, even considering the interests of the landlord class, the best thing is to destroy their unequal privilege and to do exact Justice between man and man, to acknowl-edge the rights of each. (Applause.) It cannot be good for any man to be placed, in wealth or in station, way above his fellows. Look at that child of four years of age who, by the untimely death of his father yesterday, is heir to an estate of millions of pounds a year. Take any boy, and let him grow up knowing that that enormous sum will come to him, and let him be surrounded by flunkeys and panders, and only a miracle can make a man of him. (Loud applause.) Envy the rich! I have not in any heart the slightest particle of envy for the rich. I fear poverty. I know better what it is than some of these men who talk so idly of the sufferings of poverty. I fear poverty, for it is indeed the deepest hell that man can know. (Loud applause.) It is not that you may go cold and may go hungry, but that the finest feelings of man's noblest nature are cut and lacer-ated and seared day-by-day. But I don't envy the rich; and I would not be rich.

Once, when I was a young man, it was my ambition to have an income of five million dollars a year and I wanted to go round like Monte Cristo.[28] (Laughter.) But by-and-by, when I came alongside of men who were rich, I learned absolutely to pity them. When I saw men coming up from the work-shop or the little merchant's counter into positions of enormous wealth, I saw that they were surrounded by flatterers and sycophants, that they had constantly on their minds the care of these great fortunes, that they did not dare to believe when a man came to them and took their hand and said he liked them that he was in earnest, and when a woman smiled on these men thought she had designs. (Laughter and applause.)

No. The state of equality is the best state and the highest state. Where is it, from what class is it, that have come the men who have made this great English literature? The men of whom we of the English-speaking race are most proud. Have they come from the very rich? Have they come from the ranks of landed aristocracy? No, no more than from those who were ground down by poverty.

No, they came from the ranks of the men, who, without having had too much to raise them above their fellows, had enough to give them leisure, and that is the condition all men ought to have. (Cheers.)

Not in envy, not in malice, but in Justice and in love, even to those people who think we are going to rob them, ought we to reassert, the natural inalienable rights of men, the rights of our forefathers on this island generations ago. But there are other reasons. Turn your eyes to the poor in this England of yours, to that festering mass of poverty, and misery, and vice and crime. ("Hear, hear.") Read the "Bitter Cry of Outcast London" and similar accounts.[29] What are you going to do with these people? Are they not your own flesh and blood? Those little children, are they not just such children, as was that child whom Christ set among his disciples, and said that it was better that a man had a millstone tied around his neck and he fling himself into the uttermost depths of the sea rather than to offend one of those little ones?[30] (Cheers.) What can you do for them? Charity! Charity will only degrade. ("Hear, hear.") Charity will do nothing, [needed is something] higher than charity . . . it is Justice. (Cheers.) Charity! Why, men, did you ever think of it? I speak with all reverence but so long as the laws of the universe are what they are, it is not in the power of Almighty God to relieve that suffering, that unseeing, that starvation so long as private property in land continues (Cheers.)

Our record we have. You remember when the children of Israel crossed the desert, and they cried for water and Moses struck the rock with his staff and the water gushed out?[31] What good would that water have been to the Israelites, if there had stood somebody there to say: "That rock is mine?" ("Hear, hear.") Why, think of it. If manna were to fall from heaven upon these islands of yours what good would it do people? Who would own it? Where would it fall? It would fall on the land, and necessarily it would belong to the people. I will stop here because I have exceeded my hour. (Loud cries of "go on.") Why think of it. I am talking this way because I want to shame you. Supposing you were to get to the kingdom of heaven, and you found this system existing there, and that those who had got there first had preempted it. What good would be the fruits of Paradise to you? Private property in land. Why, it is simply a form of slavery. ("Hear, hear.") Necessarily and absolutely the man who owns the land on which, and from which, another human being must live is his master, and his master even to life or to death. Under the state of things which we see here, and which exists all over the civilized world, for we are imposing it on the other side of the water, on our own virgin confinement, are there not slaves? Are not the working masses of these civilized countries slaves, just as utterly as were the chattel slaves, only they do not know their particular master. ("Hear, hear.")

When a man, without doing a thing, can draw from the earnings of the community the results of labor, when he can have a palace, and yachts, and horses, and hounds, and all the things that labor produce, is not the laborer necessarily robbed? Are not the fruits of his labor necessarily taken without any return to the man who gave the labor? Why in a paper last week, in which I read a column denunciation of myself and my proposals for "theft and confis-cation" (laughter), I turned over two or three pages and there was an article headed "The White Slaves of England," in which it stated, and stated the truth, that the condition of a large section of the English people was worse than that of any chattel slaves. And so it is; no Southern slaveholders would have worked and kept his Negroes as white men and women and children are worked and kept in this free England. (Cheers.) I will take the annals of any system of chattel slavery, and for every horror that you produce, I will produce a double horror from the files of our papers. Our missionaries do not read those papers to the heathen. (Laughter.)

Every word that is now said against the demands for equal rights in land was said in defense of chattel slavery. I could take such letters as that of the Rev. Page Hopps (hisses and applause) and by simply substituting the word "slaveowner" for "landowner" I would show you precisely what just such preachers were preaching over in my country when I was a boy; when the men who dared to assert the inalienable right, the God-given right of man to his own strength and views, were called Communists and destroyers; were told they were attacking the sacred rights of property, and that if those wild notions ever carried, nothing would be secure; when they were denounced in the church, ostracized in public, rotten-egged on the stump, and sometimes driven to their death. The men who were helping the Negroes of the South to escape from slavery were termed thieves and robbers; not the man stealers, not the men who tore the husband from the wife, the child from the mother, but the men who would put an end to that state of things. ("Hear, hear.") But the truth grew and grew, and despite possession, and despite denunciation, and the talk of thievery and confiscation, there came a time when the slaveholders determining at any cost to resist, went a little too far.

There came a time when the gun flashed from Charleston Harbor, a bit of bunting flag, and a million men spring to the deathlock – and slavery was dead[32] (cheers); died in blood, and flame and agony, died at the cost of a million of lives and millions of treasure. Why? Because the nation had done an injustice. When our Declaration rang out, asserting the equal and inalienable and God-given rights of man, our fathers did not apply it to the Negroes. That was the right of property. They had been bought with money. So careful were they not to interfere with vested interests that they would not even stop the slave trade, but provided that for a term of years it should not be in the power of the Congress of the United States to prevent men, stolen from Africa, who were brought to our shores and sold as merchandise. And at length the time came, and the nation paid for its injustice. ("Hear, hear.") And so, as all the history of the world shows, must men pay for injustice. There is no escape from it. I have never concerned myself with results. If I had asked whether this man is with me or against me I should never have stood here tonight. (Cheers.) What is coming I do not know; but this I do know, that in the history of the world and in the providence of God, the time has come when this injustice that is eating the heart out of our civilization; when this injustice that condemns little children to a life of misery and vice, that destroys the mind and soul as well as the body, cannot possibly continue. (Cheers.) The struggle has begun, and it must go on (cheers); and you men of Birmingham, you leaders of the van of Radical England[33] be true to yourselves, care nothing for denunciation, put aside all those fears and old superstitions; ask only for what is true, only what is right, and go forward. (Loud cheering.)

Questions

The Chairman: Mr. George is willing to answer questions, not to have a debate.

A gentleman in the hall asked: Whether Mr. George was a direct advocate for plain and simple confiscation, without any compensation whatever: and if not in all cases, where would he draw the line.

Mr. George: I mean to draw no line. (Cheers.) I am a believer in the absolute rights of man, I believe that every Englishman on this island has an equal right to the soil of that island. (Cheers.) It is the right that adheres to every child as it comes into the world, and I would pay nothing for the resumption of it. (Cheers.)

The Questioner asked Mr. George: Whether, if he went into a shop and bought a loaf, which be afterwards found had been stolen, would he consider that loaf his property? (A voice: "It is not land.")

Mr. George: I will answer you in a word.

The Questioner: I will speak to land if you like.

Mr. George: If you can understand English –

The Questioner (interrupting): I ask[ed] you when I have bought and paid for land by the labor of my hand and brain, whether I am entitled to it or not.

Mr. George: Yes, whenever you can show a title from any man who had a right to give you a title ("oh, oh," and cheers), from the maker of the land, or any who got the title from the maker of the land – until that time, [N]o. (Cheers.)

The Questioner: Then let me tell you, Mr. George, you may go back to your Americans in America and tell them to restore the land to the red Indians. I tell you that Englishmen believe in vested rights. (Cries of "order.")

The Chairman: The object of the meeting tonight is to give Mr. George an opportunity of expounding his views. Mr. George is willing to answer questions, and I therefore appeal to the audience to listen to the questions with perfect silence and then we shall hear Mr. George's answer and we shall be able to form our opinions.

A second questioner asked: Whether Mr. George would not approve a plan, whereby the burden of the national debt which had been incurred for the protection of the land of the landlord might not be thrown upon the shoulders of the landowners?

Mr. George: I am not here to advocate plans. Whatever plan you men of England choose to adopt, that is your own affair. One plan may be best for one country, and another for another country, but the eternal principle is the same all the world over. (Cheers.) You may, if you choose, compensate your landlords, you may make up a collection for them. (Laughter.) Let every man who thinks the landlords ought to be compensated, put his hand in his pocket. (Laughter and "hear, hear.") What I contend for is this: Not one of you, and not all of you together, have a right to say that the humblest English child shall be denied its birthright until it [is] compensated [by] the landlords.

Mr. McClelland: What Mr. George would do with the land in Ireland, where at the present moment the people were in favor of peasant proprietary?

Mr. George: I would give the land to the people (cheers), I would form no peasant proprietary. The principle is the same whether a man holds a quarter of an acre, or a quarter of a million of acres. I care nothing for the tenant-farmers of Ireland, the men I care for are the agricultural laborers. (Cheers.) Elevate the condition of the whole of society; anything short of that merely divides the people into the rich and poor. Peasant proprietary is impossible. ("Hear, hear.")

This land of England was once cut up into small estates. Only two centuries ago, historians tell us, the great majority of English farmers were owners of their own land. Where have they gone now? What is going on in the United States? The small American farmer is being exterminated; the majority of the American people do not realize that fact, but it is so. I know of farms of 60,000 acres made out of small farms. Concentration is the law of the time, and you cannot avoid it. Peasant proprietary! They have got peasant proprietary in France and there are 150,000 unemployed workmen in Paris. They have peasant proprietary in Belgium and M. de Laveleye, the very highest authority, says the tenants there are rack-rented even worse than in Ireland. Do not stand at any of these miserable halfway houses, go to the eternal principle.

A third questioner asked: Whether, if the land was held by the country, it would not be the duty of the Government to let every acre of land at the highest rent they could get, and if so, how would the poor man be one bit the better for such a condition?

Mr. George: Unquestionably it would be the duty of the Government to rent land for the very highest price it could get. If ten of us owned a horse, and only one could ride him, would it not be the fair thing, and the only fair thing, for the whole ten of us, that man, if he is willing to pay the highest price for the horse, should ride him? Certainly. The rent, then, would go back to the people. (Cheers.) There is [no] harm in rent. Rent to my mind is one of the most beautiful adaptations of Nature, one of those things in which a man who looks can see an absolute proof of a divine and beneficent intelligence. It is that we have perverted the good gifts of the Creator that the children's bread is tossed to the dogs. Rent rises and increases as society progresses. What does that mean? Simply this, that in the order of Nature the progress in civilization should be a progress towards truer and truer quality: that the common interest of all should grow larger and larger as against the individual interest. (Cheers.)

I would like to ask this audience a question for my own curiosity and infor-mation. I have been told that English audiences do not, and would not, agree with me on this point of compensation to the landowners before the resump-tion by the people of their rights. I would put it to the vote of this audience, I want to ask you, all who stand with me: Who hold that vested interests cannot stand in the way of natural rights, and that it is not necessary to buy the land from those who now hold it before the people resume their own[?] – say aye; contrary no. (Loud cries of "aye" and "no.") I believe the ayes have it. (Loud cheers.)

A fourth questioner asked Mr. George: What distinction he would draw between the landowner and the capitalist whose income came in the same way, from the earnings of the people. ("Hear, hear.")

Mr. George: Your capitalist has done something. The very fact that he is a capitalist proves that. Your landlord has done nothing whatever; he takes no more part in production than a wild animal. I do not say that there may not be landlords, there are many, who do other things, but I say that the landlord as a landlord does nothing. There was one landlord in Ireland who went to bed and stayed there for eight years.

The Questioner: But suppose I inherit those shares and have done nothing at all for them.

Mr. George: Shares! What do the shares represent? They represent actual work done.

The Questioner: Not by me.

Mr. George: By somebody. Whoever does work, the product of that work belongs to him by the law of Nature. Nature gives nothing to man without hard labor, and he can transfer that right to whoever he pleases; that which the man produces is his as against the world, to give, to sell, to bequeath, or to destroy. Let the man who has real property, real wealth do what he likes with it. What we want is the soil, the element of Nature for which all wealth proceeds. (Cheers.)

SOCIALISM AND RENT APPROPRIATION: A DIALOGUE
(Henry George and Henry M. Hyndman,
The Nineteenth Century: A Monthly Review, 1885).[34]

Mr. Hyndman: I see that you have been hard at work since you have been over here in land agitation. I think, as you know, that you expect far too much from nationalization of the land by itself.

Mr. George: Why?

H.: Because I understand you to advocate merely the confiscation of competition rents, and that, to my mind, will not benefit the laborers.

G.: I advocate the recognition of equal rights to land. As for any particular plan of doing this, I care little; but it seems to me that the only practicable way is to take rent for common purposes. Rent is, of course, fixed by competition.

H.: That is rather vague to me. I am as much in favor of nationalization and communization of the land as you are; but taking rent would not bring this about; it would leave the laborers, whom we both wish to benefit, competing against one another for subsistence wages just as they were before.

G.: That I think a mistake. It would not only give the laborers their equal share in the benefits of an enormous fund which now goes to individuals, but, by making land valuable only to the user, would break up the monopoly which forces men who have nothing but their own powers to that fierce competition which drives wages down to the lowest possible rate. The fundamental mistake of Socialists of your school, it seems to me, is in your failure to see that this competition is not a natural thing, but solely the result of the monopolization of land.

H.: I should dispute to begin with, if it were worthwhile, that all rent is necessarily competition rent. Customary rents are still far more common than competition rents. But let that pass. What we contend is that the confiscation of rent leaves the competition untouched. This you admit yourself. And if you broke up the landlord monopoly tomorrow by taxing land up to what you would call its full value – an impossibility as I believe in practice – the control by the capitalist class of the means and instruments of production would remain untouched, and the laborers would still compete under the control of that class, who would derive all the advantage from the change. The historical growth of private property in land has ended in the domination of the capitalist class or *bourgeoisie*. In England, at any rate, the landlord is a mere hanger-on of this class – a sleeping partner in the product taken from the laborer by the capitalist.

G.: You are using the term rent in one sense, and I in another. By rent I mean the value of the advantage which accrues from the use of a particular piece of land, not what may, as a matter of fact, be paid by the user to the owner. The amount paid by custom, or the amount paid under lease, may be lower or higher than the true rent. In the one case the owner gets more than is really rent, in the other cast he gets less, and the tenant or intermediate tenants get the difference.

Rent involves competition. Until two men both desire the same piece of ground, land, no matter what its capacities, can have no value. True rent is always that rent which could be obtained by free competition, nor is it desirable that it should, but it would destroy that power of withholding land from use which in so many cases forces its price far beyond the point which free competition would fix. The price which crofters and agricultural laborers pay for land, the price, many times the agricultural value, which must be paid for building sites on the outskirts of town, is in reality not rent, but blackmail. In short, were every holder of land compelled to pay its competitive value to the community, the power of withholding land from use would be gone, and there would be substituted for the present one-sided competition in which men deprived of the natural means of livelihood are forced to underbid each other, a free competition in which employer would compete with employer as fully as laborer must now compete with laborer.

The social difficulties we are both conscious of do not arise from competition, but from one-sided competition. No monopoly of capital of which it is possible to conceive would, so long as land was open to labor, drive wages to the starvation point. As for the landlord being a mere hanger-on of the capitalist, the monopoly of land is the parent of all other monopolies. Give men land and they can get capital, but shut men out from land, and they must either get someone to let them work for him or starve.

H.: Perfectly free competition for land is unknown, but I am quite content to argue the question as you put it. Without going into the genesis of capital in its modern sense, I urge that the system of production which assumes the payment of wages to "free" competitors who necessarily produce commodities for exchange, and have no control, individually or collectively, over the means and instruments of production or their own products, would not be in the least affected by the confiscation of competition rents. The men who attempted to work on the land would inevitably fall under the yoke of the capitalist class just as they do now. They would be producing for the world market under the control of the employing class just a they are now, and would be at their mercy as a class in the same way.

Moreover, even from the individual point of view, crops or buildings can be hypothecated[35] just as easily as the land. I repeat, therefore, that the capitalists as a class would be the sole gainers by the confiscation of rent and its application to the reduction of taxation or to public purposes. The workers would be no freer than they are today.

G.: I cannot see why not. The confiscation of rent would give to all laborers, as well as capitalists, a share in the income thus arising. This would necessarily be to the far greater relative gain of the laborer than of the capitalist, and would greatly increase his power of making a fair bargain. Even now an equal division of the rent of the United Kingdom would give to a large number of families more than they have at present to live upon. Further than this, the effect, as I have before pointed out, would be to make land valuable only to the user, that is to say, to make it free to labor and to keep it free. No harm whatever that I can see could come from the power of hypothecating crops and buildings or anything of that nature.

As for producing for one's own use or producing for the world market, that form of production is best for the producer which will give him most of the things he wants, and which are the real objects of his production. The effect of the recognition of equal rights to land would be necessarily to greatly diminish the proportion of wage workers, but I can see nothing wrong in that system in itself. The trouble is not in men's working for wages, but in the fact that, deprived of the natural means of employing themselves, they are forced to work for unfair wages.

You speak of the means and instruments of production as though they did not include land, which is the principal and only indispensable means and instrument of production. All other means and instruments have been produced from land. Give men land, and they may produce capital in all its forms. But give them all the capital you please, and deprive them of land, and they can produce nothing.

H.: I venture to think my argument is quite clear. You say that an equal division of the rent of the United Kingdom at the present time would give a large number of families more than they now have to live upon. The total rent of Great Britain, agricultural rent and ground rent together, apart from interest

on capital invested, is estimated by no authority worth a moment's consideration at more than £60,000,000 or £70,000,000 a year. The population of Great
Britain is 30,000,000. An equal division would therefore give £2 to £2 7s. a
head, or from £10 to £11 15s. per family. None but pauper families can possibly
subsist on that yearly sum, and they of course are kept by ratepayers at a much
higher cost.

But admitting the division to be so made, what then? Each working family
could afford, in periods of severe competition, dull times, crisis, or the like, to
accept 4s. a week less wages from the employing class by reason of this dole,
and thus the £10 a year per family would almost immediately go into the pockets
of the capitalist class. Surely that is clear.

G.: Rent is generally underestimated, many important items being omitted,
and I am inclined to think that the estimate of £300,000,000 as the true rental
value of the three kingdoms is nearer the mark than that you gave. But we
have not time to dispute as to statistics. Even £2 per head would be a most
important addition to the income of many working-class families, relieving them
from the necessity of forcing women and children into the labor market for any
pittance they can obtain.

But such a division would not be the best way of utilizing the fund. By
substituting the income thus derived for the income now raised by taxation,
which, falling upon production and consumption, represses enterprise and bears
most heavily on the poorer classes, the gain of the laborer would be much more
considerable than from a simple division of rent; and while it is true that, land
being monopolized, any general addition to the income of the working classes
would ultimately carry wages still further down, the breaking up of the land
monopoly, which the appropriation of rent for public purposes would cause,
would prevent the competition that has this effect. Here is the main point which
it seems to me you fail to appreciate, that the tendency of wages to a minimum
which will merely enable the laborer to live and reproduce is not an inherent
tendency, but results solely from the monopolization of land.

H.: The first argument is merely a contravention of my figures. As a matter
of fact, agricultural rents are falling all over England today, and will continue
to fall in the face of American competition and the rapid improvements of trans-
port. With relation to municipal rents, I am talking merely of the ground
landlords; the moment you go beyond that, you enter upon the expropriation
of the capitalist, and, even as it is, the capitalist class takes a large percentage
of the ground rents and agricultural rents as interest on mortgage. The effect
of the reduction of taxation would be precisely the same as the division of the
income from rent among the population; the working classes would be able to
accept to that degree lower wages in a period of fierce competition, and the

total benefit in the long run would go to the capitalist class. My answer to your last statement will come better later on.

G.: Agricultural rents are falling in England, but I think this fall is overestimated, and cannot long continue, for they are rising elsewhere, noticeably in the United States, where in new sections of the country half the crop is now being paid as rent. But urban rent and mineral rents, which in this country are more important than agricultural rents, are steadily rising. In the cities the rise of house rent is entirely the rise of ground rent. Buildings do not become more valuable, but land does. Where the capitalist gets interest on a mortgage on land, he is simply getting a portion of the rent.

H.: Of course, as his class gets a much greater portion of all products of labor (being more powerful under present conditions) than the landlord class.

G.: Capitalists, as capitalists, are not, and never can be, more powerful than landlords. Men can live in a rude fashion without capital, but cannot live without land. But the point I wish to make is, that where the capitalist receives rent he is to that extent a landlord. The effect of a mortgage is simply to divide proprietary rights.

H.: If you come to that, the effect of the whole system is to confiscate labor. The question is, which of the expropriating classes is dependent on the other. I say, under the present economic conditions, which result from an historical growth extending over centuries, the landlord class is dependent on the capitalist class. You say the contrary. It is at any rate impossible to argue as if conditions of society which have long since passed away still existed.

G.: It is only necessary to argue from the facts of Nature, which are the same today as they always have been. Give men land, and they can produce capital. Deprive them of land, and they cease to exist. Were all the capital of this nation destroyed tomorrow, some remnant would continue to live, and would in time reproduce capital; but destroy the land, and what would become of the nation? You are talking of a capitalist class and of a landlord class; as a matter of fact, these classes blend into each other. There are probably no landlords who are not to some extent capitalists, and few capitalists that are not to some extent landlords. If we talk of the capitalists and the landlords, it must be by considering them in the abstract.

H.: The facts of Nature are perpetually modified by man, and in an increasing degree as social forms develop. I have already said that in England the landlord is merely a hanger-on of the capitalist, and that the two classes therefore do blend. Capital, to my mind, expresses a whole series of social relations; class monopoly of the means of production on the one hand, and competition of wage earners on the other.

Again, what is the good of giving men access to land if they have to compete with other men who own much larger capital, and therefore can undersell them by sheer force of cheaper production owing to superior machinery and greater command over the forces of Nature – who can produce with less labor, that is? Man cannot live by bread alone He must exchange his agricultural products if he is to attain a decent standard of life in other respects.

Yet here the big capitalist steps in and reduces the exchange value of his raw commodities as estimated in the universal measure of value – gold. Big capitals must in the long run crush small; you admit that, I know. Agricultural land in short, whether "nationalized" or not, is just as much capital today as any factory, and is used as a factory, the wear and tear being made good in precisely the same way as the wear and tear of machinery or buildings. Louisiana is a great raw cotton and sugar factory; Minnesota and Wisconsin are great grain factories; Lancashire is a great manufacture[r of] cotton and iron factory, and so on; all carried on under the control of the capitalist class who produce for profit on the world market.

Nationalize the land as much as you please, therefore, without giving the producers the collective control of the social machinery, the means of production and distribution as well as of the exchange, and no good really will have been done. The land is only one of the means of production and under existing conditions is useless without the others. Production for profit, and competition for wages under the control of capital, will in my opinion go on equally when the land is nationalized; wages will equally tend to a minimum; and there will be as now the same phenomena, the causes of which we Socialists alone explain – overproduction, crisis, and glut, followed by periods of boom and prosperity.

_G._: It seems to me that you Socialists confuse yourselves by using terms in varying senses. Here we are discussing the relations of capital and land, the inference necessarily being that they are separate things, whereas you include land as capital, and also include as capital such things as monopoly and competition. Land is not merely one of the means of production, but the natural factor in all production, the field and material upon which alone human labor can be exerted. No matter how much Nature may be modified by man, man can never get beyond his dependence upon land until he can discover some way of producing things out of nothing. Capital is simply wealth (that is to say, the material products of human labor exerted upon land) applied to assist in further production.

It is the monopolization of land that always drives men into such competition with each other that they must take any wages they can get, and thus forces wages down, as may be seen by the fact that in new countries, where land is more easily obtained, the wages of laborers are always higher than where the monopoly of land has further progressed. Land being free, capital cannot force wages down, for capital must compete with capital; a competition which where monopoly does not exist is quicker and more intense than any other competition. Break up the land monopoly, and not only would capital become more equally diffused, but capitalists must compete against capitalists for the employment of labor; or, to put it in more absolute form, labor would have the use and assistance of capital on the lowest terms which the competition between capitalists would bring about.

H.: Whether Socialists explain the phenomena of industrial crisis or not is a point we have not yet discussed, and may come later. Nobody has ever disputed that land is necessary to the production of wealth; but land has been used very differently, and has been the basis of very varying social relations in the history of mankind, as you are perfectly aware. You say that capital is wealth applied to further production.

> A spinning jenny is a machine for spinning cotton;[36] only in certain conditions is it transformed into capital. When torn away from those conditions it is just as little capital as gold is money in the abstract or sugar the price of sugar. In the work of production men do not stand in relation to Nature alone: They only produce when they work together in a certain way, and mutually exchange their different kinds of energy. In order to produce, they mutually enter upon certain relations and conditions, and it is only by means of those relations and conditions that their relation to Nature takes place, and production becomes possible.[37]

The fact that in new countries wages are high arises from the fact that there a man is, in many cases, able to take himself out of the wage earning class altogether, to dissociate himself from his period in fact, as the Mormons did in Utah, and as other people have done in the West of America and other parts of the world; but this cannot be permanent, whether land were common property or not. We Socialists, as I have said before, are as much in favor of making land common property as you are; our only difference arises as to the means whereby such nationalization and communication should be brought about.

There are two other points I must touch upon briefly. The capitalist system of production involves class monopoly of the means of production and competition among propertyless wage earners. As to the competition between capital and capitalists, that is going on most fiercely today; the result always is that in the long run the capitalist class, as a whole, gets a greater relative proportion of the products of labor, and the working class a less relative proportion. The same would be the case if the land were nationalized, the other conditions remaining unchanged.

G.: The point at issue between us is as to what would be the effects of nationalization of the land unaccompanied by nationalization of capital. We are talking of land, of labor, and of capital, and of three corresponding classes – landlords, laborers, and capitalists. We can never reach any clear conclusion unless we attach to those words a definite meaning and exclude from what we embrace in one that which is embraced in others.

A spinning jenny is an article of wealth, a product of labor and land, and, like all other articles of wealth, may or may not be capital according as it is used. But land or labor never can be classed as capital as long as the three terms are used in contradistinction. Whatever varying social relations may exist among men land always remains the prime necessity – the only indispensable requisite for existence.

The development of exchange and the division of labor do not change the essential facts that each laborer is endeavoring to produce for himself the things which he desires, and that land is the raw material from which they must come. Make labor free to [work the] land, and it will be impossible for capital to take any undue advantage of it. Just as men in new countries can take themselves out of the wage-receiving class by going to work for themselves, so would it be in such a country as this.

The competition of capital with capital is intense today, and its effects may be seen in the lowering rate of interest. But labor being shut out from land, wages tend to a minimum, and the advantages of the improved processes of production and exchange go either to the landowner who possesses the natural element indispensable to production or to capitalists who in other ways secure a monopoly which shelters them from competition and who thus take what, if these monopolies were abolished, would go not to the laborers but to the landowners. In short, our social difficulties arise not from capital or from capitalistic production but from monopoly.

H.: As to monopoly, capitalism, and a wage earning class involve class monopoly, or we should not be arguing now. It is almost unnecessary for me to repeat that, in England, at any rate, I consider the landlord to be a mere appendage to the capitalist, and that you cannot get at the land with any advantage to the people except through capital. You seem to forget that the mass of mankind who labor would be wholly helpless on the land if they had perfect freedom to go there. Each man produces not what he himself desires to keep and to use for himself, but things which other people desire to have in exchange under the control of the capitalist class, all production now being practically conducted with a view to exchange.

As to the competition of capital lowering the rate of interest, 3% on £100,000 is the same as 30% on £10,000, and the capitalist class may be taking a very much higher amount out of the total product of labor, although the rate of interest may be very low. We are both agreed of course that labor applied to natural objects is the sole source of wealth; and that the quantity of labor socially

necessary to produce commodities is on the average the measure of their relative value in exchange.

G.: The mass of mankind, even the men of the cities, would not be wholly
helpless on the land. Man had in the beginning nothing but land, and it is from
land that all the instruments that he uses to assist production are derived. But
even admitting that man in the present state of society can make no use of land
without some capital, the effect of throwing open land to labor would be that
those who had some capital would go upon the land, thus at once relieving the
competition of wage earners and increasing the demand for their labor.

Not to prolong the discussion on these lines, it seems to me the difference
between us is simply this. We both agree that labor does not find its proper
opportunities or get is fair reward. Your contention is that, to remedy this state
of things, not merely land but also capital must be made common property;
while I contend that it is only necessary to make common property of that to
which natural rights are clearly equal, and without which men cannot exist or
produce – land.

H.: With regard to your last sentence I agree that it formulates our difference (omitting the phrase "natural rights"), and using "capital" in the sense of
means and instruments of production, and all improvements upon them.

G.: What do you mean by "capital" in that sense?

H.: I mean railways, shipping, machinery, mines, factories, and so forth. I
contend, in short, that all production today is necessarily social, and that
exchange is conducted for the benefit of individuals or a class, the products
belonging to the capitalists, not to the producers. I wish to socialize both the
means of production and the forms of exchange as well as the land

G.: This seems to me indefinite. I am quite with you as to the desirability
of carrying on for public benefit all businesses which are in their nature monopolies, such as telegraphs and railways; but it does not seem to me necessary to
go any further than this, as, where free competition is possible, the same end
will be much better served by leaving such things to individual enterprise. Even
as to railways, telegraphs, and such agencies, the assumption of them by the
community is quite a minor matter. Give the people land, and they can live
without either the railway or the telegraph, and though a railway or telegraph
may be a monopoly, its owners must in their own interest fix their charges at
such a rate as would induce people to use them.

As a matter of fact, we see everywhere, the advantages that accrue from such
improved instruments as railways and telegraphs do not go to the capitalists who
construct them, but very largely to landowners in the increased value of land. I
can understand how a society must at sometime become possible in which all
production and exchange should be carried on under public supervision and for

the public benefit, but I do not think it possible to attain that state at one leap, or to attain it now.

In the meantime, people are suffering and are starving because the element which is indispensable to existence, and to which all have naturally equal rights, has been monopolized by some. Destroy this monopoly, and the present state of things would at the very least be enormously improved. If it were then found expedient to go further on the lines of Socialism, we could do so, but why postpone the most necessary and the most important thing until all that you may think desirable could be accomplished?

H.: The capitalists as a class would meantime be strengthened. But we wish to postpone nothing. In our opinion, given the necessary political predominance of the producers, the economic forms are all ready for the nationalization of the means and instruments of production of which I have spoken. From the Company to State or Communal control is an easy, and, as you would say, a natural transition – the salaried officials and wage earners of the Companies becoming the salaried officials and wage earners of the State or the Municipalities, and would certainly have nothing to gain by making profits out of their own overwork or underpay.

The economic forms for the nationalization and communization of agricultural land for productive purposes are in my opinion not ready, except in the way I state. The suffering and starvation which we see around us now are due to the capitalist system of production, which throws people out of employment the moment production at a profit to the capitalist class ceases to be possible. This brings us, I think, to practical proposals. I, at any rate, have said all I wish to say at present on the main issue between us.

G.: Taxation supplies the form for the virtual nationalization of land, and I cannot see your reason for thinking that of itself it would not relieve labor. Take, for instance, the overcrowding of cities. That does not arise from any system of capitalistic production, but merely from the fact that people are not permitted to build houses without first paying an enormous price, and that when the houses are built a further tax is placed upon them which must necessarily fall upon the user. The effect of appropriating rent would be to at once [to] increase the number of houses by reducing the price that must be paid for their sites and abolishing the tax now imposed upon them. So, all through the agricultural districts men would be able to go upon and cultivate land from which they are now debarred, thus relieving the labor markets and producing a greater demand for the commodities which the working classes of the towns and cities produce. It would make it impossible for men to shut up mineral resources as a certain Scotchman recently shut up an iron mine employing a great number of hands, saying he could afford to keep it idle as it would "not eat anything." In short, the effect would be to stimulate production in every direction.

H.: Taxation leaves competition among wage earners untouched. The people are driven into the towns by improvements in machinery which enable the farmers to do the same amount of work with fewer hands, and therefore ought to benefit the whole community; partly also, in England at any rate, by the substitution of pasturage for arable culture. Thus, driven into the towns, they compete with their fellows. Moreover, overcrowding in cottages in the country, where the ground rent and the taxation is a mere trifle, is just as bad as it is in towns. As to shutting up the iron mine, the men could not work that mine unless they appropriated the mineowner's capital. The causes of the present universal crisis, the seventh of this century, lie, I think much deeper than you suppose, and would not be affected by any one single proposal.[38]

G.: I think, on analysis, all these evils are traceable to the fact that land, which is necessary to all, is made the property of some.

H.: That is you contention, I know; but do you not think we have argued sufficiently now to be able to speak on points of agreement rather than of difference? For instance, we are quite at one in wishing to bring about greater freedom, comfort, and happiness for the mass of mankind, who at the present moment are driven into degradation and misery by class monopoly of one sort or another.

Whichever way we look we see the adulteration of goods; overwork of women and children; science and art at the command of a privileged few; education in any high sense shut away from the mass of the people; hours of labor unduly prolonged; men forced into idleness who wish nothing better than to work for the good of themselves and their fellow creatures.

Anything, therefore, which tends to bring the workers of the world together upon a common basis, whether of nationalization of land or collective ownership of capital, must necessarily tend to the overthrow of these abuses. You know that today peasant proprietorship is being put forward as a remedy for the ills of this country. On this point, at any rate, looking across the Channel and seeing the condition of the French and other peasant proprietors all over Europe we are thoroughly of one mind, that no benefit whatever can accrue by such an extension of the rights of private property.

Here, in England, a bill is before Parliament which, I believe, will be carried, for the enfranchisement of leaseholds. This, again, will but interest a larger class in that very monopoly of land to which you and I equally object. With regard to the thousands of the unemployed whom you spoke for yesterday in front of the Royal Exchange, and in whose interests we Socialists have been working for many months past, you, I presume, would be as glad as we should be to see the Government recognize its responsibility and organize their labor alike in town and country for the benefit of the community at large.

The present depression which has extended through every civilized country is independent of despotism or republic, protection or free trade.[39] These social questions evidently lie below all forms of government and all fiscal arrangements. We hold, as you know, that these decennial crises are due to the revolt of the socialized form of production against the individual form of exchange all over the civilized world. The incapacity of the *bourgeoisie* or middle class to handle its own system is, at any rate, proclaimed in every industrial and commercial center. Whether we are right, or whether we are wrong, no doubt it makes a difference to the tactics of the immediate future, but it can make no difference in the desire which we all must have to work in common for the great end of the emancipation of our fellowmen.

The mere fact that you and I are meeting here to discuss in a friendly manner the deepest social and economic problems, however, cursorily, is a proof that men are learning to sink differences of opinion in the sincere desire to find a base of agreement in view of the silent anarchy of today, and the furious anarchy which, unless some serious and important measures are immediately taken, threatens to overwhelm the civilized world tomorrow. The equality of men and the enfranchisement of women, which today are spoken of by many as a dream, are becoming really a necessity for the advance of civilization. Only by and through an International Socialist feeling, and a brotherhood among the workers of the world, can we hope for the happy future which thousands of the noblest of our race have longed to see.

G.: With all your sentiments I heartily agree. We who seek to substitute for the present social order one in which poverty should be unknown are not the men who threaten society. They are really the dangerous men who insist that injustice must continue because it exists. Nothing but good can come from a free interchange of opinion. Every man who looks at civilized society today must feel that the order that exists, and which you have so graphically described, is not that order which the Creator has intended.

The only question between us is as to the best way of substituting for it that order of things which will give free play to the powers and full scope to the aspirations of mankind. And questions of method are as yet but secondary. The great work is to break up the "pitiable contentment of the poor," and rouse the conscience of the rich, to spread everywhere the feeling of brotherhood. And this your Socialists are doing. These are indeed worldwide questions. We on the other side of the Atlantic have the same social problems to solve that are forcing themselves upon you here. The great change in public feeling that I have observed since my visit here a year ago proves to me that you in England are indeed taking hold of these questions with a determination to solve them. In my opinion, the greatest of English revolutions has already commenced, and it means not merely revolution in England, but one which will extend over the whole civilized world.

HENRY GEORGE AND THE SINGLE TAX
(J. Bruce Glasier, *Commonweal*, June 1, 1889).[40]

The large audiences which assembled to hear Mr. George during his recent lecturing tour in Scotland and the enthusiasm with which his denunciation of private property in land was received, are cheering signs to Socialists. It was especially gratifying to observe that almost every speaker who preceded and followed Mr. George at those meetings, spoke more Socialism than Single Tax; and that just in proportion as their utterances were boldly Socialistic – and many of them were remarkably so – was the applause and enthusiasm of the meetings.

I think, however, that I am not guilty of bias in saying that when Mr. George attempted to show that land can be nationalized by the imposition of a single tax on land values, there was a palpable fall in his argument, if not from the sublime to the ridiculous, at least from the convincing to the confounding – a fall, too, that was apparent alike in the lecturer's manner of speech and in the audience's manner of receiving it. It is true Mr. George admitted that he did not believe the imposition of a single tax upon land values would do "everything," and that he defined the right of private property in the products of labor in terms that implied Socialism; yet the general tenor of his utterance when eulogizing his tax theory and when replying to questions put by Socialists, was so patently sophistical and reactionary, that one could not help agreeing with Dr. Clark when he said – speaking at the farewell meeting in the City Hall, Glasgow – that: "He (pointing to Mr. George) was the most Conservative man in Scotland."

As Mr. George's single-tax theory has no doubt charms for those whose minds have been warped by the unhealthy teaching of expedients and palliatives common to all political reformers, and for those who feel almost persuaded to become Socialists but hesitate, looking for some stepping-stone, fearing to take the bold leap, I will endeavor to show briefly the error of fact and thought upon which it is founded. A reply given to Mr. George to a question asked in Glasgow, seems to me to expose the radical defect of perception upon which the whole superstructure of his method is built. When asked if a man was justified in taking the highest price he could get for a product of his labor, Mr. George replied: "Yes! If I make a rod and line and go and fish in a stream and catch a fish, and if everybody else is equally free to go and do the same, I am certainly entitled to ask, and if possible get, what I please for the fish."

It is hard to believe that Mr. George is not conscious of the utter absurdity that underlies such a declaration. What man ever would pay to another more for a fish than it would cost him to procure one himself, *if he was equally free and able to go and procure it?* Fishers do not usually succeed in robbing fishers, any more than lawyers succeed in robbing lawyers. But fishers may succeed in robbing lawyers, and lawyers usually succeed in robbing fishers when they get them in their clutch. It is apparent on the face of it, that the only reason why men can and do make a profit off their fellows is because they have a direct or indirect power to compel their fellows to submit to their extortion. Mr. George declares that he is opposed to all monopoly; but what is monopoly but the possession of some advantage – it may be special opportunity or special knowledge – that gives some the power to get more than the labor value of what they give? When there is no special advantage or monopoly, profit and interest, which Mr. George justifies, will be impossible.

The streams of our land may be equally free to all who care or can get an opportunity of fishing in them. But if while one man residing near a stream may be free to fish in it, another residing far away from it is surely not so free; and the man who is down in the mine during the day procuring coal to boil, or salt to season the fish, can scarcely be said to be free to fish at all. Are men, therefore, to take advantage of their neighbors' different pursuits, or even of their neighbors' ignorance or misfortune? Shall the fisher make a profit off the miner, the miner off the tailor, and the tailor off the husbandman? And when the one complains to the other: "Oh, you are robbing me!" shall the other answer him: "True, but you are free to rob me or someone else if you can. You are free to turn your hand to my occupation, and then you will no longer be robbed by me!" Surely the complainer would reply: "Yes, but then I again in turn would be robbed by the man who did the work which I now do! I am not free to fish, and bake, and make clothes, and cultivate the ground all at the same time; and unless I can do everything for myself someone or another will rob me!"

I freely admit, that if by any system of land restoration, or by the imposition of any tax, we could achieve a state where men might, even at a great sacrifice, be free to persistently revolt against robbery, or be free to rob others just as much as others robbed them, the problem would be solved. For these men, unless we assume that they had become idiots, would in sheer despair declare for Communism within twenty-four hours. But alas!, no theory of mere land restoration yet expounded, no method of taxation yet proposed, gives the slightest hope of such a result. Mr. George, while admitting that under his single-tax system the land would be rented to the highest bidder, and when, as he must know, the highest bidder would be the man with the most capital – the man who today can pay the highest rents, the man who would employ as much machinery and as few *men* as possible – still clings to the belief that his system will make the soil of the country free to the workers, so that they will be able to escape from the oppression of the capitalists and be able to sell their labor for its full, and if they can for more than its full value. There is nothing, however, in his single-tax theory per se that implies that a great number of men would on its adoption be drawn from the congested labor market and be employed upon land, not to speak of being free and able to employ themselves upon it. Unless the price of food is increased, it would not pay to employ more than are now employed upon the arable portions of it. Horticulture, it is true, might be substituted for agriculture, and that undoubtedly would employ more; but horticulture is not an exclusive or even an integral principle of the single-tax economy. Horticulture could be adopted without the single tax being imposed, and, as a matter of fact, is already largely adopted in many parts of the country.

As for the converting of sheep farms and sporting lands into arable soil, would not the number of men it might be profitable to employ in this way (and this again is not a remedy dependent upon the adoption of the single tax) be counterbalanced by the thousands of domestic servants, flunkeys, gamekeepers, gillies, fencemakers, and indeed cabinetmakers, tailors, etc., who at present make a livelihood by ministering to the landlord class, and whom the dispossessed – or taxed out – landlords and landladies would require to disband? Would not even the 30,000 or 50,000 landlords themselves – who would (if the tax proved effectual) be thrown on all fours upon the labor market to earn their bread by the sweat of their brows – be a somewhat important mob of recruits to the army of unskilled labor?

There are the tramway servants, the match-factory girls, the Cradley Heath chainmakers, the shirt seamstresses, and the thousands of other victims of usury and profit making – how, I ask, are they going to get on the land? Supposing the carse[41] of Gowrie or the fields of Lincolnshire were waiting for them rent free; without capital they could not go there, and even if they could they would perish of starvation in a single week. And if they cannot go to the land and employ themselves they must submit to be employed by others, and others will only employ them on condition of making a profit off their labor – that is robbing them. And need more be said than this, that any system that gives one class of men the power to be masters over another class, that compels the poor to labor for the profit of the rich, is a system – however hypocritically it may disguise itself – that no Socialist should assist in establishing, and if established every Socialist should war against until its utter destruction was assured.

As for the tax itself, Mr. George says it would, by means of the State, be distributed directly or indirectly to all the people of the country – rich and poor alike. Just so. What the rich got of it would be robbery; for, quibble about land values as Mr. George and his disciples may, there is not an atom of wealth that can be acquired by the State by means of a tax, that does not come exclusively from the labor of those who work. Those who don't work can produce no wealth, either to be taken in the form of taxation or any other form and every iota of the land tax which the wealthy idlers received directly or indirectly would be robbery of the poor. On the other hand, that portion of the tax which the poor received back, would also, by the iron law of wages (from which as we have seen the single tax provides no escape) go to the rich – the capitalists – in the form of reduced wages paid to the workers! Thus, the single tax would ultimately become only another means of making the rich richer and the poor poorer. So also would it operate in this way so far as it made unnecessary the exaction of all other direct and indirect taxation; for, while the single tax upon land would relieve the rich capitalist of say £1,000 a year of indirect and direct taxes, the worker would only be relieved of say £5. The capitalist would in this way be actually receiving a greater share of the value of the land than the poor worker.

There is no real economic distinction between land that can bear taxation and capital. A piece of taxable land is a manufactured article, just like a wheelbarrow or a steam engine, the difference being really, economically speaking, one of degree. It is true we cannot carry an acre of land upon our back or move it to and fro, and therefore, instead of carrying it or moving it, we have to carry or move ourselves to it; and it is just the labor involved in carrying ourselves and placing our factories and our cities upon any piece of land that gives it whatever value it possesses, apart from actual cultivation, which is a species of manufacture. We cannot, for example, bring our meadows and our mountains to our railways and stations; we have to take our railways and stations to our meadows and mountains. Our meadows and mountains are thus modified by labor, and those who happen to own land thus modified by the labor of others and thus increased in value, can of course charge for that extra value in the form of rent, and become thereby robbers of labor. This value, which Mill termed "unearned increment," and which Mr. George proposes to tax, is a value given solely by labor; it is a value too that accrues to other articles besides land.[42] It is a value, for instance, that accrues to an old violin, or a bottle of wine, the value in fact of the care and trouble of maturing and preserving an article. Human labor has also this "unearned increment," for by eight hour's labor today a mechanic or a millworker can do treble or quadruple the value of work that a mechanic or millworker could do fifty years ago. This increased value of labor, like the increased value of land, is not given by the efforts of any one man, but by the efforts of the workers and thinkers of the present and past generations; and therefore, no man can claim that the products of his own labor are exclusively his own. We cannot escape from Communism.

"No man made the land!," cries Mr. George. But no man ever *made* anything. No man ever made art, education, science, or the knowledge of how to cook, make clothes, or do anything that distinguishes us from brutes. That skill and knowledge is no more made by any one man, or generation of men, than the soil of the country. No man of himself ever made a machine, the raw and even largely the manufactured material was provided him. No man ever of himself, and from his own created knowledge, invented a machine; the science, the skill of workmanship, was preserved for him and supplied to him by society. To society, to the past and present generations of men, he owes everything that distinguishes him from a bushman or a brute; and shall he be allowed to take all things from society and give nothing to society in return? Shall he, in fact, be permitted to be a Communist in *taking* from the community and an Individualist in *giving* to it?

Mr. George advises Socialists not to concern themselves so much about the monopoly of capital. The entire capital of the nation, he tells us, does not amount to more than four years national production. The stored product of four years labor!, and Mr. George actually thinks that is not of very great account! Why, if every worker in the community had his portion of that capital, the value of four years of his own labor, what would the landlords and capitalist[s] have? Nothing! The workers would own every stick and stone, every atom of material in the country. Then the capitalists would know, what Mr. George apparently does not know, what it is to be without, not four years, but maybe not four weeks or even four days stored labor to fall back upon. And that is the plight of the great portion of our workers today, and it is just because of that plight that they are unable to live without hiring themselves to the capitalists who have appropriated their stored labor, or to resist being plundered in the future. It is because of that plight that even did the workers appropriate the entire soil of the country, and squat upon it rent free, they would be unable to subsist upon it without the assistance of the capitalists – unless they appropriated the capital of the country, or at least a mighty large portion of it, as well.

As Socialists, we are all for Land Nationalization, or Land Communization, but we do not believe it possible to nationalize or communize the land by a single tax upon it, or any other mere fiscal adjustment. Land Nationalization is impossible without Socialism, and any attempt to really nationalize the land would be resisted by capitalists and landlords alike, as they now resist Socialism. Let us then dally no further with the matter, but boldly teach our right to obtain, and boldly set ourselves to obtaining, all the wealth and all the means of producing wealth which our labor and the labor of the past generations of workers (whose inheritors we are) has produced. Let us appease no superstition by hypocrisy, no prejudice by sophistry, and no fear by untruth, but lift boldly up the banner of revolution and all true and good men will be drawn irresistibly to it.

HENRY GEORGE'S FALLACIES
(H. Quelch, *Justice*, July 13, 1889).[43]

Many persons, including some Socialists, have frequently questioned the wisdom of our hostility to Mr. Henry George's theories, and have counseled a friendly neutrality if not active sympathy towards those theories on the ground that after all, they were Socialistic in character, and would be a step in the right direction. The debate at St. James's Hall must have given a rude shock to those persons, and to all who have ever given Mr. George credit for Socialistic leanings. Certainly no bourgeois economist could have given utterance to a more glowing panegyric on competition than that which flowed from the lips of Mr. George on that occasion.

To competition he attributed all the achievements of science, all the triumphs of invention, all the progress and development of mankind. It was competition, he said, which had produced, among other vast results, those ocean liners, those greyhounds of the sea, which bridged over the thousands of miles of water which divided the people of the Eastern from those of the Western world. To destroy competition would be to destroy the very source of activity and progress, and humanity would become stunted, would wither and die. Competition, he might have added, produces slums and sweating, shoddy clothing, bosh butter, and all the other innumerable cheap and nasty things that enter into the daily life of the people. Competition for existence between individuals means industrial war, general degradation for the advantage of the individual. Competition, involving the cheapening of labor, reduces the purchasing power of the laborer, increases the demand for the most inferior commodities, and results in the production of the shoddy, scamped articles with which the markets are now glutted, and of that adulteration which has been lauded as its legitimate outcome.

The substitution of cooperation for competition would supply a greater incentive to progress than any which is afforded by competition, by making each individual interested in the success of all, and by that healthy emulation which would induce each one to strive to win the approbation of his fellows and the advancement of the general well-being rather than a material advantage for himself at the expense of his fellows.

Now, as to Mr. George's remedy. We have not the slightest objection to taxing out the landlords, and it is from no love for them that we have felt it to be our duty to oppose Mr. George. A tax on land values was one of the proposals put forward by the S.D.F.[44] in connection with the County Council elections, and we do not object if, so long as landlords exist they are taxed twenty shillings in every pound of their income. But we are opposed to the single tax, because we believe its advocacy in the form put forward by Mr. George is misleading and mischievous.

It is sometimes urged by those who support Mr. George's theory that the taking of rent, which they put at [£]150 millions, or more than double what we believe the proper estimate to be, the State would have a fund at its disposal for building dwellings, providing places of recreation, organizing the unemployed, and many other social purposes. This, however, forms no part of Mr. George's proposal. Indeed, judging by his earnest denunciations of Governmental interference the other evening he would be strongly opposed to anything of the kind. He simply advocates a single tax, to be used for the abolition of all other taxation, to be levied on and ultimately to absorb rent. This he contends would free land to the laborer, and that is all that is necessary. A tax levied on rent would make it exceedingly undesirable for the landlords to hold it from use, they would therefore loosen their grip; deer forests and manorial parks would disappear, the laborer would have free access to the land, which under his hand would blossom as the rose, and bring forth abundantly all the good things of life. But it by no means follows that because a rent tax would make it incumbent upon the landlords to let their land, therefore the laborer would be able to get at it.

Mr. George seems oblivious of the fact that landlords are an avaricious set, and that many of them are, from their point of view, by no means well-off. They are far more interested in getting rent than they are in holding the land, consequently they now let for the highest possible rent they can get; they could do no more if their rent were taxed. But does this benefit the laborer? Is the land that is most highly rented the most accessible to him? Not at all. Neither does the exaction of rent stimulate cultivation. On the contrary most of the districts which have been cleared of men to make way for deer, have been turned into deer forests because the landlord could get a higher rent for them than for farms, and hundreds of acres of arable land have been turned into pasture because it pays better to pasture cattle than to grow grain.

Suppose we had this single tax, what would be the result? Mr. George, with anarchistic horror of collective control, is most anxious that the landlords should retain the position of rent collectors and managers, therefore it would be necessary to allow them to retain a percentage of the rent in order to induce them to continue to pose as landlords. If they were going to reap no advantage from holding the land they would throw it up and then that hideous thing State management would supervene. But we will suppose them still in the position of landlords, either for fun or profit, and having to pay the whole or the greater part of the rent to the Government. How does that make the land more accessible to the laborer?

The landlords will be compelled then, even if they are not now, to let the land at the highest possible rent, and the person who can pay that would assuredly not be the penniless laborer, but the capitalist farmer. It would not be the starving proletaires who fight at the dock gates for the right to earn a crust, and whom Mr. George is anxious to get back on the land, who would benefit by this change, but the capitalist, the banker, the usurer, the manufacturer, the merchant. They could well afford to pay two or three pounds an acre for a farm or for the luxury of a park or a deer forest, and would pay this rent all the more cheerfully seeing that its payment would relieve them of all other taxation.

But, urges Mr. George, the land at the margin of cultivation, that land which yields no economic rent, would at once be made accessible to the laborer, by the imposition of the single tax. Why? We confess we should like Mr. George to explain this. If the landlords could gain nothing by holding this land, neither could they gain anything by letting the laborers have it free, and if they should be so silly as to continue to hold land for which they had to pay rent while deriving no benefit from it, they would assuredly stick to that which cost them nothing, and which might be used to minister to their pleasures.

Mr. George ascribes the want of employment, the poverty and the evils that flow therefrom, which exists in our midst today, to the divorcement of the laborer from the land. The remedy is to put the laborer back on the land. Even if this were correct the remedy would not be effected by the single tax. The single tax would relieve the capitalist of all taxation, but, as we have shown it would not put a single laborer back on the land. But neither would it be possible in any other way to put the laborer back on the land in the individualistic sense advocated by Mr. George, and for him to argue for it is to ignore all the conclusions to be drawn from his own writings. He shows, both in *Progress and Poverty*, and in *Social Problems*, that Society is becoming more and more organized, that capital is rolling up into larger and larger masses, and that labor is becoming more interdependent and more subdivided, and yet he thinks to reverse all this economic development and go back to primitive individual cultivation of the land by peasant cultivators, as tenants of landlords who are simply state rent collectors! It won't work.

He sneers at capital; he says: "Give man land and he will produce his own capital. The land was here, and man was here before capital. Labor produced capital, therefore capital is the servant of labor." That by no means follows. Labor created capital, it is true, but in so doing it created a Frankenstein, which, under competition, enslaves, exploits, degrades, and crushes its creator. Capital, the machine, the creature of labor it is, as [William] Morris so forcibly puts it, which:

> Fast and faster, our iron master,
> The thing we made, for ever drives,
> Bids us grind treasure and fashion pleasure,
> For other hopes and other lives.[45]

Let the laborer get on the land, says George, and competition in the labor market will cease; a man will not work for less than he could earn by his own exertion on the land. That is true, but he can get land by paying a rack rent for it now, and he could no more afford to pay a rack rent under the single-tax system than he can now. He cannot compete against the capitalist now, and he could not do so then. As to the idea that the reduction of taxation effected by this taking of rent would benefit the workers we have before shown that to be perfectly groundless. The capitalists of course would benefit, but upon what compulsion must they share any such benefit with their wage slaves?

Competition, so much admired by Mr. George would still control their wages. Therefore, we oppose this single-tax scheme as being a middle-class dodge and urge our fellow workers to set their faces against that and any other burden shifting and go straight for collective control of all the means of production and to Agitate, Educate, Organize for such measures as will hasten the Social Revolution which can alone secure the emancipation of the workers of the world.

NOTES

1. Henry George, "Nationalization," *Irish World and American Industrial Liberator* (Aug. 5, 1882): ILD. The letter was mailed from London and dated June 18, 1882.
2. This lecture appears on pages 38–52.
3. For James Finton Lalor see footnote no. 64, page 101.
4. In Latin: The earth He hath given to the children of men.
5. See page 53 for nearly the same quotation.
6. Henry George, "The Nationalization of Land," London *Times* (Sep. 6, 1882): PDM.
7. Alfred Russel Wallace (1823–1913) was an English scientist who simultaneously and independently of Charles Darwin hypothesized a theory of natural selection and evolution for animal species. He was also very active in social reform, especially as an advocate of land nationalization.
8. The title of Wallace's 1882 book is *Land Nationalisation*.
9. A reference to William the Conqueror. See footnote no. 29, page 99.
10. George's two arrests in Ireland in 1882 created a stir.

11. The Duke of Sutherland was notorious during the land clearances in Scotland in the early nineteenth century. According to Julia Bastian, "in 1807 the Duke of Sutherland ordered his agents to clear some 15,000 inhabitants from his land and to burn their homes." See chapter 6 in Wenzer, *An Anthology of Single Land Tax Thought*, 159.

12. The following sentence was deleted by the editor: "Vote to Mr. George, who is about to return to America, and to the Chairman [who] closed the proceedings."

13. This transcribed speech given in the Lecture Theatre of the Midland Institute on June 23, 1884 is taken, in part, from Wenzer, *An Anthology of Henry George's Thought*, pp. 86–99.

14. Thomas Walker was a British friend and supporter of George who nevertheless differed on points of theory.

15. I have not been able to determine the "butcher of mankind."

16. See footnote no. 30, page 251.

17. Winfield S. Hancock was also a noted Union general.

18. The *Alabama*, *Florida*, as well as *Shenandoah*, were famous Confederate ships that wreaked havoc on the Union's merchant marine.

19. In this instance, the Established Church refers to the Church of England.

20. The Reform Bill of 1832 eliminated fifty-six "rotten boroughs" in England that were sparsely populated parliamentary districts easily controlled by a single faction. Apparently they also existed in Ireland.

21. Sir James Caird (1816–1892) was an agriculturalist, author, and Liberal M.P.

22. Possibly a reference to the steel manufacturer Andrew Carnegie (1835–1919) or to the already mentioned but untraceable Mr. Winans.

23. William III (1650–1702). He was the son of William II of Orange (Holland), who ruled England jointly with his wife Mary II.

24. The Civil List in general is a yearly allowance granted to a royal family, a legislative body controlling the funds. The accession of William III in 1689 inaugurated the Civil List. In the nineteenth century, the phrase referred to money granted only for royal household uses.

25. In England beginning in the 1100s and reaching its height in the late seventeenth century, enclosure was the growing practice of fencing off lands formerly considered to be under common rights. This activity was part of the transition from the feudal system to free cultivation, and caused a population shift of the poor from the countryside to the cities. It did foster more efficient husbandry and stockraising.

26. See footnote no. 29.

27. To "spike a gun" is an old expression probably meaning to render something useless.

28. Protoganist in the novel of the same name by Alexander Dumas (père), (1844–1845).

29. Refers to a series of articles in the *Pall Mall Gazette* during October, 1883.

30. In the Christian Scriptures, St. Luke 17:2.

31. In the Hebrew Scriptures, Numbers 20: 11.

32. The Confederate attack on Fort Sumter on April 12, 1861 was the first military action of the Civil War.

33. Radical was used to refer to the advanced reform wing of the Liberal Party in the late nineteenth century. Prior to this time a radical also referred to someone who had espoused the ideas of William Godwin, David Ricardo, David Hume, and John Stuart Mill, or thereafter parliamentary representation reform, the Chartist movement, anti-Corn Law agitation, and the Manchester school of economics.

34. Henry George and Henry M. Hyndman, "Socialism and Rent Appropriation," *The Nineteenth Century: A Monthly Review* XVII, 96 (Feb. 1885), pp. 369–380: PDM. The authors desire to say that from pressure of time, they could only carry out their suggestion of a joint article in this form by dictating to a shorthand writer whom they called in a few hours before Mr. George's departure from London. They hope that any literary blemishes may be attributed to this circumstance (H.G. and H.H.).

35. To hypothecate: a pledge to a creditor as security.

36. James Hargreaves (?–1778) invented the spinning jenny about 1764.

37. This section was within quotation marks and without a citation.

38. A reference to the major business slump in the mid-1880s.

39. Ibid.

40. J. Bruce Glasier, "Henry George and the Single Tax," *Commonweal: The Official Journal of the Socialist League* (June 1, 1889): ILD.

41. A carse is a low-lying alluvial land along the course of a Scottish river.

42. See footnote no. 73, page 101.

43. H. Quelch, "Henry George's Fallacies," *Justice: The Organ of Social Democracy* (July 13, 1889): ILD.

44. S.D.F: The Social Democratic Federation. See footnote no. 73. page 102.

45. William Morris (1834–1896) was an outspoken English writer, poet, printer, artist, and socialist. I have not been able to locate the source of these lines.

BIBLIOGRAPHY

This bibliography contains relevant works in English relating to Henry George and the Single Tax, and includes graduate papers. It is hoped that this list will serve as a touchstone for additional inquiry. For articles from *The American Journal of Economics and Sociology*, *AJES* has been used as an abbreviation. Thanks must be extended to Edward Dodson for his assistance in their compilation.

Allen, H. W. (1936). *Prosperity in the Year 2000 A.D.* Boston: The Christopher Publishing House.

Andelson, R. V. (Ed.) (1979). *Critics of Henry George.* Rutherford, NJ: Fairleigh Dickinson Press.

Argyll, Duke of; Campbell, G. D. (1894). *Property in Land: A Passage-at-Arms between the Duke of Argyll and Henry George.* New York: Sterling.

Argyll, Duke of; Campbell, G. D. (1884). *The Peer and the Prophet.* London: Reeves.

Aslanbeigui, N., & Wick, A. (Apr., 1990). "Progress: Poverty or Prosperity? Joining the Debate Between George and Marshall on the Effects of Economic Growth on the Distribution of Income." *AJES, 49*(2), 239–256.

Barker, C. A. (1991). *Henry George.* New York: Robert Schalkenbach Foundation.

Beggs, G. H. (1967). *The Fairhope Single Tax Corporation: An Analysis of the Efforts of a Single Tax Colony to Apply the Ideas of Henry George.* Ph.D. diss., University of Arizona.

Bell, S. (1968). *Rebel, Priest and Prophet: A Biography of Dr. Edward McGlynn.* New York: Robert Schalkenbach Foundation.

Benestad, J. B. (July, 1985). "Henry George and the Catholic View of Morality and the Common Good: George's Overall Critique of Pope Leo X's Classic Encyclical, Rerum Novarum." *AJES, 44*(3), 365–378.

Benestad, J. G. (Jan., 1986). "George's Proposals in the Context of Perennial Philosophy." *AJES, 45*(1), 115–123.

Bengough, J. W. (1908). *The Whole Hog Book.* Boston: American Free Trade League.

Benz, G. A. (1969). *The "Single Tax" as a Means of Support for a Local Government.* Edmond, Oklahoma. Ph.D. diss., University of Oklahoma.

Bernstein, W. S. (1978). *Lewis Henry Morgan, John Wesley Powell, and Henry George: A Study in the Relation Between Nineteenth Century Intellectual Thought and Social Reform.* M.A. thesis, Brown University.

Birnie, A. (1939). *Single-Tax George.* London: T. Nelson & Sons.

Bonaparte, T. (1985). *Henry George: His Impact Abroad and the Relevancy of His Views on International Trade.* New York: Pace University.

306

Bonaparte, T. (Jan., 1987). "Henry George's Impact at Home and Abroad: He Won the Workers of Marx's Adopted Country But Through Leninism Marxism Has Won Half the World." *AJES*, *46*(1), 109–124.

Bonaparte, T. (Apr., 1989). "George on Free Trade, At Home and Abroad: The American Economist and Social Philosopher Envisioned a World Unhindered in Production and Exchange." *AJES*, *48*(2), 245–255.

Borcherding, T. E., Dillon, P., & Willett, T. D. (Apr., 1998). "Henry George: Precursor to Public Choice Analysis: Some Conclusions from a Lifetime's Study of the Relationship Between Ethics and Economics." *AJES*, *57*(2), 173–182.

Bradley, P. (July, 1980). "Henry George, Biblical Morality and Economic Ethics." *AJES*, *39*(3), 209–215.

Bramwell, G. (1895). *Nationalisation of Land: A Review of Mr. George's Progress and Poverty*. London: Liberty and Property Defense League.

Brann, H. A. (1887). *Henry George and His Land Theories*. New York: Catholic Publication Society.

Briggs, G. (1950). *Comment on Henry George's Definitions*. San Diego, CA: Henry B. Cramer.

Briggs, M. C. (1891). *Regress and Slavery vs. Progress and Poverty*. New York: Hunt & Eaton.

Brooks, N. (1899). *Henry George in California*. New York.

Browne, M. N., & Powers, B. (Oct., 1988). "Henry George and Comparable Worth: Hypothetical Markets as a Stimulus for Reforming the Labor Market." *AJES*, *47*(4), 461–471.

Buurman, G. B. (Oct., 1986). "Henry George and the Institution of Private Property in Land: A Property Rights Approach." *AJES*, *45*(4), 489–502.

Buurman, G. (1971). A Comparison of the Single Tax Proposals of Henry George and the Physiocrats. M.A. thesis, Western Washington State College.

Candeloro, D. L. (1981). *Louis F. Post: Carpetbagger, Singletaxer, Progressive*. Ann Arbor, MI: University Microfilms.

Cantwell, H. (1901). *The Philosophy of Henry George*. St. Louis, MO: Kenmore Press.

Cathrein, V. (1889). *The Champions of Agrarian Socialism: A Refutation of Emile de Laveleye and Henry George*. Buffalo, NY: P. Paul & Bros.

Clancy, R. (1950). *The Story of the Georgist Movement*. London: Land & Liberty Press.

Clancy, R. (1954). *A Seed Was Sown: The Life, Philosophy, and Writings of Oscar H. Geiger*. New York: Henry George School of Social Sciences.

Coleman, J. (1887). *George and Democracy*. Georgetown College, DC: By the author.

Collier, C. (1976). Henry George's System of Economics: Analysis and Criterion. Ph.D. diss., Duke University.

Cord, S. B. (1985). *Henry George: Dreamer or Realist?* New York: Robert Schalkenbach Foundation.

Croft, A. (1952). The Speaking Career of Henry George: A Study in Ideas and Persuasion. M.A. thesis, Northwestern University.

Crump, A. (1884). *An Exposure of the Pretensions of Mr. Henry George as Set Forth in His Book Progress and Poverty*. London: E. Wilson.

D'A Jones, P. (Apr., 1987). "Henry George and British Labor Politics." *AJES*, *46*(2), 245–256.

D'A Jones, P. (Oct., 1988). "Henry George and British Socialism." *AJES*, *47*(4), 472–491.

Davidson, J. (1899). *Concerning Four Precursors of Henry George and the Single Tax*. London: Labor Leader Pub. Dept.

de Mille, A. G. (1950). *Henry George: Citizen of the World*. Chapel Hill: University of North Carolina Press.

Dewey, J. (1927). *An Appreciation of Henry George*. New York: Robert Schalkenbach Foundation.

Dewey, J. (1927). *John Dewey on Henry George and What Some Others Say*. New York: Robert Schalkenbach Foundation.

Dewey, J. (Ed.) (1928). *Significant Paragraphs from Progress and Poverty*. New York: Robert Schalkenbach Foundation.

Douglas of Barlock. (1937). *Social Science Manual, Guide to the Study of Henry George's Progress and Poverty*. London: Henry George Foundation of Great Britain.

Dixwell, G. (1882). *Progress and Poverty: A Review of the Doctrine of Henry George*. Cambridge, MA: J. Wilson & Son.

Dixwell, G. (1882). *Premises of Free Trade Examined*. Cambridge, MA: J. Wilson & Son.

Dudden, A. P. (1971). *Joseph Fels and the Single Tax Movement*. Philadelphia: Temple University Press.

Dwyer, T. M. (Oct., 1982). "Henry George's Thought in Relation to Modern Economics." *AJES*, *41*(4), 363–374.

Easterly, J. (1976). Louis F. Post 1849–1928: The Henry George Man as Progressive and Reformer. Ph.D. diss., Duke University.

Easterly, J. (1970). Louis F. Post: Popularizer and Propagandist for Henry George and the Single Tax, 1849–1928. M.A. thesis, Duke University.

Ely, R. T. (1880). *Land, Labor, and Taxation*. Baltimore: Cushing & Co.

Faidy, J. (1903). *The Political Economy of Henry George*. Cedar Rapids, IA: Why.

Fillebrown, C. (1917). *Thirty Years of Henry George*. Boston: By the author.

Fillebrown, C. (1917). *The Catholic Church and Henry George*. Boston: By the author.

Fillebrown, C. (1917). *Henry George and His Single Tax: An Appreciation*. Boston: By the author.

Fillebrown, C. (1960). *Henry George and the Economists*. Boston.

Flaherty, J. (1985). *Henry George: Motivating the Managerial Mind*. New York: Pace University.

Flattery, H. (1887). *The Pope and the New Crusade*. New York: Thomas R. Knox & Co.

Foksch, M. D. (Jan., 1980). "Theoretical Background of Henry George's Value Theory." *AJES*, *39*(1), 95–104.

Foldvary, F. (Jan., 1996). "A Review Article: George and Democracy in the British Isles." *AJES*, *55*(1), 125–127.

Fuller, III, A. B. (Jan., 1983). "Selected Elements of Henry George's Legitimacy as an Economist" Aaron. *AJES*, *42*(1), 45–61.

Gaffney, M., & Harrison, F. (1994). *The Corruption of Economics*. London: Shepheard-Walwyn.

Geiger, G. R. (1933). *The Philosophy of Henry George*. New York: MacMillan Co.

Geiger, G. R. (1939). *Henry George: A Biography*. London: Henry George Foundation of Great Britain.

Genovese, F. C. (1985). *Henry George and the Labor Unions*. New York: Pace University.

Genovese, F. C. (Jan., 1991). "Henry George and Organized Labor." *AJES*, *50*(1), 113–127.

Genovese, F. C. (Oct., 1984). "An Economic Classic and Plutology: The 'Science of Wealth' Reminds Economists That Their Goal Should Be Well-being for All." *AJES*, *43*(4), 455–467.

George, H. (1999). *Our Land and Land Policy*. Edited by K. C. Wenzer, East Lansing: Michigan State University Press. First essay originally published in 1871.

George, H. (1992). *Progress and Poverty*. New York: Robert Schalkenbach Foundation. First published in 1879.

George, H. (1982). *The Land Question*. New York: Robert Schalkenbach Foundation (First published as The Irish Land Question in 1881). Included are other works dating from 1884 to 1891.

George, H. (1881). *The Land Question: What It Involves and How Alone It Can Be Settled*. New York: D. Appleton & Co.

George, H. (1992). *Social Problems*. New York: Robert Schalkenbach Foundation. First published in 1883.

308

George, H. (1992). *Protection or Free Trade?* New York: Robert Schalkenbach Foundation. First published in 1886.

George, H. (1889). *Verbatim Report of the Debate in St. James Hall, July 2, 1889.* (with H. M. Hyndman). London: Justice.

George, H. (1988). *A Perplexed Philosopher.* New York: Robert Schalkenbach Foundation. First published in 1892.

George, H. (1988). *The Science of Political Economy.* New York: Robert Schalkenbach Foundation. First published posthumously in 1898.

George, H. (1930). *Causes of Business Depressions.* New York: Robert Schalkenbach Foundation.

George, H. (1930). *Gems from Henry George.* London: Henry George Foundation of Great Britain.

George, H. (1942). *More Progress and Less Poverty: A Businessman Reviews Henry George.* New York: Robert Schalkenbach Foundation.

George, H. (1947). *The Wisdom of Henry George: Excerpts from Social Problems.* Girard, KS: Haldeman-Julius Pubs.

George, Jr., H. (1981). *Henry George.* Edited by Daniel Aaron. New York: Chelsea House.

Green, C. (1934). *The Profits of the Earth.* Boston: Christopher Pub. House.

Green, J. J. (1956). The Impact of Henry George's Theories on American Catholics. Ph.D. diss., Notre Dame University.

Green, J. J. (1948). First Impact of the Henry George Agitation on Catholics in the United States. M.A. thesis, Notre Dame University.

Grigg, K. (1983). *Sun Yat Sen: The Third Alternative for the Third World.* Melbourne: Henry George Foundation.

Gronlund, L. (1887). *Insufficiency of Henry George's Theory.* New York: New York Labor News.

Hackner, W. (1887). *Socialism and the Church, or, Henry George versus Archbishop Corrigan.* New York: Catholic Pub. Society.

Harriss, C. L. (1985). *Taxation: Today's Lessons from Henry George.* New York: Pace University.

Harriss, C. L. (July, 1989). "Guidance from an Economic Classic: The Centennial of Henry George's Protection or Free Trade." *AJES, 48*(3), 351–356.

Hawks, C. (1981). M. Herbert Quick: Iowan. Ph.D. diss., University of Iowa.

Hazleton, R. L. (1973). Henry George's Social Economics. Ph.D. diss., University of Utah.

Heath, S. (1952). *Progress and Poverty Reviewed and Its Fallacies Exposed.* New York: Science of Society Foundation.

Hellman, R. (1987). *Henry George Reconsidered.* New York: Carlton Press.

Higgins, E. (1887). *Fallacies of Henry George Exposed and Refuted.* Cincinnati, OH: Press of Keating & Co.

Horner, J. (Oct., 1997). "Henry George on Thomas Robert Malthus: Abundance vs. Scarcity." *AJES, 56*(4), 595–607.

Horner, J. H. (Apr., 1993). "Seeking Institutionalist Signposts in the Work of Henry George: Relevance Often Overlooked." *AJES, 52*(2), 247–255.

Horton, J., & Chisholm, T. (July, 1991). "The Political Economy of Henry George." *AJES, 50*(3), 375–384.

Howell, S. (1970). Scholars of the Urban-Industrial Frontier: 1880–1889. Ph.D. diss., Vanderbilt University.

Hubbard, E. (1907). *Little Journeys to the Homes of Great Reformers.* E. Aurora, NY: The Roycrofters.

Hughes, B. F. *The Basis of Interest (A Reply to Mr. Lowery).* New York: Kraus Reprint.

Inkster, I. (July, 1990). "Henry George, Protectionism and the Welfare of the Working Class: The Economist Offered a Basically Conventional Approach to Protectionism, Different from Today's Issues." *AJES*, *49*(3), 375–384.

Johannsen, O. B. (July, 1987). "Henry George and His Philosophy: He Sought Equality of Opportunity to Use the Earth's Resources as Well as the End of Land Monopoly." *AJES*, *46*(3), 379–382.

Johnson, E. (1910). The Economics of Henry George's *Progress and Poverty*. Ph.D. diss., University of Chicago.

Johnson, M. S. (Oct., 1995). "An Address by a Georgist Sympathizer: Practical Issues in Georgist Thought." *AJES*, *54*(4), 481–488.

Jones, L. B. & T. H. (Apr., 1994). "Huxley's Critique of Henry George: An Expanded Perspective." *AJES*, *53*(2), 245–255.

Jones, P. (1991). *Henry George and British Socialism*. New York: Garland Pubs.

Jonsson, P. O. (Oct., 1997). "On Henry George, the Austrians, and Neoclassical Choice Theory: A New Look at the Similarities Between George and the Austrians." *AJES*, *56*(4), 577–594.

Jordan, D. S. (1899). *The True Basis of Economics*. New York: Doubleday & McClure Co.

Jorgensen, E. O. (1936). *Did Henry George Confuse the Single Tax?* Elkhart, IN: James A. Bell Co.

Jorgensen, E. O. (1925). *False Education in Our Colleges and Universities*. Chicago: Manufacturers and Merchants Federated Tax League.

Kamerschen, D. R. (Oct., 1987). "Some Surviving Elements in the Work of Henry George." *AJES*, *46*(4), 489–493.

Kelly, J. M. (July, 1981). "New Barbarians: The Continuing Relevance of Henry George." *AJES*, *40*(3), 299–308.

Kindleberger, C. (1987). *Henry George's Protection or Free Trade*. Williamstown, MA: Williams College.

Lawrence, E. (1957). *Henry George in the British Isles*. East Lansing: Michigan State University Press.

Lewandowski, E. (1980). The Great American Paradox: Tom L. Johnson and the Controversy Surrounding His Role in History. M.A. thesis, Ohio State University.

Lindner, E. W. (1985). The Redemptive Politic of Henry George: Legacy to the Social Gospel Christianity. New York City. Ph.D. diss., Union Theological Seminary.

Lissner, W. (Jan., 1979). "On the Centenary of Progress and Poverty." *AJES*, *38*(1), 1–16.

Lissner, W., & Lissner, D. (Eds.) (1991). *George and the Scholars*. New York: Robert Schalkenbach Foundation.

Lissner, W. (1992). *George and Democracy in the British Isles*. New York: Robert Schalkenbach Foundation.

Love, J. (1899). *Japanese Notions of European Political Economy*. Philadelphia: By Kuya Shihosho.

Love, J. (1897). *A Correspondence Between an Amateur and a Professor of Economics*. Philadelphia: J. B. Lippincott.

Lowrey, D. (1892). *The Basis of Interest. A Criticism of the Solution Offered by Henry George*. Philadelphia: American Academy of Political and Social Sciences.

Madison, C. A. (1947). *Critics and Crusaders: A Century of American Protest*. New York: Henry Holt & Co.

Mallock, W. H. (1884). *Property and Progress or, a Brief Inquiry into Contemporary Social Agitation in England*. New York: G. P. Putnam's Sons.

Martin, T. L. (Oct., 1989). "Protection or Free Trade: An Analysis of the Ideas of Henry George on International Commerce and Wages." *AJES*, *48*(4), 489–501.

310

Marx, K., & Engels, F. (1969). *Letters to Americans: 1848–1895.* New York: International Publishers.

McMillion, J. L. (1975). Henry George on Land and Liberty. M.A. thesis, Bowling Green State University.

Miller, B. J. (1887). *Trade Organization in Politics.* New York: The Baker & Taylor Co.

Miller, B. J. (1886). *Progress and Robbery: Two Americans Answer to Henry George, the Demi-Communist.* New York: Cherouny.

Moffat, R. (1885). *Mr. Henry George, the Orthodox.* London: Remington & Co.

Moss, L. S. (Oct., 1997). "This Special Henry George Issue: Editor's Introduction." *AJES, 56*(4), 365–388.

Muirhead, J. (1935). *Land and Unemployment.* London: Oxford University Press.

Murray, J. E. (Apr., 1996). "Henry George and the Shakers: Evolution of Communal Attitudes Towards Land Ownership." *AJES, 55*(2), 245–256.

Nitoche, C. G. (1981). Albert Jay Nock and Frank Chodorov: Case Studies in Recent American Individualist and Anti-Statist Thought. Ph.D. diss., University of Maryland, College Park.

Nock, A. J. (1939). Henry George. New York: W. Morrow & Co.

Noyes, R. (Ed.) (1996). *Now the Synthesis, The New Social Control.* London: Shepheard-Walwyn.

Nuesse, C. J. (Apr., 1985). "Henry George and 'Rerum Novarum:' Evidence is Scant that the American Economist was a Target of Leo X's Classic Encyclical." *AJES, 44*(2), 241–254.

O'Donnell, E. T. (1997). "Though Not an Irishman Henry George and the American Irish." *AJES, 56*(4), 407–419.

Oser, J. (1974). *Henry George.* New York: Twayne Pubs.

Padover, S. K. (1960). *The Genius of America: Men Whose Ideas Shaped Our Civilization.* New York: McGraw-Hill Book Co.

Pedder, D. C. (1908). *Henry George and His Gospel.* London: A. C. Fifield.

Perehman, M. (1997). "Henry George and Nineteenth-Century Economics: The Village Economy Meets the Railroad." *AJES, 56*(4), 441–449.

Petrella, F. (Apr., 1988). "Henry George and the Classical Scientific Research Program: The Economics of Republican Millennialism." *AJES, 47*(2), 239–256.

Petrella, F. (July, 1988). "Henry George and the Classical Scientific Research Program: George's Modification of It and His Real Significance for Future Generations." *AJES, 47*(3), 371–384.

Petrella, F. (Apr., 1981). "Henry George, the Classical Model and Technological Changes." *AJES, 40*(2), 191–206.

Petrella, F. (July, 1984). "Henry George's Theory of State's Agenda: The Origins of His Ideas on Economic Policy in Adam Smith's Moral Theory." *AJES, 43*(3), 269–286.

Portner, S. (1963). *Louis F. Post: His Life and Times.* Ann Arbor, MI: University Microfilms.

Post, L. F. (1930). *The Prophet of San Francisco: Personal Memoirs and Interpretation of Henry George.* New York: Vanguard Press.

Post, L. F. (1976). *Henry George's 1886 Campaign.* Westport, CT: Hyperion Press.

Post, L. F. (1912). *Taxation of Land Values; An Explanation.* Chicago: The Public.

Post, L. F. (1899). *The Single Tax.* Cedar Rapids, IA: F. Vierth.

Post, L. F. (1912). *Outlines of Lectures on the Taxation of Land Values.* Chicago: The Public.

Quinby, L. (1925). *Henry George's Progress and Poverty.* Los Angeles, CA: Chimes Press.

Rafalko, R. J. (Jan., 1988). "Henry George and the Contemporary Debate over Industrial Protectionism." *AJES, 47*(1), 111–123.

Rafalko, R. J. (July, 1989). "Henry George's Labor Theory of Value: He Saw the Entrepreneurs and Workers as Employers of Capital and Land, and Not the Reverse." *AJES, 48*(3), 311–320.

Ralston, J. (1945). *Confronting the Land Question*. Bayside, NY: American Association for Scientific Taxation.

Ralston, J. (1909). *Open Letter Concerning Tax Reform*. Cincinnati, OH: Joseph Fels Fund of America.

Ralston, J. (1931). *What's Wrong With Taxation*. San Diego, CA: Ingram Institute.

Ramhurst, R. (1953). The Single Tax and Its Practical Modifications. M.A. thesis, University of Arizona.

Rather, L. (1978). *Henry George – Printer to Author*. Oakland, CA: Rather Press.

Redfearn, D. (1992). *Tolstoy: Principles for a New World Order*. London: Shepheard-Walwyn.

Reeves, C. E. (Jan., 1965). "Henry George's Speaking in the Land Reform Movements: The West Coast 'Training Phase.'" *AJES*, *24*(1), 51–68.

Ring, H. F. (1887). *The Case Plainly Stated*. New York: Henry George.

Rose, E. J. (1968). *Henry George*. New Haven, CT: College & University Press.

Rutherford, R. C. (1887). *Henry George versus Henry George*. New York: D. Appleton & Co.

Saldji, V. (1959). *Is Progress and Poverty Outdated?* London: Land & Liberty Press.

Salter, W. (1884). *Progress and Poverty: Rev. Dr. Salter's Sermon at the Congregational Church*. Burlington[?], IA.

Samuels, W. J. (Jan., 1983). "Henry George's Challenge to the Economics Profession." *AJES*, *42*(1), 63–66.

Schwartzman, J. (Oct., 1997). "The Death of Henry George: Scholar or Statesman?" *AJES*, *56*(4), 391–405.

Schwartzman, J. (Jan., 1986). "Henry George and the Ethics of Economics." *AJES*, *45*(1), 101–114.

Schwartzman, J. (Jan., 1990). "Henry George and George Bernard Shaw: Comparison and Contrast: The Two 19th Century Intellectual Leaders Stood for Ethical Democracy vs. Socialist Statism." *AJES*, *49*(1), 113–127.

Schubart, R. D. (1984). Ralph Borsodi: The Political Biography of a Utopian Decentralist, 1886–1977. Ph.D. diss., State University of New York at Binghamton.

Scott, W. (1898). *Henry George and His Economic System*.

Scudder, M. L. (1884). *The Labor-Value Fallacy*. Chicago: Jansen, McClurg & Co.

Shafer, R. E. *The Philosophy of Henry George*.

Shapiro, A. H. (Oct., 1988). "Moses – Henry George's Inspiration." *AJES*, *47*(4), 493–501.

Shearman, T. G. (1889). *Henry George's Mistakes*. New York: Henry George.

Shearman, T. G. (1895). *National Taxation*. New York: G.P. Putnam's Sons, 1895.

Shearman, T. G. (1883). *Free Trade*. New York: The American Free Trade League.

Shearman, T. G. (1910). *Shortest Road to the Single Tax*. Cedar Rapids, IA: Frank Vierth.

Shearman, T. G. (1889). *Objections to the Single Tax*. New York: Henry George.

Shearman, T. G. (1892). *Taxation and Revenue: The Free Trade View*. New York: Appleton & Co.

Sheilds, C. H. (1914). *Single Tax Exposed*. Seattle.

Siemens, R. P. (Jan., 1995). "Henry George: An Unrecognized Contributor to American Social Theory." *AJES*, *54*(1), 107–127.

Siemens, R. P. (Apr., 1995). "Henry George and Social Theory: Consequences of Inattention to His Contributions." *AJES*, *54*(2), 249–256.

Silagi, M. (Jan., 1994). "Henry George and Europe: The Ending of the Venture in Georgist Tax Reform in Hungary." *AJES*, *53*(1), 111–127.

Silagi, M. (July, 1993). "Henry George and Europe: Several Statutes and a Constitutional Revision Won by George's Followers under A. Damaschke." *AJES*, *52*(3), 369–384.

Silagi, M. (Jan., 1993). "Henry George and Europe: Failure of the Early Efforts to Organize Germany's Land Reformers." *AJES*, *52*(1), 119–127.

Silagi, M. (Oct., 1992). "Henry George and Europe: Michael Flürscheim Publicized George's Ideas in Germany." *AJES*, *51*(4), 495–501.

Silagi, M. (Apr., 1986). "Henry George and Europe: The Far-Reaching Effect of the Ideas of the American Social Philosopher at the Turn of the Century." *AJES*, *45*(2), 201–213.

Silagi, M. (Jan., 1989). "Henry George and Europe: As Dissident Economist and Path-breaking Philosopher, He was a Catalyst for British Social Reform." *AJES*, *48*(1), 113–122.

Silagi, M. (July, 1986). "Henry George and Europe: As Social Philosopher, He Was Seen as Synthesizing Jefferson, the Enlightenment and Mother Earth." *AJES*, *45*(3), 373–384.

Silagi, M. (Apr., 1991). "Henry George and Europe: His Followers Awakened the British Conscience." *AJES*, *50*(2), 243–255.

Silagi, M. (Oct., 1987). "Henry George and Europe: Ireland, the First Target of His Efforts to Spread His Doctrines Internationally, Disappointed Him." *AJES*, *46*(4).

Stabile, D. R. (July, 1995). "Henry George's Influence on John Bates Clark." *AJES*, *54*(3), 373–384.

Stebbins, G. (1887). *Progress from Poverty. Review and Criticism of Henry George's Progress and Poverty and Protection or Free Trade*. Chicago: C. H. Kerr & Co.

Suit, W. W. (1988). Tom Loftin Johnson, Businessman, Reformer (Ohio Progressive). Ph.D. diss., Kent State University.

Sutton, P. M. (1887). *Work's Dialogue with Henry George*. Marshalltown, IA: Marshall Printing Co.

Swinerton, S. (1993). The Jeffersonian Civil Revolution of 1886: Henry George's First New York City Mayoral Campaign. M.A. thesis, Northeast Missouri State University.

Teilhac, E. (1936). *Pioneers of American Economic Thought in the Nineteenth Century*. New York: The MacMillan Co.

Thomas, J. L. (1983). *Alternative America: Henry George, Edward Bellamy, Henry Demarest Lloyd, and the Adversary Tradition in America*. Cambridge, MA: Belknap Press.

Tipple, J. O. (1960). *Andrew Carnegie/Henry George: The Problems of Progress*. Cleveland, OH: H. Allen.

Tribe, H. F. (1991). Disciple of "Progress and Poverty:" Robert Crosser and Twentieth Century Reform: Robert Crosser, Congressman, Ohio. Ph.D. diss., Bowling Green State University.

Toynbee, A. (1883). *Progress and Poverty: A Criticism of Mr. Henry George*. London: Kegan Paul.

Tucker, B. (1896). *Henry George, Traitor*. New York: By the author.

Wedgwood, J. (1908). *Henry George for Socialists*. London: Independent Labour Party.

Wenzer, K. C. (1997). *An Anthology of Henry George's Thought. Vol. I, The Henry George Centennial Trilogy*. Rochester, NY: University of Rochester Press.

Wenzer, K. C. (1997). *An Anthology of Tolstoy's Spiritual Economics. Vol. II, The Henry George Centennial Trilogy*. Rochester, NY: University of Rochester Press.

Wenzer, K. C. (1999). *An Anthology of Single Land Tax Thinkers. Vol. III, The Henry George Centennial Trilogy*. Rochester, NY: University of Rochester Press.

Wenzer, K. C. (Ed.) (1999). *Land-Value Taxation: The Equitable and Efficient Source of Public Finance*. Armonk, NY: M. E. Sharpe Publishers.

Wenzer, K. C., & West, T. R. (2000). *The Forgotten Legacy of Henry George*. Waterbury, CT: Emancipation Press.

Wenzer, K. C., & Sawyer, R. A. (2000). *Henry George and the Single Tax: A Catalogue of the Collection in the New York Public Library*. New York: Robert Schalkenbach Foundation. First published in 1926.

Wenzer, K. C. (Oct., 1997). "Tolstoy's Georgist Spiritual Political Economy (1897–1910): Anarchism and Land Reform." *AJES*, *56*(44), 639–667.

Wray, C. R. (1958). The Theories and Tactics of Henry George: 1886–1888. M.A. thesis, Notre Dame University.

Wunderlich, G. (July, 1982). "The USA's Land Date Legacy from the 19th Century: A Message From the Henry George – Francis A. Walker Controversy over Farm Land Distribution." *AJES*, *41*(3), 269–280.

Wynns, P. L. (1965). Henry George, the Single Tax, and the Economic Rent of Land During the Twentieth Century. M.A. thesis, Florida State University.

Yanosky, R. W. (1993). Seeing the Cat: Henry George and the Rise of the Single Tax Movement, 1879–1890. Ph.D. diss., University of California at Berkeley.

Yardley, A. (Ed.) (1905). *Addresses at the Funeral of Henry George*. Chicago: The Public Pub. Co.

Young, A. N. (1916). *The Single Tax Movement in the United States*. Princeton, NJ: Princeton University Press.

Zimard, S. (1900). *Modern Social Movements; Descriptive Summaries and Bibliographies*. New York: The H. W. Wilson Co.